Parametric Modeling

with Siemens NX 2020

Randy H. Shih
Oregon Institute of Technology

SDC
PUBLICATIONS

SDC Publications
P.O. Box 1334
Mission, KS 66222
913-262-2664
www.SDCpublications.com
Publisher: Stephen Schroff

Examination Copies
Books received as examination copies are for review purposes only and may not be made available for student use. Resale of examination copies is prohibited.

Electronic Files
Any electronic files associated with this book are licensed to the original user only. These files may not be transferred to any other party.

Trademarks
Siemens NX and NX I-deas are registered trademarks of *Siemens PLM Inc*. Microsoft Windows are either registered trademarks or trademarks of *Microsoft Corporation*. All other trademarks are trademarks of their respective holders.

The author and publisher of this book have used their best efforts in preparing this book. These efforts include the development, research and testing of the material presented. The author and publisher shall not be liable in any event for incidental or consequential damages with, or arising out of, the furnishing, performance, or use of the material.

ISBN-13: 978-1-63057-380-5
ISBN-10: 1-63057-380-9

Printed and bound in the United States of America.

Preface

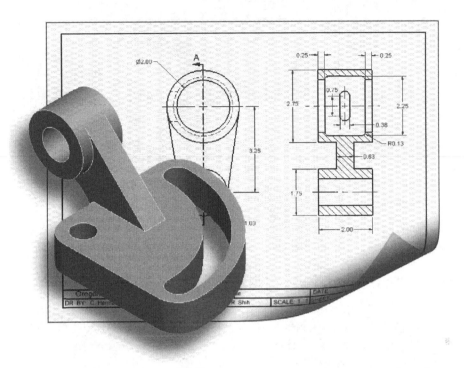

The primary goal of *Parametric Modeling with Siemens NX* is to introduce the aspects of designing with **Solid Modeling** and **Parametric Modeling**. This text is intended to be used as a practical training guide for students and professionals. This text uses *Siemens NX* as the modeling tool and the chapters proceed in a pedagogical fashion to guide you from constructing basic solid models to building intelligent mechanical designs, creating multi-view drawings and assembly models. This text takes a hands-on, exercise-intensive approach to all the important *Parametric Modeling* techniques and concepts. This textbook contains a series of ten tutorial style lessons designed to introduce beginning CAD users to **Siemens NX**. This text is also helpful to *NX* users upgrading from a previous release of the software. The solid modeling techniques and concepts discussed in this text are also applicable to other parametric feature-based CAD packages. The basic premise of this book is that the more designs you create using *Siemens NX*, the better you learn the software. With this in mind, each lesson introduces a new set of commands and concepts, building on previous lessons. This book does not attempt to cover all of *Siemens NX*'s features, only to provide an introduction to the software. It is intended to help you establish a good basis for exploring and growing in the exciting field of **Computer Aided Engineering**.

Acknowledgments

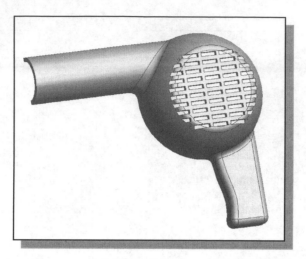

This book would not have been possible without a great deal of support. First, special thanks to two great teachers, Prof. George R. Schade of University of Nebraska-Lincoln and Mr. Denwu Lee, who showed me the fundamentals, the intrigue, and the sheer fun of Computer Aided Engineering.

The effort and support of the editorial and production staff of SDC Publications is gratefully acknowledged. I would especially like to thank Stephen Schroff for his support throughout this project. And I would also like to thank Mr. Mark H. Lawry of *Siemens PLM Inc.* for his encouragement, suggestions and support.

I am grateful that the Manufacturing and Mechanical Engineering and Technology Department of Oregon Institute of Technology has provided me with an excellent environment in which to pursue my interests in teaching and research.

Finally, truly unbounded thanks are due to my wife HsiuLing and our daughter Casandra for their understanding and encouragement throughout this project.

Randy H. Shih
Klamath Falls, Oregon
Spring, 2020

Table of Contents

Preface
Acknowledgments

Chapter 1
Introduction - Getting Started

Chapter 2
Parametric Modeling Fundamentals

Chapter 3
Constructive Solid Geometry Concepts

Chapter 4
Model History Tree

Chapter 5
Parametric Constraints Fundamentals

Chapter 6
Geometric Construction Tools

Chapter 7
Parent/Child Relationships

Chapter 8
Part Drawings and Associative Functionality

Chapter 9
Datum Features and Auxiliary Views

Chapter 10
Introduction to 3D Printing

Chapter 11
Symmetrical Features in Designs

Chapter 12
Advanced 3D Construction Tools

Chapter 13
Basic Sheet Metal Designs

Chapter 14
Assembly Modeling - Putting It All Together

Chapter 15
Advanced Assembly Modeling and Animation

Appendix A

Index

Notes:

Chapter 1
Introduction - Getting Started

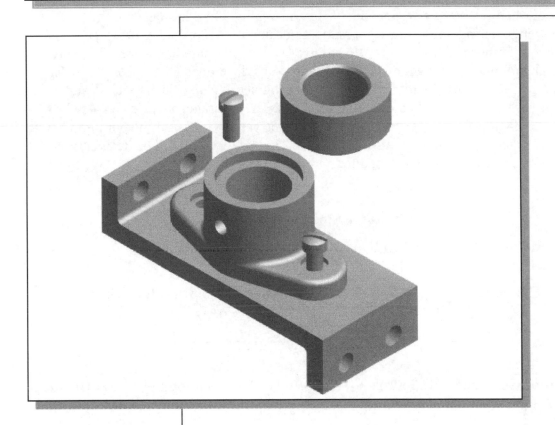

Learning Objectives

♦ **Development of Computer Geometric Modeling**

♦ **Feature-Based Parametric Modeling**

♦ **Startup Options and Units Setup**

♦ **Siemens NX Screen Layout**

♦ **User Interface & Mouse Buttons**

♦ **Siemens NX Online Help**

Introduction

The rapid changes in the field of **Computer Aided Engineering** (CAE) have brought exciting advances in the engineering community. Recent advances have made the long-sought goal of **concurrent engineering** closer to a reality. CAE has become the core of concurrent engineering and is aimed at reducing design time, producing prototypes faster, and achieving higher product quality. *Siemens NX* is an integrated package of mechanical computer aided engineering software tools developed by *Siemens PLM Inc*. *Siemens NX* is a tool that facilitates a concurrent engineering approach to the design and stress-analysis of mechanical engineering products. The computer models can also be used by manufacturing equipment such as machining centers, lathes, mills, or rapid prototyping machines to manufacture the product. In this text, we will be dealing only with the solid modeling modules used for part design and part drawings.

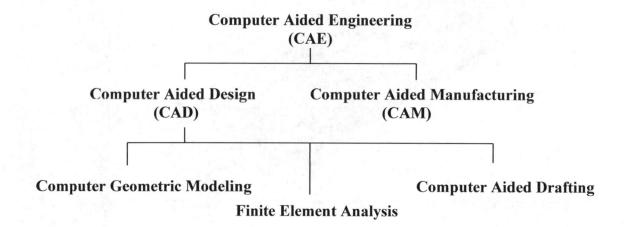

Development of Computer Geometric Modeling

Computer geometric modeling is a relatively new technology, and its rapid expansion in the last fifty years is truly amazing. Computer-modeling technology has advanced along with the development of computer hardware. The first-generation CAD programs, developed in the 1950s, were mostly non-interactive; CAD users were required to create program-codes to generate the desired two-dimensional (2D) geometric shapes. Initially, the development of CAD technology occurred mostly in academic research facilities. The Massachusetts Institute of Technology, Carnegie-Mellon University, and Cambridge University were the leading pioneers at that time. The interest in CAD technology spread quickly and several major industry companies, such as General Motors, Lockheed, McDonnell, IBM, and Ford Motor Co., participated in the development of interactive CAD programs in the 1960s. Usage of CAD systems was primarily in the automotive industry, aerospace industry, and government agencies that developed their own programs for their specific needs. The 1960s also marked the beginning of the development of finite element analysis methods for computer stress analysis and computer aided manufacturing for generating machine toolpaths.

The 1970s are generally viewed as the years of the most significant progress in the development of computer hardware, namely the invention and development of **microprocessors**. With the improvement in computing power, new types of 3D CAD programs that were user-friendly and interactive became reality. CAD technology quickly expanded from very simple **computer aided drafting** to very complex **computer aided design**. The use of 2D and 3D wireframe modelers was accepted as the leading edge technology that could increase productivity in industry. The developments of surface modeling and solid modeling technologies were taking shape by the late 1970s, but the high cost of computer hardware and programming slowed the development of such technology. During this period, the available CAD systems all required room-sized mainframe computers that were extremely expensive.

In the 1980s, improvements in computer hardware brought the power of mainframes to the desktop at less cost and with more accessibility to the general public. By the mid-1980s, CAD technology had become the main focus of a variety of manufacturing industries and was very competitive with traditional design/drafting methods. It was during this period of time that 3D solid modeling technology had major advancements, which boosted the usage of CAE technology in industry.

The introduction of the *feature-based parametric solid modeling* approach, at the end of the 1980s, elevated CAD/CAM/CAE technology to a new level. In the 1990s, CAD programs evolved into powerful design/manufacturing/management tools. CAD technology has come a long way, and during these years of development, modeling schemes progressed from two-dimensional (2D) wireframe to three-dimensional (3D) wireframe, to surface modeling, to solid modeling and, finally, to feature-based parametric solid modeling.

The first-generation CAD packages were simply 2D **computer aided drafting** programs, basically the electronic equivalents of the drafting board. For typical models, the use of this type of program would require that too many views of the objects be created individually as they would be on the drafting board. The 3D designs remained in the designer's mind, not in the computer database. Mental translations of 3D objects to 2D views are required throughout the use of these packages. Although such systems have some advantages over traditional board drafting, they are still tedious and labor intensive. The need for the development of 3D modelers came quite naturally, given the limitations of the 2D drafting packages.

The development of three-dimensional modeling schemes started with three-dimensional (3D) wireframes. Wireframe models are models consisting of points and edges, which are straight lines connecting between appropriate points. The edges of wireframe models are used, similar to lines in 2D drawings, to represent transitions of surfaces and features. The use of lines and points is also a very economical way to represent 3D designs.

The development of the 3D wireframe modeler was a major leap in the area of computer geometric modeling. The computer database in the 3D wireframe modeler contains the locations of all the points in space coordinates, and it is typically sufficient to create just one model rather than multiple views of the same model. This single 3D model can then be viewed from any direction as needed. Most 3D wireframe modelers allow the user to create projected lines/edges of 3D wireframe models. In comparison to other types of 3D modelers, the 3D wireframe modelers require very little computing power and generally can be used to achieve reasonably good representations of 3D models. However, because surface definition is not part of a wireframe model, all wireframe images have the inherent problem of ambiguity. Two examples of such ambiguity are illustrated.

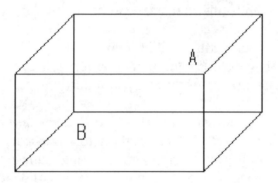

Wireframe Ambiguity: Which corner is in front, A or B?

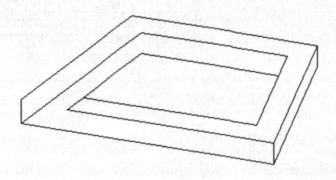

A non-realizable object: Wireframe models contain no surface definitions.

Surface modeling is the logical development in computer geometry modeling to follow the 3D wireframe modeling scheme by organizing and grouping edges that define polygonal surfaces. Surface modeling describes the part's surfaces but not its interiors. Designers are still required to interactively examine surface models to ensure that the various surfaces on a model are contiguous throughout. Many of the concepts used in 3D wireframe and surface modelers are incorporated in the solid modeling scheme, but it is solid modeling that offers the most advantages as a design tool.

In the solid modeling presentation scheme, the solid definitions include nodes, edges, and surfaces, and it is a complete and unambiguous mathematical representation of a precisely enclosed and filled volume. Unlike the surface modeling method, solid modelers start with a solid or use topology rules to guarantee that all of the surfaces are stitched together properly. Two predominant methods for representing solid models are **constructive solid geometry** (CSG) representation and **boundary representation** (B-rep).

The CSG representation method can be defined as the combination of 3D solid primitives. What constitutes a "primitive" varies somewhat with the software but typically includes a rectangular prism, a cylinder, a cone, a wedge, and a sphere. Most solid modelers also allow the user to define additional primitives, which are shapes typically formed by the basic shapes. The underlying concept of the CSG representation method is very straightforward: we simply **add** or **subtract** one primitive from another. The CSG approach is also known as the machinist's approach, as it can be used to simulate the manufacturing procedures for creating the 3D object.

In the B-rep representation method, objects are represented in terms of their spatial boundaries. This method defines the points, edges, and surfaces of a volume, and/or issues commands that sweep or rotate a defined face into a third dimension to form a solid. The object is then made up of the unions of these surfaces that completely and precisely enclose a volume.

By the 1980s, a new paradigm called *concurrent engineering* had emerged. With concurrent engineering, designers, design engineers, analysts, manufacturing engineers, and management engineers all work together closely right from the initial stages of the design. In this way, all aspects of the design can be evaluated and any potential problems can be identified right from the start and throughout the design process. Using the principles of concurrent engineering, a new type of computer modeling technique appeared. The technique is known as the *feature-based parametric modeling technique*. The key advantage of the *feature-based parametric modeling technique* is its capability to produce very flexible designs. Changes can be made easily and design alternatives can be evaluated with minimum effort. Various software packages offer different approaches to feature-based parametric modeling, yet the end result is a flexible design defined by its design variables and parametric features.

Feature-Based Parametric Modeling

One of the key elements in the *Siemens NX* solid modeling software is its use of the **feature-based parametric modeling technique**. The feature-based parametric modeling approach has elevated solid modeling technology to the level of a very powerful design tool. Parametric modeling automates the design and revision procedures by the use of parametric features. Parametric features control the model geometry by the use of design variables. The word *parametric* means that the geometric definitions of the design, such as dimensions, can be varied at any time during the design process. Features are predefined parts or construction tools for which users define the key parameters. A part is described as a sequence of engineering features, which can be modified and/or changed at any time. The concept of parametric features makes modeling more closely match the actual design-manufacturing process than the mathematics of a solid modeling program. In parametric modeling, models and drawings are updated automatically when the design is refined.

Parametric modeling offers many benefits:

- **We begin with simple, conceptual models with minimal detail; this approach conforms to the design philosophy of "shape before size."**

- **Geometric constraints, dimensional constraints, and relational parametric equations can be used to capture design intent.**

- **The ability to update an entire system, including parts, assemblies and drawings after changing one parameter of complex designs.**

- **We can quickly explore and evaluate different design variations and alternatives to determine the best design.**

- **Existing design data can be reused to create new designs.**

- **Quick design turn-around.**

The feature-based parametric modeling technique enables the designer to incorporate the original **design intent** into the construction of the model, and the individual features control the geometry in the event of a design change. As features are modified, the system updates the entire part by re-linking the individual features of the model.

Siemens NX is a digital product development system that is designed to help companies transform the product lifecycle. With the industry's broadest suite of integrated, fully associative CAD/CAM/CAE applications, Siemens NX can be used to cover the full range of development processes in product design, manufacturing and simulation.

Siemens NX is the ninth release, with many added features and enhancements, of the *NX* software produced by Siemens PLM Inc. *Siemens PLM Inc.* is also considered the industry leader in product lifecycle management (PLM), which empowers businesses to make unified, information-driven decisions at every stage in the product lifecycle.

Getting Started with Siemens *NX*

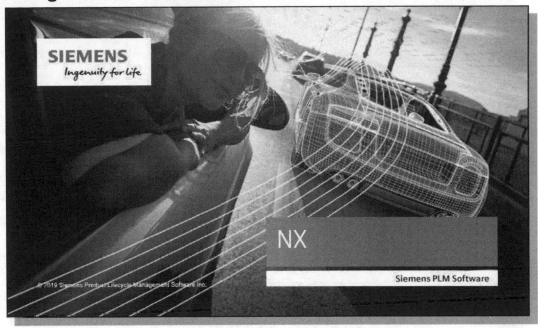

Siemens NX is composed of several application software modules (these modules are called *applications*), all sharing a common database. In this text, the main concentration is placed on the solid modeling modules used for part design. The general procedures required in creating solid models, engineering drawings, and assemblies are illustrated.

How to start *Siemens NX* depends on the type of workstation and the particular software configuration you are using. With most *Windows* systems, you may select **Siemens NX** on the *Start* menu or select the **Siemens NX** icon on the desktop. Consult your instructor or technical support personnel if you have difficulty starting the software. The program takes a while to load, so be patient.

The tutorials in this text are based on the assumption that you are using **Siemens *NX's*** default settings. If your system has been customized for other uses, contact your technical support personnel to restore the default software configuration.

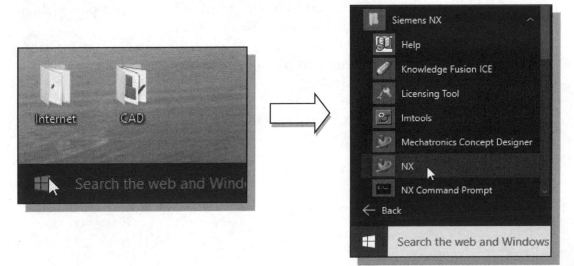

The Siemens NX Main Window

Once the program is loaded into the memory, the *Siemens NX Main Window* appears at the center of the screen.

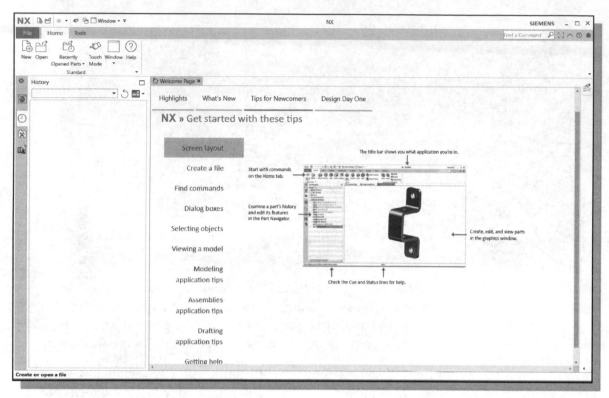

Note that very few options are available at the current state of the main window. This is because *Siemens NX* uses three **work area concepts** called the **"No Part"** state, **"Applications,"** and **"Task environments."**

No Part State: When you start *Siemens* NX, it opens in a default environment called the "No Part" state. This environment is similar to standing in the lobby of an office building, in the sense that you have entered the building, but you have not yet entered a specific room in the building. From the "No Part" state, you can access all the other rooms in the building or, in this case, all the other applications in *Siemens* NX.

Applications: To begin work, first open a part file. Then, you can start the application you want to use, such as Modeling or Drafting.

Task Environments: From some applications, you can enter what is called a Task Environment. For example, from the Modeling application you can enter the Sketcher. When done using the Sketcher, you can return to the Modeling application.

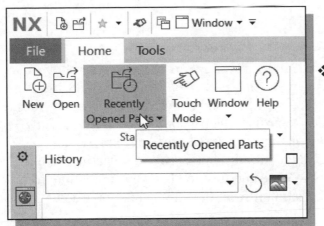

❖ Note that only general setup options are available under the No-Part state.

❖ We will next switch to the *Siemens* NX Application work area.

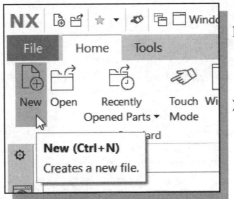

1. Select the **New** icon with a single click of the left-mouse-button in the *What to Do* dialog box.

➢ The **File New** dialog box will appear on the screen with three option tabs: **Model**, **Drawing** and **Simulation**. Note that the default is set to Model and we can create a new model using any of the pre-defined templates.

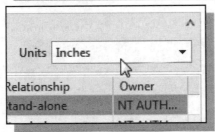

2. Select the **Inches** units as shown in the window. When starting a new CAD file, the first thing we should do is to choose the units we would like to use.

3. Select **Blank** in the *Template list*. Note that the *Blank template* contains no pre-defined settings.

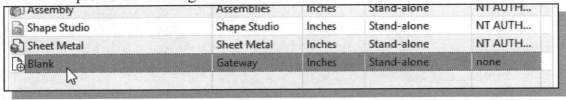

Assembly	Assemblies	Inches	Stand-alone	NT AUTH...
Shape Studio	Shape Studio	Inches	Stand-alone	NT AUTH...
Sheet Metal	Sheet Metal	Inches	Stand-alone	NT AUTH...
Blank	Gateway	Inches	Stand-alone	none

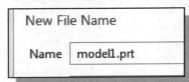

4. In the *New File Name* section, note that **model1.prt** as the default *File Name*.

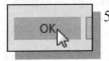

5. Click **OK** to accept the settings and proceed with the new file command.

Siemens NX Screen Layout

Upon opening an existing part or creating a new part in the *Siemens* NX *no part* state, we now entered the *Siemens* NX **Gateway** application. The *Siemens* NX Gateway allows us to perform the general functions, such as open existing part files, create new part files, save part files, plot drawings and screen layouts, import and export various types of files, just like in the no part state. It also provides controls to *view display operations, screen layout and layer functions, WCS manipulation, object information and analysis,* and access to online help. Gateway is the gateway to all other interactive applications and is the first application we enter in *Siemens* NX. We can return to Gateway at any time from the other applications in *Siemens* NX by selecting it from the Application pull-down menu.

The default *Siemens NX* Application screen contains the *pull-down* menus, the *Standard* toolbar, the *View* toolbar, the *Utility* toolbar, the *Graphics* area, the *Selection toolbar*, the *Snap point* toolbar and the *Resource Bars*. A line of quick text appears next to the icon as you move the *mouse cursor* over different icons. You may resize the *Siemens NX* window by clicking and dragging at the edges of the window, or relocate the window by clicking and dragging at the window title area.

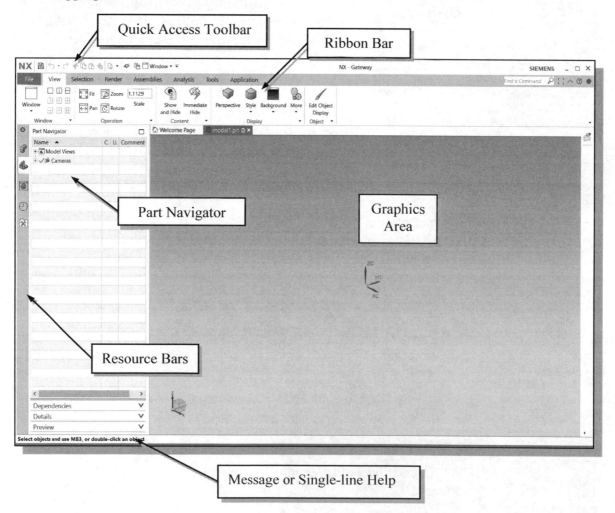

- **Quick Access Toolbar**

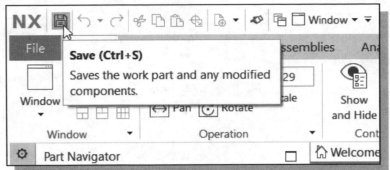

The *Quick Access Toolbar* at the top of the main window contains operations that you can use for all modes of the system.

- **Ribbon Bar**

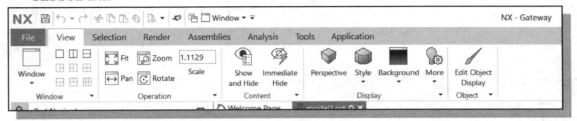

The *Ribbon* bar at the top of the screen window organizes commands in groups on tabs. Commands can be hidden or displayed as preferred.

- **Zoom Toolbar**

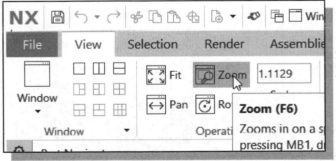

The *Zoom* toolbar allows us quick access to frequently used zoom-related commands, such as Zoom and Zoom Scale.

- **Additional Tools**

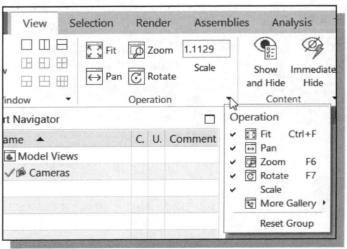

The arrow next to each toolbar icon can be used to access the additional tools that are available.

- **Message and Status Bar**

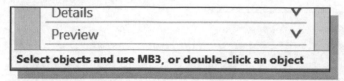

The *Message and Status Bar* area shows a single-line help when the cursor is on top of an icon. This area also displays information pertinent to the active operation.

- **Part Navigator**

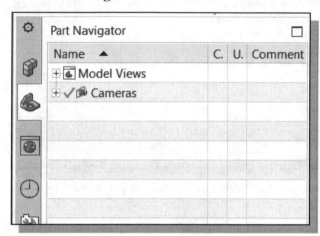

The *Part Navigator* area shows information regarding the current active model. This area is used by the Resource Bars options described below.

- **Resource Bars**

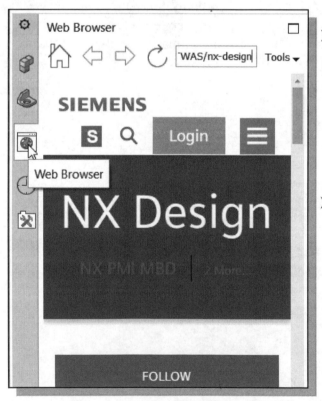

➢ The *Resource Bar* provides three groups of resources: **Navigators**, **Explorers** and **Palettes** for multiple functions such as managing access to features and editing, and providing alternate access to functions in the context menu.

➢ On your own, in the **web browser**, read the general descriptions of the available options for using the resource bars.

Mouse Buttons

Siemens NX utilizes the mouse buttons extensively. In learning *Siemens NX*'s interactive environment, it is important to understand the basic functions of the mouse buttons. It is highly recommended that you use a mouse or a tablet with *Siemens NX* since the package uses the buttons for various functions.

- **Left mouse button (MB1)**
  The **left-mouse-button** is used for most operations, such as selecting menus and icons, or picking graphic entities. One click of the button is used to select icons, menus and form entries, and to pick graphic items.

- **Middle mouse button/wheel (MB2)**
  The middle mouse button/wheel can be used to end a command (single click) or it can also be used to Rotate (hold down the wheel button and drag the mouse) or Zoom real time (turn the wheel). The software also utilizes the **middle-mouse-button** the same as the **ENTER** key and is often used to accept the default setting to a prompt or to end a process.

- **Right mouse button (MB3)**
  The **right-mouse-button** is used to bring up additional available options.

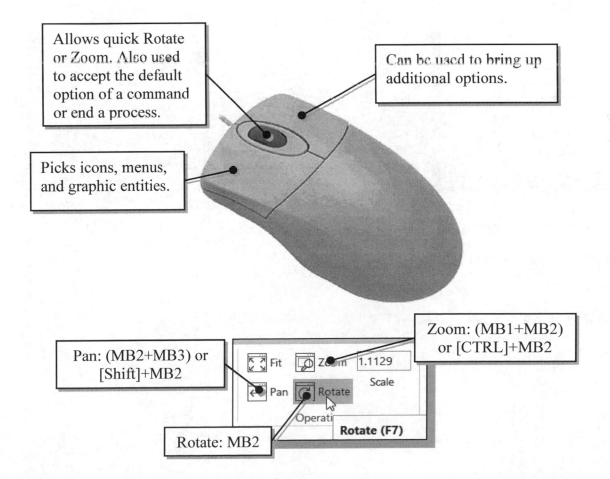

Allows quick Rotate or Zoom. Also used to accept the default option of a command or end a process.

Can be used to bring up additional options.

Picks icons, menus, and graphic entities.

Pan: (MB2+MB3) or [Shift]+MB2

Zoom: (MB1+MB2) or [CTRL]+MB2

Rotate: MB2

Fit   Zoom   1.1129

Pan   Rotate   Scale

Operati   Rotate (F7)

[Esc] - Canceling Commands

The **[Esc]** key is used to cancel a command in *Siemens NX*. The **[Esc]** key is located near the top-left corner of the keyboard. Sometimes, it may be necessary to press the **[Esc]** key twice to cancel a command; it depends on where we are in the command sequence. For some commands, the **[Esc]** key is used to exit the command.

Online Help

❖ Several types of on-line help are available at any time during a *Siemens NX* session. *Siemens NX* provides many on-line help functions, such as:

• The **Help** menu: Click on the **Help** option in the pull-down menu to access the *Siemens NX Help* **menu system**.

• Resource Bar: The resource bar provides quick access to the help menu.

• Help quick key: Press the **[F1]** key to access the *On-On Context Help* **system**.

Leaving Siemens NX

➢ To leave *Siemens NX*, use the left-mouse-button and click on **File** at the top of the *Siemens NX* screen window, then choose **Exit** from the pull-down menu.

Creating a CAD Files Folder

It is a good practice to create a separate folder to store your CAD files. You should not save your CAD files in the same folder where the *Siemens NX* application is located. It is much easier to organize and back up your project files if they are in a separate folder. Making folders within this folder for different types of projects will help you organize your CAD files even further. When creating CAD files in *Siemens NX*, it is strongly recommended that you *save* your CAD files on the hard drive.

➢ To create a new folder in the *Windows* environment:

1. On the *desktop* or under the *My Documents* folder in which you want to create a new folder...

2. *Right-click* once to bring up the option menu, then select **New→ Folder**.

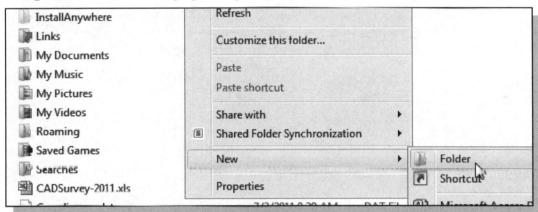

3. Type a name, such as **NX-Projects**, for the new folder, and then press [**ENTER**].

Notes:

Chapter 2
Parametric Modeling Fundamentals

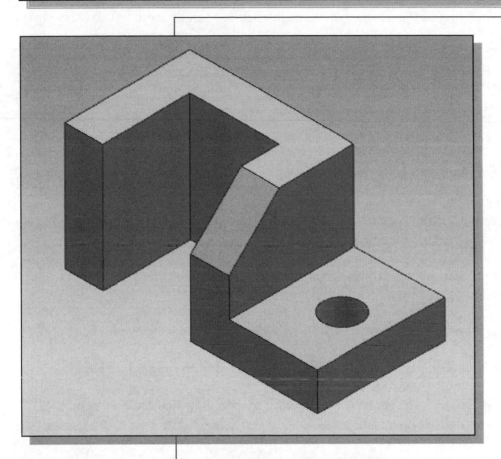

Learning Objectives

- ♦ **Create Simple Extruded Solid Models**
- ♦ **Understand the Basic Parametric Modeling Procedure**
- ♦ **Create 2D Sketches**
- ♦ **Understand the "Shape before Size" Approach**
- ♦ **Use the Dynamic Viewing Commands**
- ♦ **Create and Edit Parametric Dimensions**

Introduction

The **feature-based parametric modeling** technique enables the designer to incorporate the original **design intent** into the construction of the model. The word ***parametric*** means the geometric definitions of the design, such as dimensions, can be varied at any time in the design process. Parametric modeling is accomplished by identifying and creating the key features of the design with the aid of computer software. The design variables, described in the sketches and described as parametric relations, can then be used to quickly modify/update the design.

In *Siemens NX*, the parametric part modeling process involves the following steps:

1. Set up *Units* and *Part name*.

2. **Determine the type of the base feature, the first solid feature, of the design. Note that *Extrude*, *Revolve*, or *Sweep* operations are the most common types of base features.**

3. **Create a rough two-dimensional sketch of the basic shape of the base feature of the design.**

4. **Apply/modify constraints and dimensions to the two-dimensional sketch.**

5. **Transform the parametric two-dimensional sketch into a 3D solid.**

6. **Add additional parametric features by identifying feature relations and complete the design.**

7. **Perform analyses/simulations, such as finite element analysis (FEA) or cutter path generation (CNC), on the computer model and refine the design as needed.**

8. **Document the design by creating the desired 2D/3D drawings.**

The approach of creating two-dimensional sketches of the three-dimensional features is an effective way to construct solid models. Many designs are in fact the same shape in one direction. Computer input and output devices we use today are largely two-dimensional in nature, which makes this modeling technique quite practical. This method also conforms to the design process that helps the designer with conceptual design along with the capability to capture the ***design intent***. Most engineers and designers can relate to the experience of making rough sketches on restaurant napkins to convey conceptual design ideas. *Siemens NX* provides many powerful modeling and design-tools, and there are many different approaches to accomplishing modeling tasks. The basic principle of **feature-based modeling** is to build models by adding simple features one at a time. In this chapter, the general parametric part modeling procedure is illustrated; a very simple solid model with extruded features is used to introduce the *Siemens NX* user interface.

The Adjuster Design

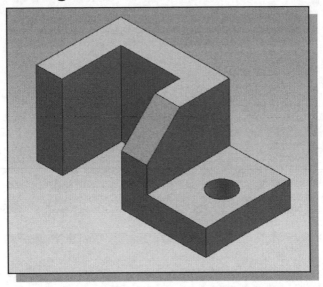

Step 1: Starting Siemens NX and Units Setup

1. Select the **Siemens NX** option on the *Start* menu or select the **Siemens NX** icon on the desktop to start *Siemens NX*. The *Siemens NX* main window will appear on the screen.

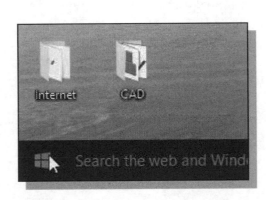

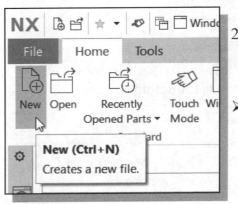

2. Select the **New** icon with a single click of the left-mouse-button (MB1) in the *Standard toolbar area*.

➢ The File New dialog box will appear on the screen with three option tabs: ***Model***, ***Drawing*** and ***Simulation***. Note that the default is set to Model and we can create a new model using any of the pre-defined templates.

3. Select the **Inches** units as shown in the window. When starting a new CAD file, the first thing we should do is to choose the units we would like to use.

4. Select **Blank** in the *Template list*. Note that the *Blank template* contains no pre-defined settings.

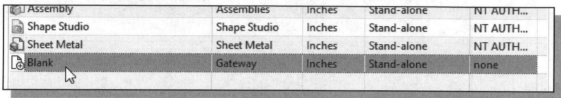

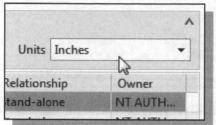

5. In the *New File Name* section, enter **Adjuster.prt** as the *File Name*.

6. Click **OK** to accept the settings and proceed with the new file command.

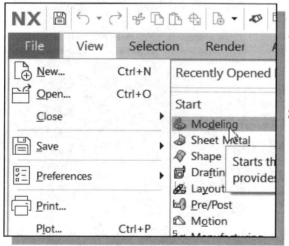

7. Select the **File** tab with a single click of the left-mouse-button (MB1) in the *Ribbon* bar area.

8. Hit [**Ctrl+M**] or pick **Modeling** in the *Applications* list as shown in the figure.

➢ The above procedure shows the switching of the three Siemens NX **work areas**; we have gone from the **"Gateway"** state to the **"Applications,"** and **"Task environments" state**. In the later chapters, the procedure to use the pre-defined template is also illustrated.

➢ Note that the Siemens NX Modeling application allows us to perform both the tasks of **parametric modeling** and **Free-form Synchronous modeling**.

● **Free-Form Synchronous Modeling**
We can also perform more complex-shape modeling tasks, such as the creation of complex surfaces and solid models. For example, creating *swept features* along 3D curves; *lofted features* using conic methods; and *meshes* of points and curves.

Siemens NX Application Screen Layout

➢ The *Siemens NX* modeling screen layout is quite similar to the *Siemens NX Gateway Application* screen layout (refer to the figure in page 1-12 for more details). The new items on the screen include the *Sketch* toolbar, *Form Feature* toolbar, the *Feature Operation* toolbar, and the *Curve* toolbar.

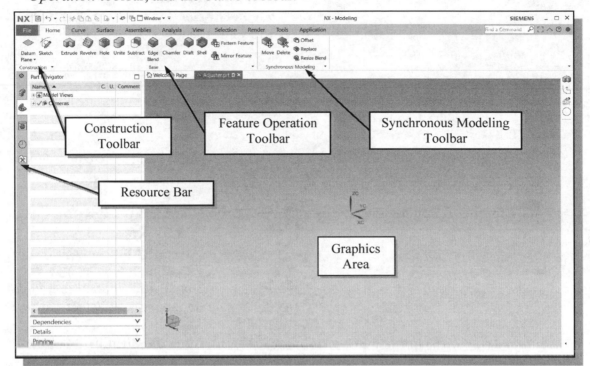

Construction Toolbar

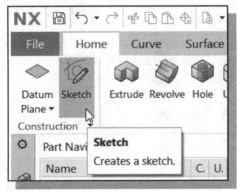

This toolbar contains tools that allow us to quickly access Datum features and switch to Sketch options.

Features Toolbar

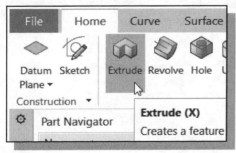

This toolbar contains tools that allow us to quickly create 3D features.

Step 2: Define/Set up the First Solid Feature

- For the *Adjuster* design, we will create an extruded solid as the first feature.

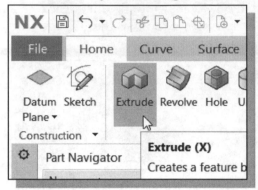

1. In the *Feature Toolbar* (toolbar next to the *Sketch* toolbar in the *Ribbon Bar*), select the **Extrude** icon as shown.

- The *Extrude Feature Options Dialogue box*, which contains applicable construction options, is displayed as shown in the below figure.

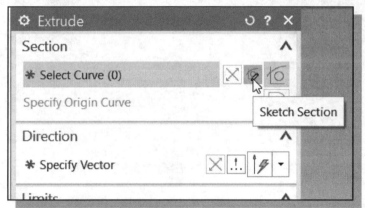

2. On your own, move the cursor over the icons and read the brief descriptions of the different options available. Note that the default extrude option is set to **Select Curve**.

3. Click the **Sketch Section** button to begin creating a new *2D sketch*.

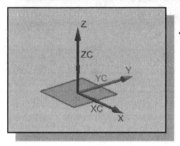

❖ Note the default sketch plane is aligned to **the XC-YC plane** of the displayed *Work Coordinate System*.

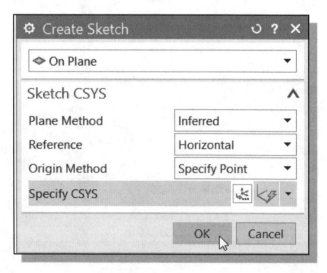

4. Click **OK** to accept the default setting of the *Sketch Plane*.

Work Plane – It is an XY CRT, but an XYZ World

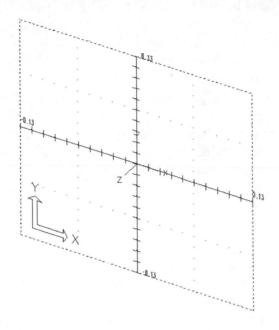

Design modeling software is becoming more powerful and user friendly, yet the system still does only what the user tells it to do. When using a geometric modeler, we therefore need to have a good understanding of what its inherent limitations are. We should also have a good understanding of what we want to do and what to expect, as the results are based on what is available.

In most 3D geometric modelers, 3D objects are located and defined in what is usually called the **absolute space** or **global space**. Although a number of different coordinate systems can be used to create and manipulate objects in a 3D modeling system, the objects are typically defined and stored using the absolute space. The absolute space is usually a **3D Cartesian coordinate system** that the user cannot change or manipulate.

In most engineering designs, models can be very complex, and it would be tedious and confusing if only one coordinate system, the absolute coordinate system, was available. Practical 3D modeling systems allow the user to define **Work Coordinate Systems (WCS)**, also known as **Local Coordinate Systems (LCS)** or **User Coordinate Systems (UCS)** relative to the absolute coordinate system. Once a local coordinate system is defined, we can then create geometry in terms of this more convenient system.

The basic concepts of coordinate systems remain the same for most CAD systems, but the actual usage of a particular type of coordinate system may vary greatly from one CAD system to another. Siemens NX has two primary coordinate systems: the **Absolute Coordinate System (ACS)** and the **Work Coordinate System (WCS)**. The ACS is fixed in space, where the WCS is a mobile system to facilitate geometry construction in different orientations. The WCS can be located and oriented anywhere in model space.

Although 3D objects are generally created and stored in 3D space coordinates, most of the 3D geometry entities can be referenced using 2D Cartesian coordinate systems. Typical input devices such as a mouse or digitizers are two-dimensional by nature; the movement of the input device is interpreted by the system in a planar sense. The same limitation is true of common output devices, such as CRT displays and plotters. The modeling software performs a series of three-dimensional to two-dimensional transformations to correctly project 3D objects onto a 2D picture plane.

The *Siemens NX* **sketch plane** is a special construction tool that enables the planar nature of 2D input devices to be directly mapped into the 3D coordinate systems. The *sketch plane* is part of the Siemens NX work coordinate system that can be aligned to the absolute coordinate system, an existing face of a part, or a reference plane. By default, the *sketch plane* is aligned to the XY plane of the default WCS; note that the default WCS is aligned to the absolute coordinate system (ACS).

Think of a sketch plane as the surface on which we can sketch the 2D sections of the parts. It is similar to a piece of paper, a white board, or a chalkboard that can be attached to any planar surface. The first profile we create is usually drawn on one of the work planes of the work coordinate system (WCS). Subsequent profiles can then be drawn on sketch planes that are defined on **planar faces of a part**, **work planes aligned to part geometry**, or **work planes attached to a coordinate system** (such as the XY, XZ, and YZ planes of WCS).

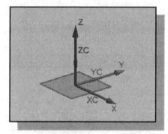

❖ The default sketch plane is aligned to the XY plane of the Work Coordinate System.

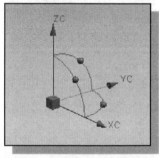

❖ The *work coordinate system* (WCS) is a mobile system to assist geometric constructions in different orientations. The WCS can be located and oriented anywhere in model space.

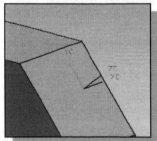

❖ Note that most solid modeling operations do not require manipulation of the WCS. Features are generally added relative to existing geometry of the solid model.

Creating Rough Sketches

Quite often during the early design stage, the shape of a design may not have any precise dimensions. Most conventional CAD systems require the user to input the precise lengths and locations of all geometric entities defining the design, which are not available during the early design stage. With *parametric modeling*, we can use the computer to elaborate and formulate the design idea further during the initial design stage. With *Siemens NX*, we can use the computer as an electronic sketchpad to help us concentrate on the formulation of forms and shapes for the design. This approach is the main advantage of *parametric modeling* over conventional solid-modeling techniques.

As the name implies, a ***rough sketch*** is not precise at all. When sketching, we simply sketch the geometry so that it closely resembles the desired shape. Precise scale or lengths are not needed. *Siemens NX* provides us with many tools to assist us in finalizing sketches. For example, geometric entities such as horizontal and vertical lines are set automatically. However, if the rough sketches are poor, it will require much more work to generate the desired parametric sketches. Here are some general guidelines for creating sketches in *Siemens NX*:

- **Create a sketch that is proportional to the desired shape.** Concentrate on the shapes and forms of the design.

- **Keep the sketches simple.** Leave out small geometry features such as fillets, rounds and chamfers. They can easily be placed using the Fillet and Chamfer commands after the parametric sketches have been established.

- **Exaggerate the geometric features of the desired shape.** For example, if the desired angle is 85 degrees, start by creating an angle that is 50 or 60 degrees (and make adjustment later). Otherwise, *Siemens NX* might assume the intended angle to be a 90-degree angle.

- **Draw the geometry so that it does not overlap.** To create a 3D feature from a 2D sketch, the 2D geometry used should define a clear boundary. *Self-intersecting* geometry shapes are not allowed, as it cannot be converted into a solid feature.

- **The sketched geometric entities should form a closed region.** To create a solid feature, such as an extruded solid, a closed region is required so that the extruded solid forms a 3D volume.

➢ **Note:** The concepts and principles involved in *parametric modeling* are very different, and sometimes they are totally opposite, to those of conventional computer aided drafting. In order to understand and fully utilize *Siemens NX's* functionality, it will be helpful to take a *Zen* approach to learning the topics presented in this text: **Temporarily forget your knowledge and experiences of using conventional Computer Aided Drafting systems.**

Sketch Settings

➢ Prior to creating a 2D sketch, we will first examine the settings in the *Sketch Settings* command. The *Sketch Settings* command provides tools for controlling the behavior of sketching tools during the creation of 2D sketches.

1. In the **Task** *pull-down menu*, aligned to the top of the main window, select the **Sketch Settings** command as shown.

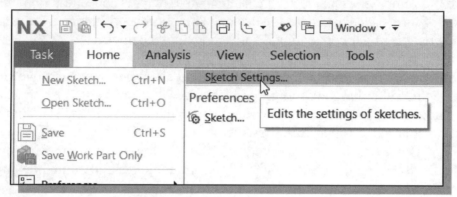

2. Turn off the **Continuous Auto Dimensioning** option by using the left-mouse-button, *MB1*, as shown.

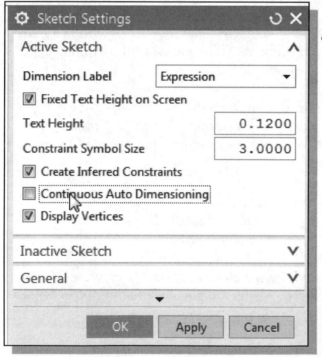

   • Note the different options under the *Sketch Settings* command are used during the creation of 2D sketches.

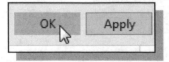

3. Click **OK** to accept the settings and exit the *Sketch Style* command.

Step 3: Creating a Rough 2D Sketch

➢ The *Sketch Curve* toolbar provides tools for creating and editing of the basic 2D geometry, construction tools such as lines, circles and editing tools such as trim, extend.

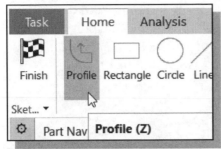

1. Activate the **Profile** tool in the *Sketch Curve toolbar* as shown.

   • The **Profile** tool allows us to create lines and/or arcs that are joined together.

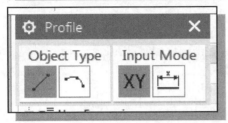

➢ Notice, by default, the **Line** option is activated. The **line** option allows us to create line segments that are joined together.

2. As you move the cursor, you will see a digital readout next to the cursor. The readout gives you the cursor location relative to the Work Coordinate System. Pick a location that is toward the right side of the WCS as the first point of a line.

3. Move the cursor around and you will notice the readout, next to the cursor, providing the length and angle of the line.

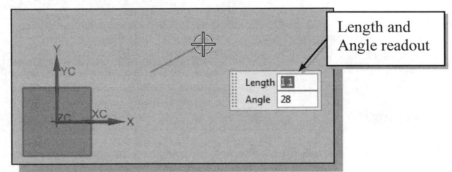

4. Move the graphics cursor directly above the previous point and create a vertical line as shown below (**Point 2**). Notice the geometric constraint symbol displayed.

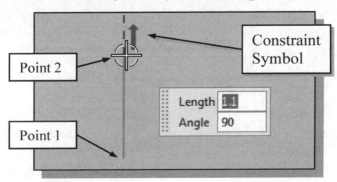

Geometric Constraint Symbols

Siemens NX displays different visual clues, or symbols, to show you alignments, perpendicularities, tangencies, etc. These constraints are used to capture the *design intent* by creating constraints where they are recognized. *Siemens NX* displays the governing geometric rules as models are built. To prevent constraints from forming, hold down the [**Alt**] key while creating an individual sketch curve. For example, while sketching line segments with the Line command, endpoints are joined with a *coincident constraint*, but when the [**Alt**] key is pressed and held, the inferred constraint will not be created.

↑	**Vertical**	indicates a line is vertical
→	**Horizontal**	indicates a line is horizontal
- - -	**Dashed line**	indicates the alignment is to the center point or endpoint of an entity
∖∖	**Parallel**	indicates a line is parallel to other entities
⊥	**Perpendicular**	indicates a line is perpendicular to other entities
	Coincident	indicates the cursor is at the endpoint of an entity
◎	**Concentric**	indicates the cursor is at the center of an entity
⌀	**Tangent**	indicates the cursor is at tangency points to curves
	Midpoint	indicates the cursor is at the midpoint of an entity
	Point on Curve	indicates the cursor is on curves
=	**Equal Length**	indicates the length of a line is equal to another line
⌒	**Equal Radius**	indicates the radius of an arc is equal to another arc

5. Complete the sketch as shown below, creating a closed region ending at the starting point (Point 1.) Do not be overly concerned with the actual size of the sketch. Note that all line segments are sketched horizontally or vertically.

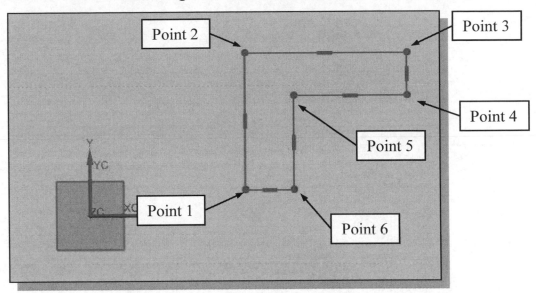

6. Inside the graphics window, click twice with the **middle-mouse-button (MB2)**, or press the [**Esc**] key twice, to end the Sketch Line command.

Step 4: Apply/Modify Constraints and Dimensions

➢ As the sketch is made, *Siemens NX* automatically applies some of the geometric constraints (such as horizontal, parallel, and perpendicular) to the sketched geometry. We can continue to modify the geometry, apply additional constraints, and/or define the size of the existing geometry. In this example, we will illustrate adding dimensions to describe the sketched entities.

1. Move the cursor to the right side of the *Sketch Curve* toolbar, note the different tools available.

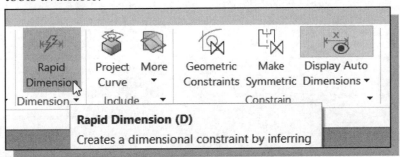

2. Activate the **Rapid Dimension** command by left-click once on the icon as shown. The Rapid Dimension command allows us to quickly create and modify dimensions.

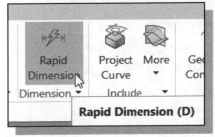

3. The message "*Select object to Dimension or select dimension to edit*" is displayed in the *Message* area. Select the **Top horizontal line** by left-clicking once on the line.

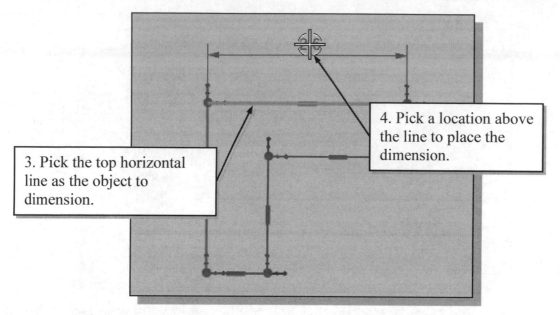

3. Pick the top horizontal line as the object to dimension.

4. Pick a location above the line to place the dimension.

4. Move the graphics cursor above the selected line and left-click to place the dimension. (Note that the value displayed on your screen might be different than what is shown in the figure above.)

❖ The **Rapid Dimension** command will create a length dimension if a single line is selected. The current length of the line is displayed, in the **Edit box**, next to the dimension as shown in the below figure.

5. Enter **2.5** as the desired length of the line. (Press the **Enter** key once after you entered the new value.)

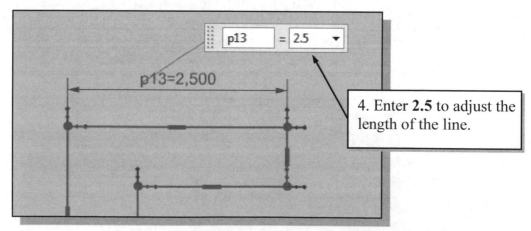

4. Enter **2.5** to adjust the length of the line.

➢ *Siemens NX* will now update the profile with the new dimension value. Note that in parametric modeling, dimensions are used as **control variables**.

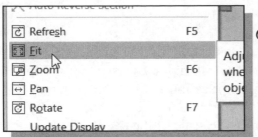

6. Right-mouse-click once to bring up the option menu and select the **Fit** option, or use hotkey [**Ctrl+F**], to resize the display window.

❖ Note that we are returned to the **Rapid Dimension** command, the *Fit* command is executed without interrupting the current active command.

7. The message "*Select object to Dimension or select dimension to edit*" is displayed in the *Message* area. Select the top-horizontal line as shown below.

8. Select the bottom-horizontal line as shown below.

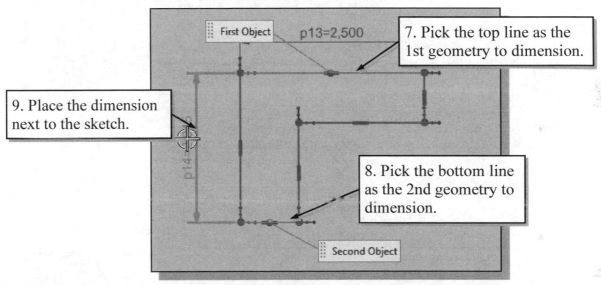

First Object   p13=2,500   7. Pick the top line as the 1st geometry to dimension.

9. Place the dimension next to the sketch.

8. Pick the bottom line as the 2nd geometry to dimension.

Second Object

9. Pick a location to the left of the sketch to place the dimension.

10. Enter **2.5** as the desired length of the line.

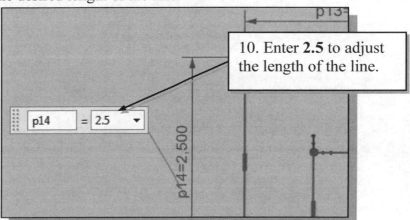

10. Enter **2.5** to adjust the length of the line.

p14 = 2.5

p14=2,500

❖ When two parallel lines are selected, the **Rapid Dimension** command will create a dimension measuring the distance between them.

❖ **Quick Zoom function by Turning the mouse wheel**

Turning the **mouse wheel** can adjust the scale of the display. Turning forward will reduce the scale of the display, using the cursor location as the center reference, making the entities display smaller on the screen. Turning backward will magnify the scale of the display.

11. The message "*Select object to Dimension or select dimension to edit*" is displayed in the *Message* area. Select the bottom-horizontal line as shown below.

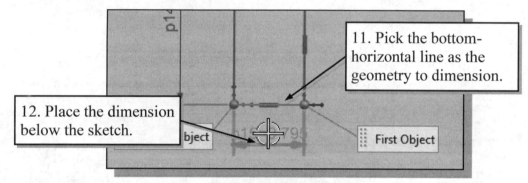

11. Pick the bottom-horizontal line as the geometry to dimension.

12. Place the dimension below the sketch.

12. Pick a location below the sketch to place the dimension.

13. Enter **0.75** as the desired length of the line.

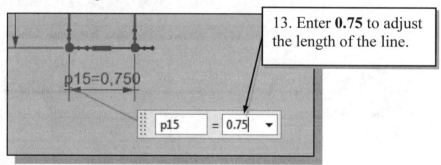

13. Enter **0.75** to adjust the length of the line.

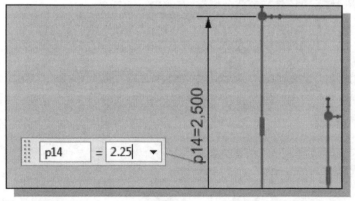

14. Pick the vertical dimension, by double-clicking on the dimension.

15. On your own, enter any number and observe the adjustment done by Siemens NX.

16. Click once on the Close button to end the **Rapid Dimension** command.

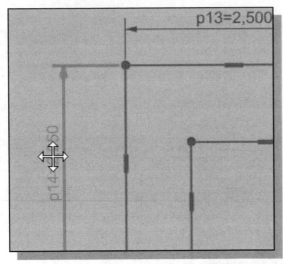

17. To reposition any of the dimensions, simply **press and drag** with the **left-mouse-button (MB1)** on the dimension as shown in the figure.

18. On your own, repeat the above steps, and create and modify dimensions so that the sketch appears as shown in the below figure.

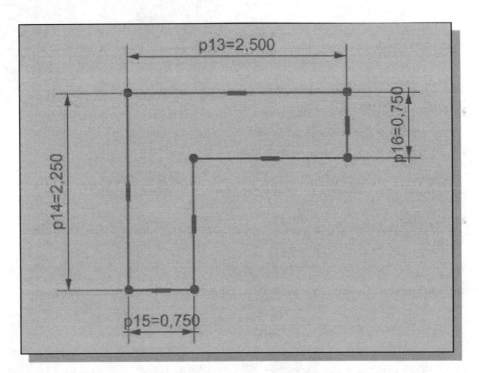

❖ The **Rapid Dimension** command is also known as the **Smart Dimension** command; dimensions are created based on the selection of the types and orientations of objects. You might want to read through this section again and consider the powerful implication of this command. Also consider this question: How would you create an *Angle dimension* using the *Rapid Dimension* command?

View Functions

- *Siemens NX* provides View related user interface called *View Toolbar* that enables convenient viewing of the entities in the graphics window.

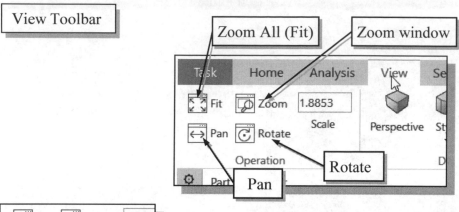

View Toolbar

Zoom All (Fit)   Zoom window

Pan

Rotate

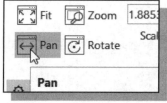

1. Click on the **Pan** icon, located in the *View Toolbar* as shown. The icon is the picture of a hand.

> The Pan command enables us to move the view to a different position. This function acts as if you are using a video camera.

Dynamic Viewing Functions

- *Siemens NX* also provides a special user interface called *Dynamic Viewing* through the use of the three mouse buttons.

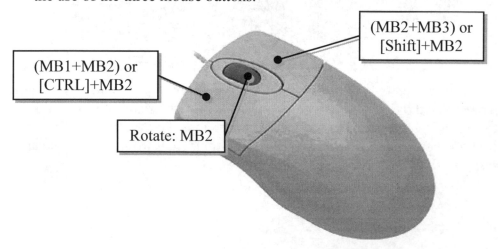

(MB2+MB3) or [Shift]+MB2

(MB1+MB2) or [CTRL]+MB2

Rotate: MB2

1. On your own, use the dynamic Zoom and Pan options by the mouse buttons to reposition the sketch near the center of the screen.

Step 5: Completing the Base Solid Feature

Now that the 2D sketch is completed, we will proceed to the next step: create a 3D part from the 2D section. Extruding a 2D section is one of the common methods that can be used to create 3D parts. We can extrude planar faces along a path. We can also specify a height value and a tapered angle. In *Siemens NX*, each face has a positive side and a negative side, and the current face we're working on is set as the default positive side. This positive side identifies the positive extrusion direction and it is referred to as the face's **normal**.

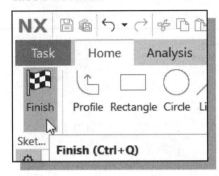

1. In the Home tab, select **Finish Sketch** by clicking once with the left-mouse-button (**MB1**) on the icon.

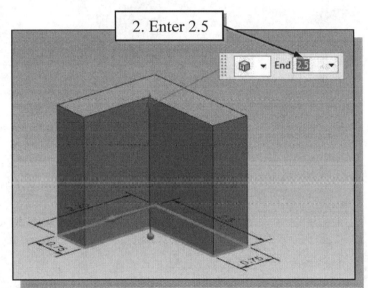

2. In the *Extrude* pop-up window, enter **2.5** as the extrusion distance. Notice that the sketch region is automatically selected as the extrusion section.

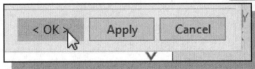

3. Click on the **OK** button to proceed with creating the 3D part.

> Note that all dimensions disappeared from the screen. All parametric definitions are stored in the *Siemens NX* **database** and any of the parametric definitions can be displayed and edited at any time.

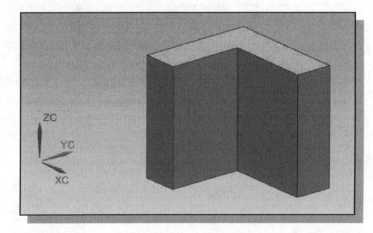

Display Orientations

❖ *Siemens NX* provides many ways to display views of the three-dimensional design. Several options are available that allow us to quickly view the design to track the overall effect of any changes being made to the model. We will first orient the model to display in the *Front view*, by using the *Option menu*.

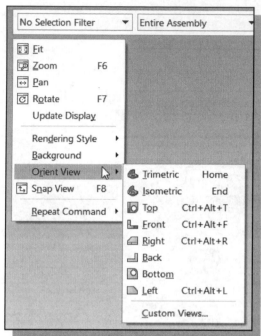

1. Inside the graphic window, right-click once to bring up the *Option menu*. Select the *Orient View List* as shown.

2. Select **Front View** in the *Orient View List* to change the display to the front view. (Note that [**Ctrl+Alt+F**] can also be used to activate this command.)

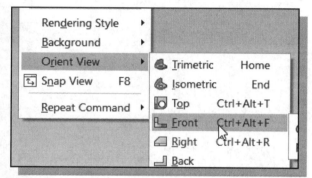

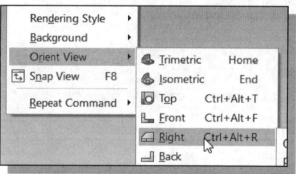

3. Inside the graphic window, right-click once to bring up the *Option menu*. Select **Right View** in the *Orient View List* to change the display to the right side view. (Note that the quick-key [**Ctrl+Alt+R**] can also be used to activate this command.)

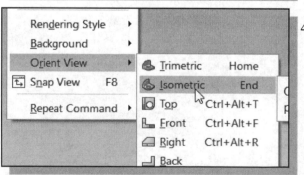

4. Inside the graphic window, right-click once to bring up the *Option menu*. Select **Isometric View** in the *Orient View List* to change the display to the isometric view. (Note that the [**End**] key can also be used to activate this command.)

❖ Notice the other view-related commands that are available under the *Orient View Drop-Down menu*.

Dynamic Viewing – Icons, Mouse buttons and Quick keys

➢ The *3D Rotate* icon can be found in the View toolbar or use the right-click option menu. *3D Rotate* enables us to manipulate the view of 3D objects by clicking and dragging with the left-mouse-button:

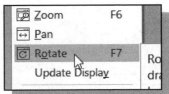

1. Inside the graphic window, right-click once to bring up the *Option menu*. Click on the **Rotate** icon in the *list*.

2. Drag with the left-mouse-button (**MB1**) for free rotation of the 3D model.

3. Press the [**F7**] key (toggle switch) once to exit the 3D rotate command.

➢ We can also use the mouse buttons to access the *Dynamic Viewing* functions.

❖ **3D Dynamic Rotation – The Middle-Mouse-Button (MB2)**

Hold and drag with the middle-mouse-button to rotate the display interactively.

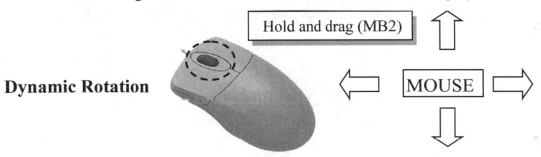

Dynamic Rotation

❖ **Panning – (1) The Middle and Right Mouse Buttons**

Hold and drag with the middle and right mouse buttons (MB2+MB3) to pan the display. This allows you to reposition the display while maintaining the same scale factor of the display.

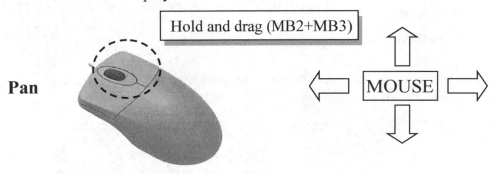

Pan

(2) [Shift] + [Middle-mouse-button (MB2)]
Hold the [**Shift**] key and the [**middle-mouse-button(MB2)**] to pan the display.

❖ **Zooming – (1) The Left and Middle Mouse Buttons**

Hold and drag with the left and middle mouse buttons (MB1+MB2) vertically on the screen to adjust the scale of the display. Moving upward will magnify the scale of the display, making the entities display larger on the screen. Moving downward will reduce the scale of the display.

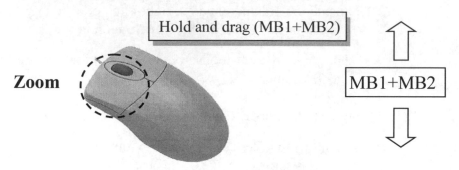

(2) [Shift] + [Middle mouse button (MB2)]

Hold the [**Shift**] key and the [**middle-mouse-button (MB2)**] to pan the display.

(3) Turning the mouse wheel

Turning the mouse wheel can also adjust the scale of the display. Turning forward will reduce the scale of the display, making the entities display smaller on the screen. Turning backward will magnify the scale of the display.

➢ Note that the *dynamic viewing* functions are also available through the **Right-Mouse-Click Option Menu**.

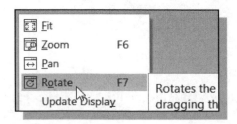

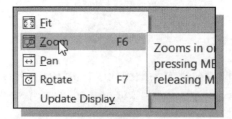

Display Modes

- The display in the graphics window has two basic display-modes: wireframe and shaded display. To change the display mode in the active window, right-click once to bring up the *Option menu* and choose the **Rendering Style** option as shown.

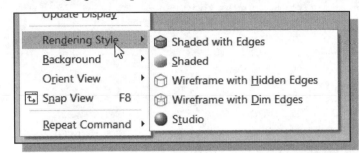

1. Inside the graphic window, right-click once to bring up the *Option menu*. choose the **Rendering Style** option as shown.

2. Pick the **Wireframe with Dim Edges** option as shown in the figure.

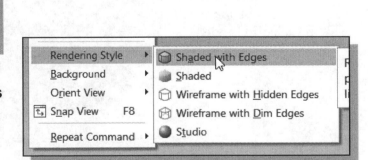

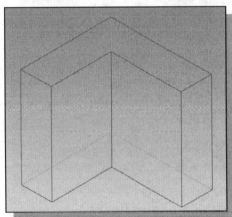

❖ **Wireframe with Dim Edges**

This display mode generates an image of the 3D object with all the back lines shown as lighter edges.

3. Pick the **Shaded with Edges** option as shown.

❖ **Shaded with Edges**

This display mode option generates a shaded image of the 3D object with the edges highlighted.

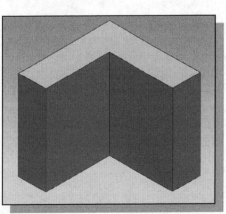

➢ On your own, use the different options to familiarize yourself with the 3D viewing/display commands.

Step 6-1: Adding an Extruded Feature

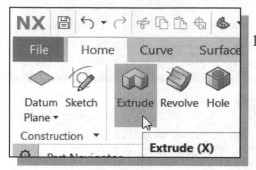

1. In the *Feature Toolbar* (toolbar next to the *Sketch* toolbar in the *Ribbon Bar*), select the **Extrude** icon as shown.

2. In the *Message* area, the message "*Select planar face to sketch or select section geometry*" is displayed. *Siemens NX* expects us to identify a planar surface where the 2D sketch of the next feature is to be created. Move the graphics cursor on the 3D part and notice that *Siemens NX* will automatically highlight feasible planes and surfaces as the cursor is on top of the different surfaces. Pick the back vertical face of the 3D solid object, with the new *work coordinate* aligned to the upper left corner, as shown.

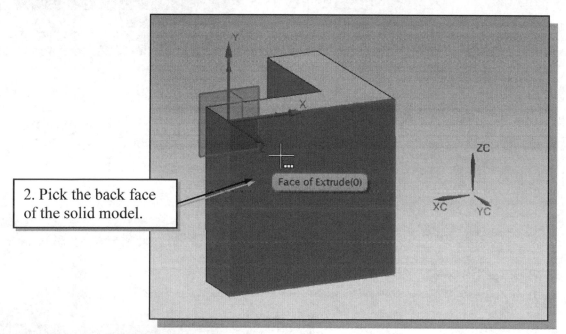

2. Pick the back face of the solid model.

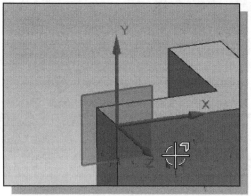

➢ Note that the sketch plane is aligned to the selected face. *Siemens NX* automatically establishes a Work-Coordinate-System (WCS) and records its location with respect to the part on which it was created.

- Next, we will create another 2D sketch, which will be used to create another extrusion feature that will be added to the existing solid object.

3. On your own, create a 2D sketch consisting of horizontal/vertical lines as shown in the below figure. Note that we are intentionally not aligning the new sketch to any of the corners of the existing 3D model.

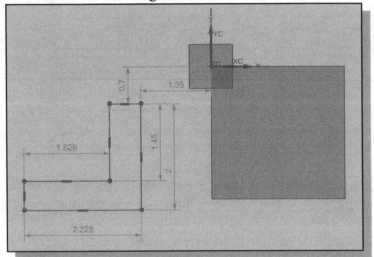

- Note that since we did not modify the **Sketch Settings**, both the *Constraints* and *dimensions* are automatically applied to the sketch.

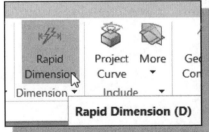

4. Activate the **Rapid Dimension** command by left clicking once on the icon as shown.

5. Create and modify the four size dimensions to describe the size of the sketch as shown. (Note that Siemens NX will maintain the right amount of dimensions to the sketch without over-constraining the sketch.)

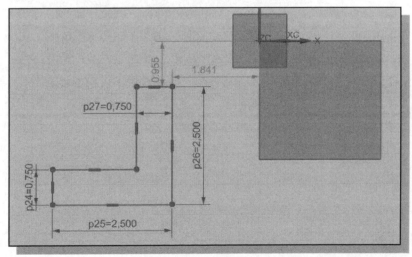

6. Modify the two unmodified location dimensions to **0.0**; this will align the position of the sketch relative to the top corner of the solid model as shown.

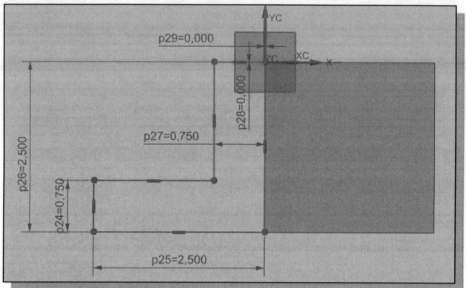

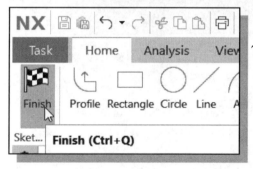

7. Select **Finish Sketch** by clicking once with the left-mouse-button (**MB1**) on the icon.

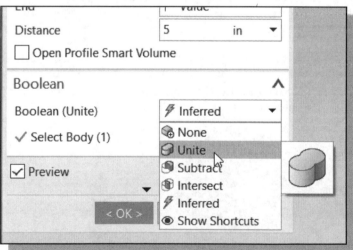

8. In the *Extrude* pop-up window, select **Unite** as the *Extrude Boolean* option.

➢ The **Unite** option will **ADD** the new feature to the existing solid. Note that the default option was set to **Inferred**, which means Siemens NX will determine the option based on the type of sketches created.

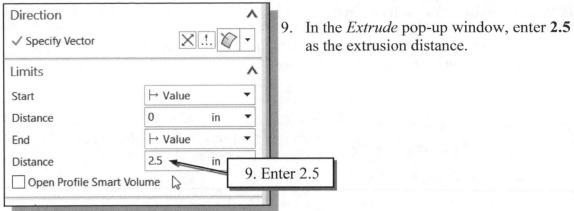

9. In the *Extrude* pop-up window, enter **2.5** as the extrusion distance.

10. Use the **Reverse Direction** icon to toggle the extrusion direction so that the extrusion appears as shown in the figure.

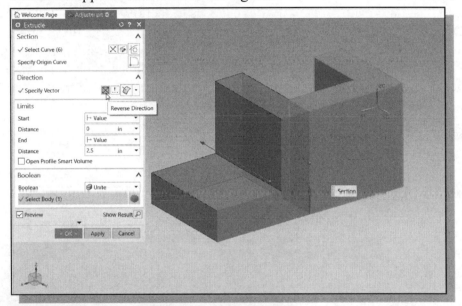

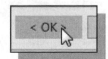

11. Click on the **OK** button to proceed with creating the 3D feature.

Step 6-2: Adding a Cut Feature

- Next, we will create and profile a circle, which will be used to create a **cut** feature that will be added to the existing solid object.

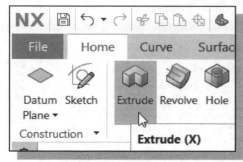

1. In the *Feature Toolbar* (toolbar next to the *Sketch* toolbar in the *Ribbon Bar*), select the **Extrude** icon as shown.

2. In the *Message* area, the message "*Select planar face to sketch or select section geometry*" is displayed. *Siemens NX* expects us to identify a planar surface where the 2D sketch of the next feature is to be created. Move the graphics cursor on the 3D part and notice that *Siemens NX* will automatically highlight feasible planes and surfaces as the cursor is on top of the different surfaces. Pick the horizontal face of the 3D solid object, with the Work Coordinate system aligned to the left corner as shown.

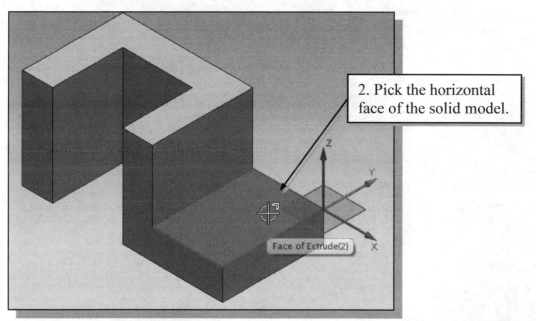

2. Pick the horizontal face of the solid model.

Face of Extrude(2)

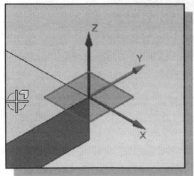

- Note that the sketch plane is aligned to the selected face. *Siemens NX* automatically establishes a *Work-Coordinate-System* (WCS) and records its location with respect to the part on which it was created.

3. Select the **Circle** command by clicking once with the **left-mouse-button (MB1)** on the icon in the *Sketch Curve* toolbar.

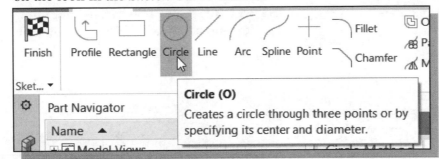

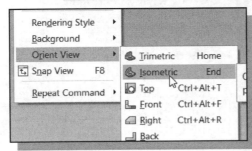

4. Reset the display to the *Isometric view* display as shown.

5. Create a circle of arbitrary size on the horizontal face of the solid model as shown.

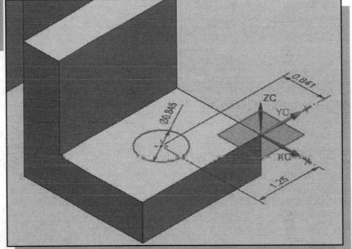

6. On your own, modify the dimensions of the sketch as shown in the figure. (Hint: there are two locational dimensions and one size dimension.)

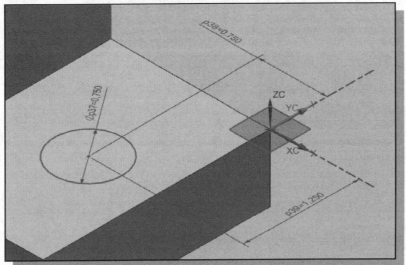

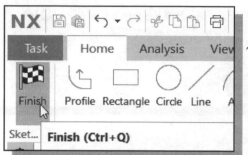

7. Select **Finish Sketch** by clicking once with the left-mouse-button (**MB1**) on the icon.

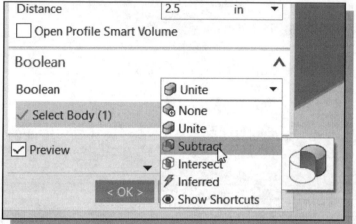

8. In the *Extrude* pop-up window, select **Subtract** as the *Extrude Boolean* option.

➢ The **Subtract** option will **Remove** the volume of the new feature to the existing solid. Note that the default option was set to **Unite**, which will add the feature to the existing solid model.

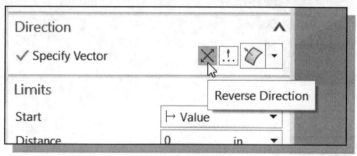

9. In the Direction option box, use the **Reverse Direction** icon to toggle the extrusion direction so that the extrusion cut through the bottom section of the model.

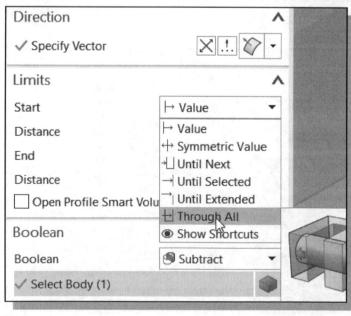

10. In the *Limits* option box, choose the **Through All** option to set the *End* extrusion limit.

11. Click on the **OK** button to proceed with creating the 3D feature.

Step 6-3: Adding another Cut Feature

- Next, we will create and profile a triangle, which will be used to create another **cut** feature that will be added to the existing solid object.

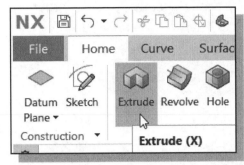

1. In the **Feature Toolbar** (toolbar next to the *Sketch* toolbar in the *Ribbon Bar*), select the **Extrude** icon as shown.

2. In the *Message* area, the message "*Select planar face to sketch or select section geometry*" is displayed. *Siemens NX* expects us to identify a planar surface where the 2D sketch of the next feature is to be created. Move the graphics cursor on the 3D part and notice that *Siemens NX* will automatically highlight feasible planes and surfaces as the cursor is on top of the different surfaces. Pick the horizontal face of the 3D solid object, with the Work Coordinate system aligned to the left corner as shown.

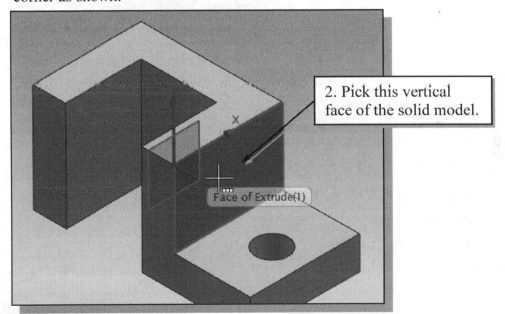

2. Pick this vertical face of the solid model.

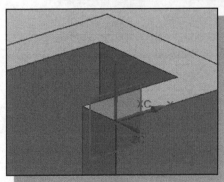

- Note that the sketch plane is aligned to one of the corners of the selected face. *Siemens NX* automatically establishes a *Work-Coordinate-System* (WCS) and records its location with respect to the part on which it was created.

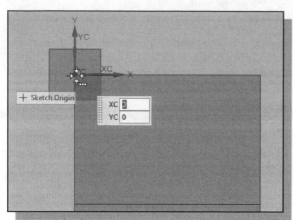

3. Use the **Profile** command and start the first line segment at the top left corner as shown.

4. Place the second point aligned to and **below** the mid-point of the left edge as shown.

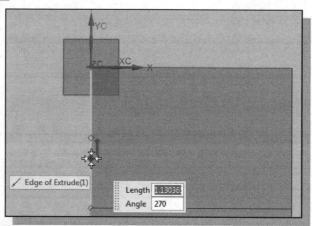

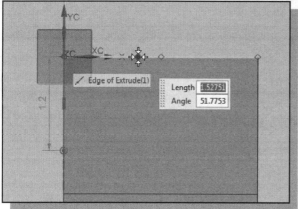

5. Place the second point aligned to and below the mid-point of the left edge as shown.

6. Complete the profile by connecting to the first point as shown.

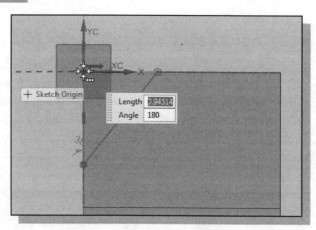

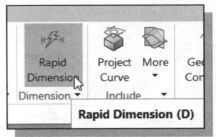

7. Activate the **Rapid Dimension** command by clicking on the icon as shown.

8. Create and modify the linear dimension on top as shown.

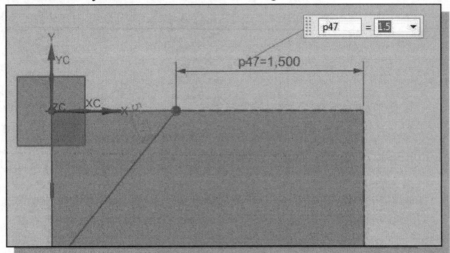

9. Select the two line segments and place the angular dimension inside the triangle to create the associated angle dimension.

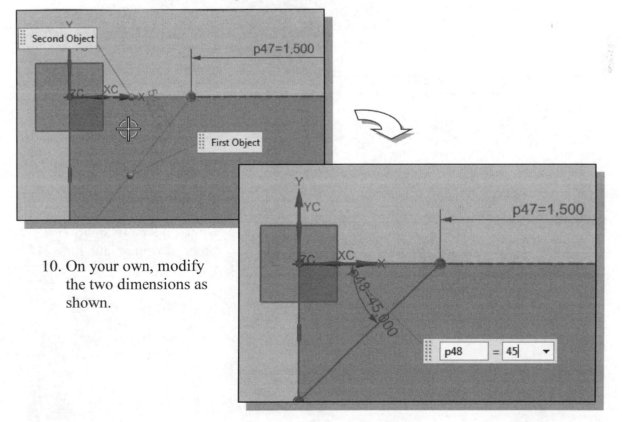

10. On your own, modify the two dimensions as shown.

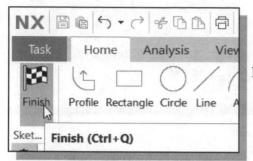

11. Select **Finish Sketch** by clicking once with the left-mouse-button (**MB1**) on the icon.

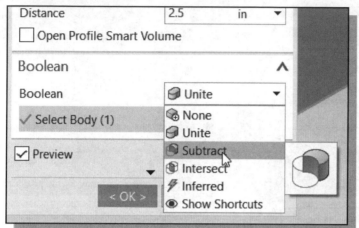

12. In the *Extrude* pop-up window, select **Subtract** as the *Extrude Boolean* option.

➢ The **Subtract** option will **Remove** the volume of the new feature to the existing solid. Note that the default option was set to **unite**, which will add the new feature to the existing solid model.

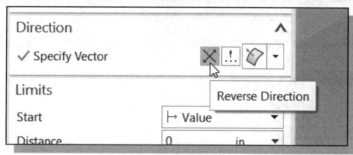

13. In the Direction option box, use the **Reverse Direction** icon to toggle the extrusion direction so that the extrusion cut through the bottom section of the model.

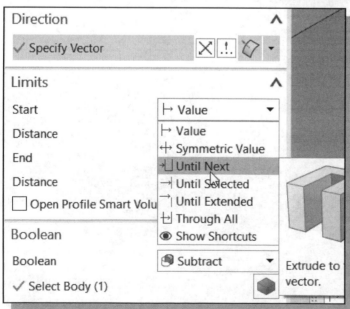

14. In the *Limits* option box, choose the **Until Next** option to set the *End* extrusion limit.

15. Click on the **OK** button to proceed with creating the 3D feature.

- On your own, use the dynamic viewing functions and examine the completed solid model.

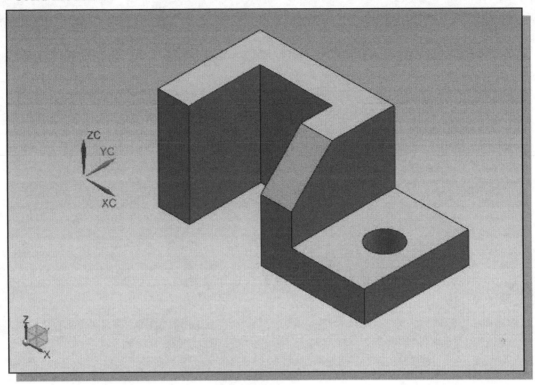

Save the Model and Exit Siemens NX

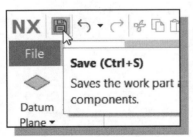

1. Select **Save** in the *quick access* menu, or you can also use the "**Ctrl+S**" key combination (hold down the "Ctrl" key and hit the "S" key once) to save the part.

❖ You should form a habit of saving your work periodically, just in case something might go wrong while you are working on it. In general, one should save one's work at an interval of every 15 to 20 minutes. One should also save before making any major modifications to the model.

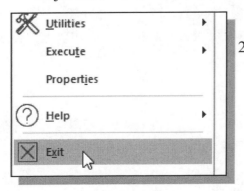

2. Select **Exit** in the *File* pull-down menu to exit Siemens NX.

Review Questions:

1. What is the first thing we should set up in *Siemens NX* when creating a new model?

2. Describe the general *parametric modeling* procedure.

3. List two of the geometric constraint symbols used by *Siemens NX*.

4. What was the first feature we created in this lesson?

5. Describe the steps required to define the orientation of the sketching plane.

6. How do we change the size of 2D geometry in the 2D sketcher mode?

Exercises: (All dimensions are in inches.)

1. Plate Thickness: **.25**

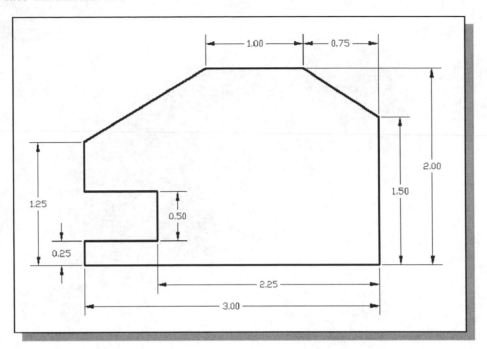

2. Plate Thickness: **.5**

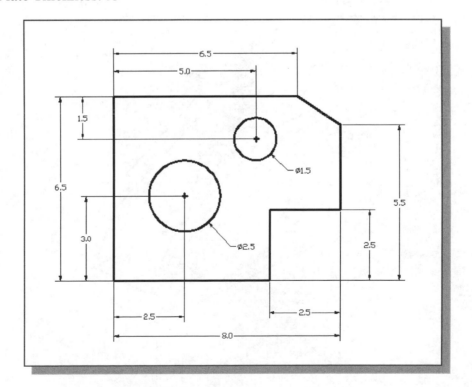

3.

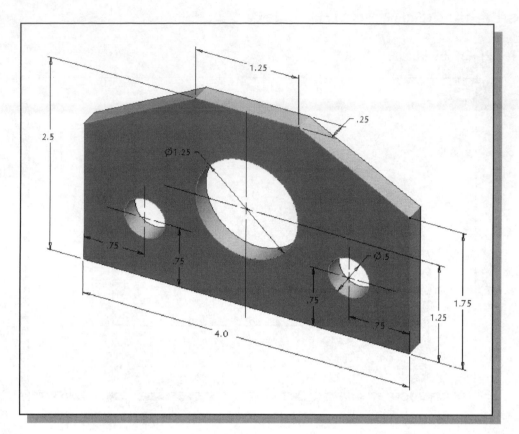

4.

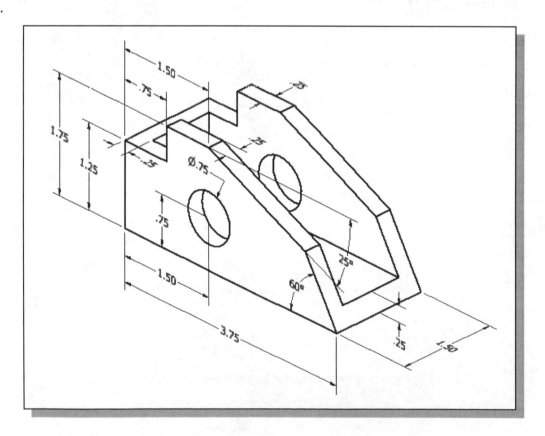

5.

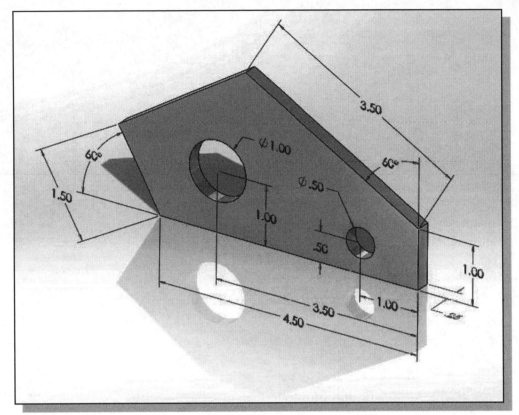

6.

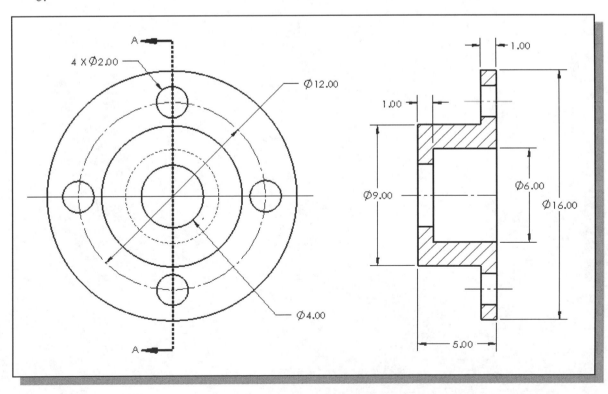

Notes:

Chapter 3
Constructive Solid Geometry Concepts

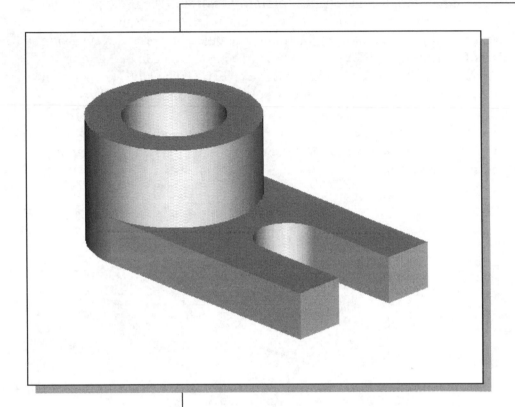

Learning Objectives

- ♦ **Understand Constructive Solid Geometry Concepts**
- ♦ **Create a Binary Tree**
- ♦ **Understand the Basic Boolean Operations**
- ♦ **Set up GRID and SNAP Intervals**
- ♦ **Understand the Importance of Order of Features**
- ♦ **Create Placed Features**
- ♦ **Use the Different Extrusion Options**

Introduction

In the 1980s, one of the main advancements in **solid modeling** was the development of the **Constructive Solid Geometry** (CSG) method. CSG describes the solid model as combinations of basic three-dimensional shapes (**primitive solids**). The basic primitive solid set typically includes Rectangular-prism (Block), Cylinder, Cone, Sphere, and Torus (Tube). Two solid objects can be combined into one object in various ways using operations known as **Boolean operations**. There are three basic Boolean operations: **UNITE (Union)**, **SUBTRACT (Difference)**, and **INTERSECT**. The *UNITE* operation combines the two volumes included in the different solids into a single solid. The *SUBTRACT* operation subtracts the volume of one solid object from the other solid object. The *INTERSECT* operation keeps only the volume common to both solid objects. The CSG method is also known as the **Machinist's Approach**, as the method is parallel to machine shop practices.

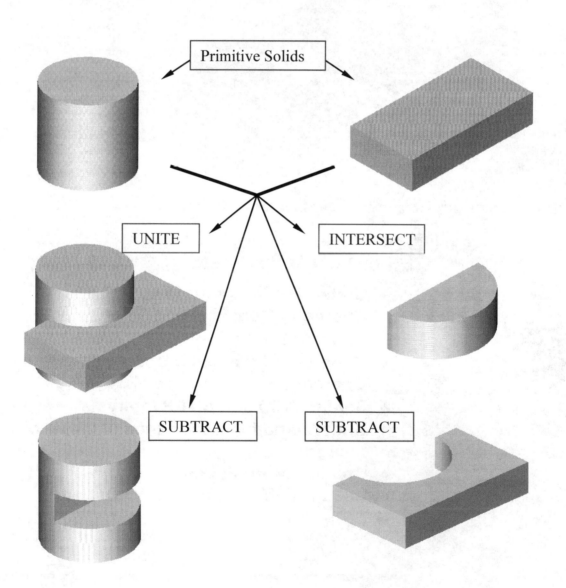

Binary Tree

The CSG is also referred to as the method used to store a solid model in the database. The resulting solid can be easily represented by what is called a **binary tree**. In a binary tree, the terminal branches (leaves) are the various primitives that are linked together to make the final solid object (the root). The binary tree is an effective way to keep track of the *history* of the resulting solid. By keeping track of the history, the solid model can be rebuilt by relinking through the binary tree. This provides a convenient way to modify the model. We can make modifications at the appropriate links in the binary tree and relink the rest of the history tree without building a new model.

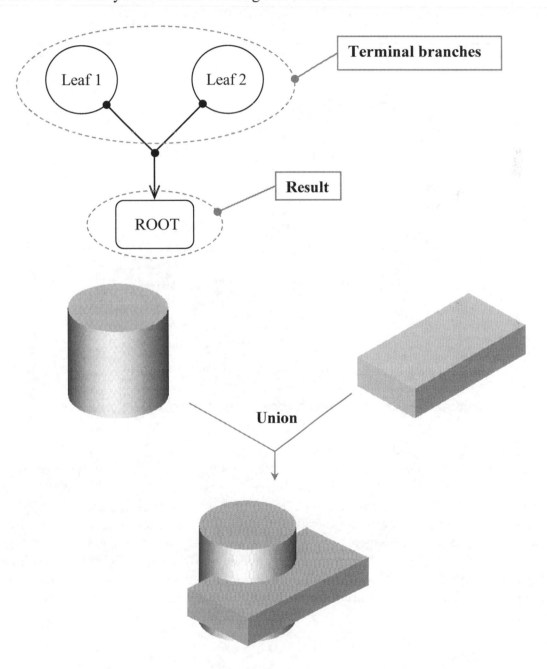

The *Locator* Design

The CSG concept is one of the important building blocks for feature-based modeling. In *Siemens NX*, the CSG concept can be used as a planning tool to determine the number of features that are needed to construct the model. It is also a good practice to create features that are parallel to the manufacturing process required for the design. With parametric modeling, we are no longer limited to using only the predefined basic solid shapes. In fact, any solid features we create in *Siemens NX* are used as primitive solids; parametric modeling allows us to maintain full control of the design variables that are used to describe the features. In this lesson, a more in-depth look at the parametric modeling procedure is presented. The equivalent CSG operation for each feature is also illustrated.

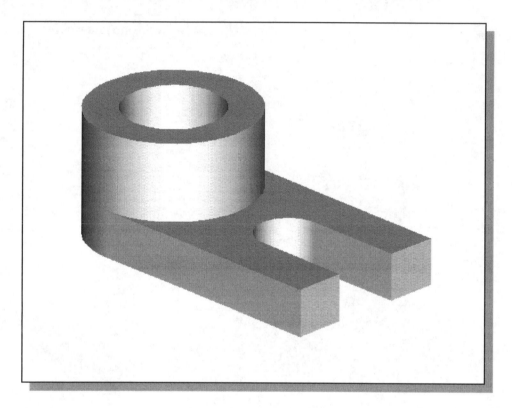

➢ Before going through the tutorial, on your own, make a sketch of a CSG binary tree of the **Locator** design using only two basic types of primitive solids: cylinder and rectangular prism. In your sketch, how many *Boolean operations* will be required to create the model? What is your choice of the first primitive solid to use, and why? Take a few minutes to consider these questions and do the preliminary planning by sketching on a piece of paper. Compare the sketch you make to the CSG binary tree steps shown on page 3-5. Note that there are many different possibilities in combining the basic primitive solids to form the solid model. Even for the simplest design, it is possible to take several different approaches to creating the same solid model.

Modeling Strategy - CSG Binary Tree

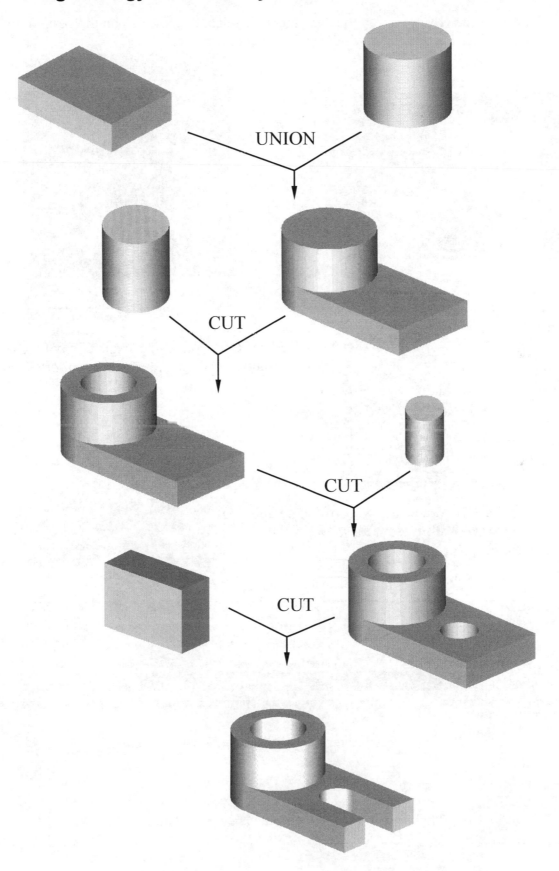

Starting Siemens NX

1. Select the **NX** option on the *Start* menu or select the **NX** icon on the desktop to start *NX*. The *NX* main window will appear on the screen.

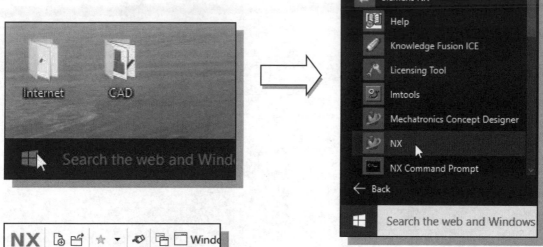

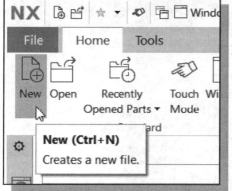

2. Select the **New** icon with a single click of the left-mouse-button (MB1) in the *Standard toolbar area*.

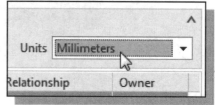

3. In the *File New* window, confirm the units are set to **Millimeters** as shown in the figure.

4. Select the **Blank** template file as shown.

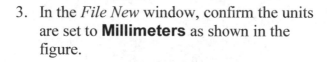

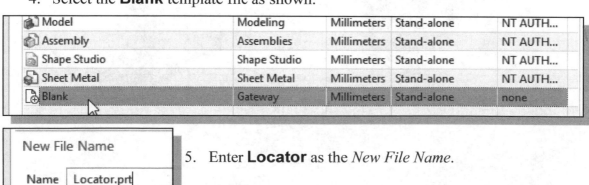

5. Enter **Locator** as the *New File Name*.

6. Click **OK** to proceed with the *New File* command.

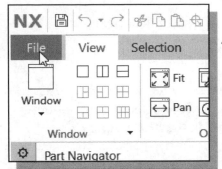

7. Select the **File** tab with a single click of the left-mouse-button (MB1) in the *Ribbon Bar area*.

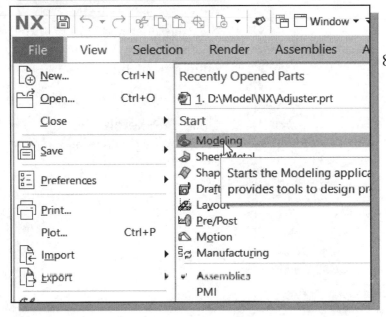

8. Pick **Modeling** in the *Applications* list as shown in the figure.

Base Feature

In *parametric modeling*, the first solid feature is called the **base feature,** which usually is the primary shape of the model. Depending upon the design intent, additional features are added to the base feature.

Some of the considerations involved in selecting the base feature are:

*   **Design intent** – Determine the functionality of the design; identify the feature that is central to the design.

*   **Order of features** – Choose the feature that is the logical base in terms of the order of features in the design.

*   **Ease of making modifications** – Select a base feature that is more stable and is less likely to be changed.

➢ For the ***Locator*** design, a rectangular block will be created first as the base feature.

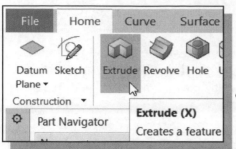

1. In the *Feature Toolbar* (toolbar next to the *Sketch* toolbar in the *Ribbon Bar*), select the **Extrude** icon as shown.

- The *Extrude Feature Options Dialogue box*, which contains applicable construction options, is displayed as shown in the below figure.

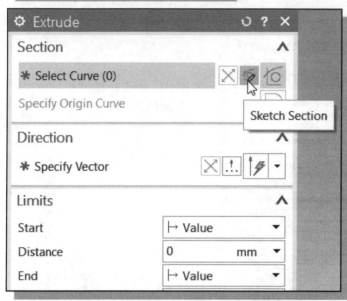

2. On your own, move the cursor over the icons and read the brief descriptions of the different options available.

3. Click the **Sketch Section** button to begin creating a new *2D sketch*.

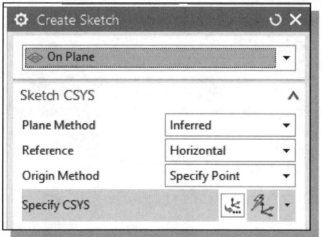

➢ The default sketch plane option is set to sketch on the **Inferred Plane,** which is aligned to the XY plane of the displayed Work Coordinate System.

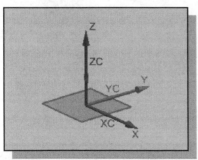

4. Click **OK** to accept the default setting of the *Sketch Plane*.

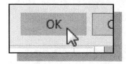

Sketch Settings

➢ Prior to creating a 2D sketch, we will first examine the settings in the *Sketch Settings* command. The *Sketch Settings* command provides tools for controlling the behavior of sketching tools during the creation of 2D sketches.

1. Click on the **Task** *pull-down menu* of the main window, select the **Sketch Settings** command as shown.

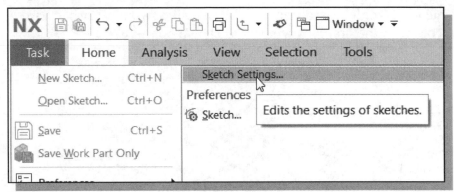

2. Turn off the **Continuous Auto Dimensioning** option by using the left-mouse-button, *MB1,* as shown.

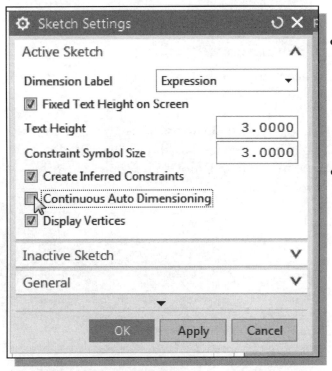

- In this chapter, we will concentrate on the basics of solid modeling. The concepts and usage of dimensions and constraints will be further examined in chapter 4 and 5.

- Note the different options under the *Sketch Settings* command are used during the creation of 2D sketches.

3. Click **OK** to accept the settings and exit the *Sketch Style* command.

Use the Rectangle Command

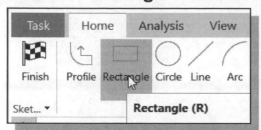

1. Select the **Rectangle** command by clicking once with the **left-mouse-button** on the icon in the *Sketch Curve* toolbar.

2. Create a rectangle of arbitrary size by selecting two locations toward the lower left corner of the graphics window as shown below.

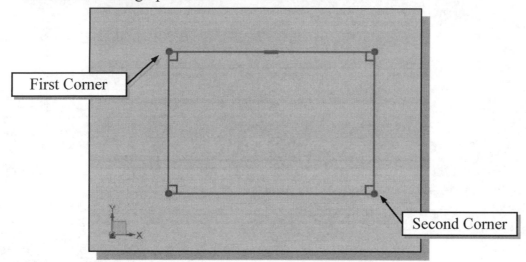

3. Inside the graphics window, click once with the **middle-mouse-button** (MB2) to end the Rectangle command.

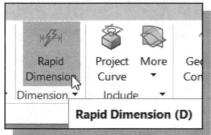

4. Activate the **Rapid Dimension** command by left-clicking once on the icon as shown. The Rapid Dimension command allows us to quickly create and modify dimensions

5. The message "*Select object to Dimension or select dimension to edit*" is displayed in the *Message* area. Select the **Top horizontal line** by left-clicking once on the line.

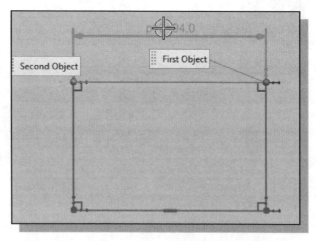

6. Move the graphics cursor above the selected line and left-click to place the dimension. (Note that the value displayed on your screen might be different than what is shown.)

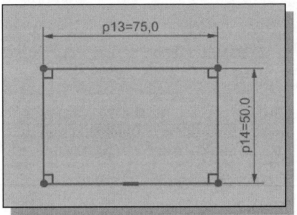

7. On your own, create and modify the dimensions of the sketched rectangle to **75mm** by **50mm** as shown.

8. Click close to end the **Rapid Dimension** command.

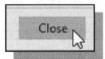

Completing the Base Solid Feature

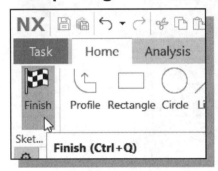

1. Select **Finish Sketch** by clicking once with the left-mouse-button (**MB1**) on the icon.

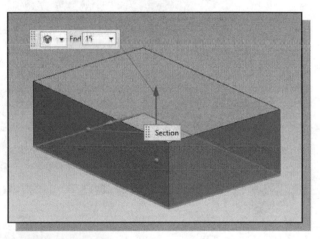

2. In the *Extrude* pop-up window, enter **15** as the extrusion distance. Notice that the sketch region is automatically selected as the extrusion section.

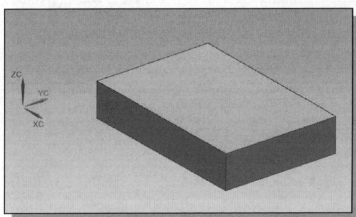

3. Click on the **OK** button to proceed with creating the 3D part. Use the *Dynamic Viewing* options to view the created part. Change the display to *isometric view* as shown before going to the next section.

Creating the Next Solid Feature

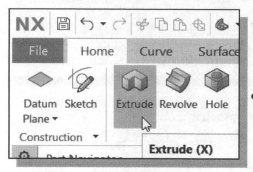

1. In the *Feature Toolbar* (toolbar next to the *Sketch* toolbar in the *Ribbon Bar*), select the **Extrude** icon as shown.

   • The *Extrude Feature Options Dialogue box*, which contains applicable construction options, is displayed as shown in the below figure.

2. Use the **dynamic rotate** function to rotate and pick the **bottom face** as the sketch plane, and align the work coordinate system as shown below.

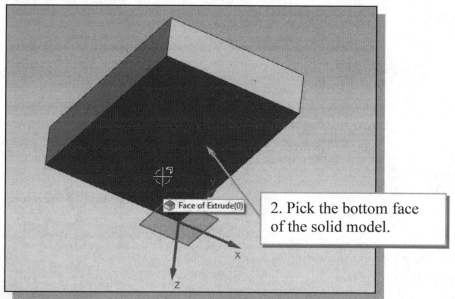

2. Pick the bottom face of the solid model.

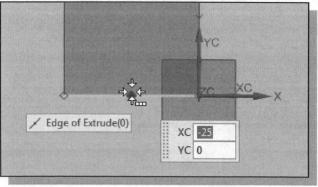

3. Select the **Circle** command by clicking once with the **left-mouse-button (MB1)** on the icon in the *Sketch Curve* toolbar.

➢ We will align the center of the circle to the midpoint of the base feature.

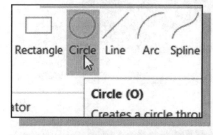

4. Move the cursor along the shorter edge of the base feature and pick the midpoint of the edge when the **midpoint** of the line is displayed with the purple constraint symbol.

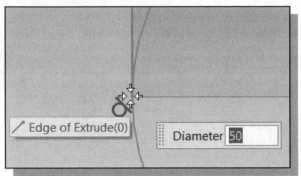

5. Move the cursor on top of the left corner and click once with the right-mouse button to display the option list and select the **select from list** option.

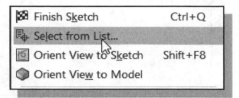

6. Move the cursor on top of the items in the list to identify the associated entities. Select the endpoint associated with the left edge and create a circle as shown.

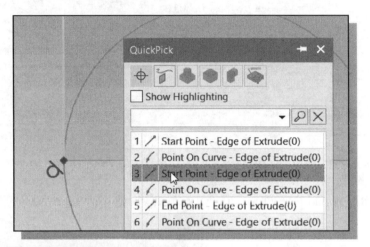

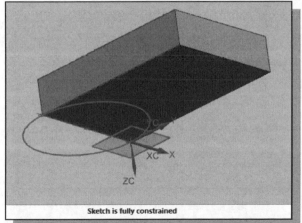

7. End the **Circle** command by clicking once with the middle-mouse-button (**MB2**) inside the graphics window.

- Note that no dimension is needed for the circle; the circle is **fully defined** with the alignment to the midpoint and corner of existing geometry.

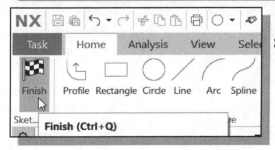

8. Select **Finish Sketch** by clicking once with the left-mouse-button (**MB1**) on the icon.

9. In the *Extrude* dialog box, enter **40** as the extrusion end limit and set the *Boolean operation* option to **Unite.** Click on the **Flip** button to reverse the direction of extrusion (set upward direction) as shown below.

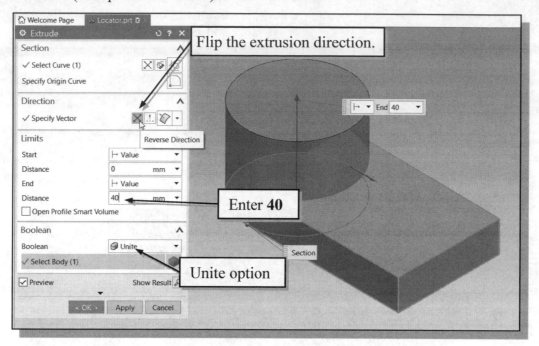

10. Click on the **OK** button to proceed with the *Unite* operation.

• The two features are joined together into one solid part; the *CSG-Union* operation was performed.

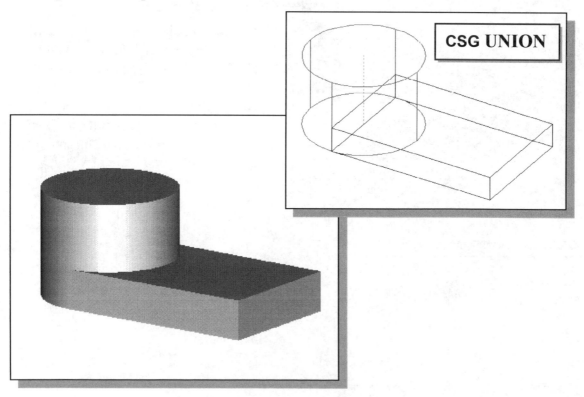

Creating a Cut Feature

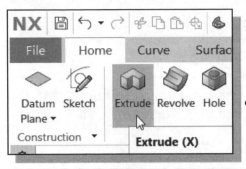

1. In the *Feature Toolbar* (toolbar next to the *Sketch* toolbar in the *Ribbon Bar*), select the **Extrude** icon as shown.

*   We will create a circular cut as the next solid feature of the design. We will align the sketch plane to the top of the last cylinder feature.

2. Pick the top face of the cylinder, with the work coordinate system aligned at the center as shown.

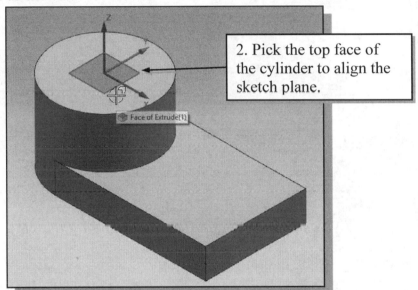

2. Pick the top face of the cylinder to align the sketch plane.

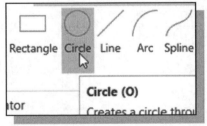

3. Select the **Circle** command by clicking once with the **left-mouse-button (MB1)** on the icon in the *Sketch Curve* toolbar.

4. Select the **Center** point of the top face of the 3D model by clicking on the circle center point with the left-mouse-button as shown. (Notice the displayed *concentric* constraint symbol.)

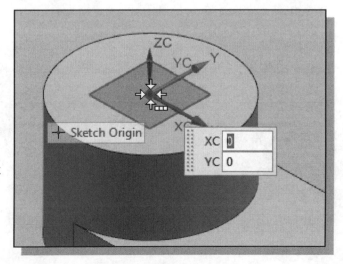

5. Sketch a circle of arbitrary size inside the top face of the cylinder as shown below.

6. Use the middle-mouse-button (**MB2**) to end the Circle command.

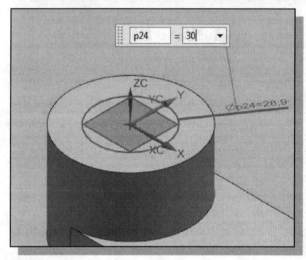

7. On your own, create and modify the size of the circle so that it is set to **30mm**.

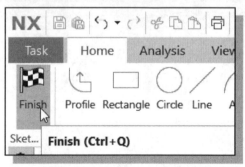

8. Select **Finish Sketch** by clicking once with the left-mouse-button (**MB1**) on the icon.

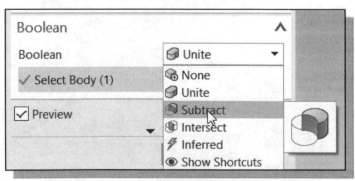

9. In the *Extrude* dialog box, set the *Boolean operation* option to **Subtract.**

10. Select **Through All** as the *End extrude limit* and click on the **Flip** button to reverse the direction of extrusion (set to downward direction) as shown below.

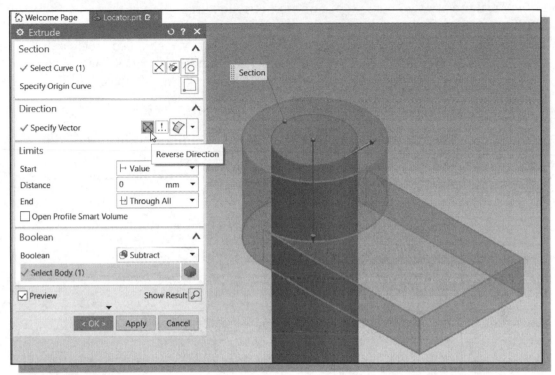

11. Click on the **OK** button to proceed with the *Subtract* operation

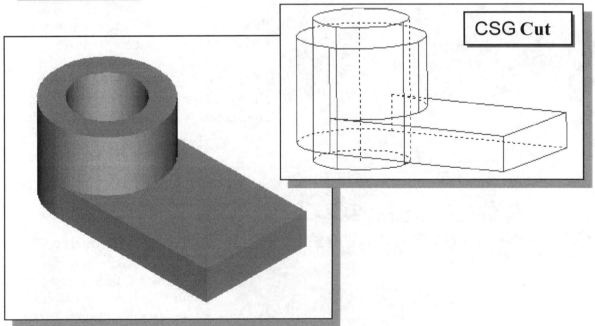

- The circular volume is removed from the solid model; the *CSG-Subtract* operation resulted in a single solid.

Creating a Placed Feature

- In *NX*, there are two basic types of geometric features: **placed features** and **sketched features**. The last cut feature we created is a *sketched feature*, where we created a rough sketch and performed an extrusion operation. We can also create a hole feature, which is a placed feature. A *placed feature* is a feature that does not need a sketch and can be created automatically. Holes, fillets, chamfers, and shells are all placed features.

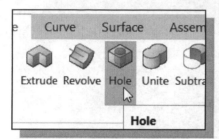

1. In the *Feature Operation* toolbar (the toolbar that is located to the right side of the *Form Feature Toolbar*), select the **Hole** command by releasing the left-mouse-button on the icon.

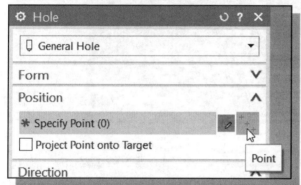

2. In the *Holes* window, notice the default hole type is set to **General Hole** as shown.

❖ Note that by default, the position of the hole is specified using the **Point** command.

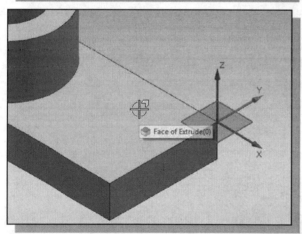

3. Pick a location inside the horizontal surface of the base feature as shown.

❖ Note the Point command is activated automatically once the sketch plane has been selected.

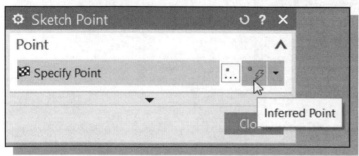

- By default, the **Inferred Point** option is activated, which means the clicked location will be used as the center point location.

4. Click **Close** to end the *sketch point* command.

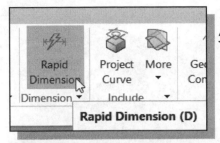

5. Activate the **Rapid Dimension** command by left-clicking once on the icon as shown. The Rapid Dimension command allows us to quickly create and modify dimensions.

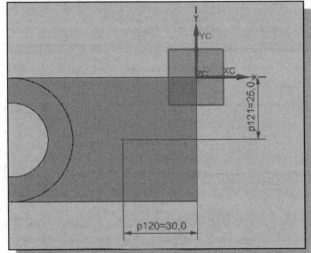

6. On your own, create the two dimensions, **25mm** and **30mm**, as shown.

7. Click **Close** to end the *Rapid dimension* command.

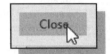

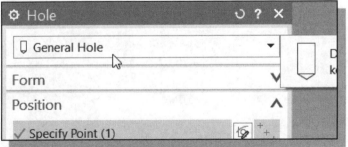

8. Select **Finish Sketch** by clicking once with the left-mouse-button (**MB1**) on the icon.

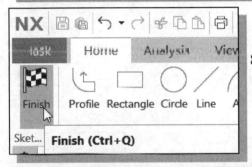

9. Confirm the *Hole type* is set to **General Hole** as shown.

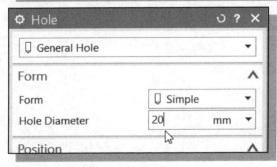

10. In the *Form and dimensions* options, enter **20 mm** as the diameter of the hole as shown in the figure.

11. Set the **Depth Limit** to **Through Body** as shown in the figure.

- Note that the *hole feature* is previewed in the graphics window as shown.

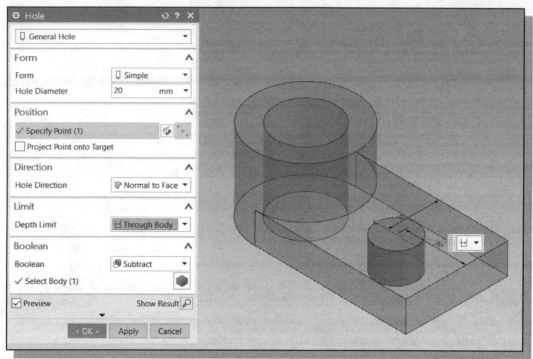

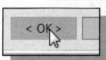 12. Click **OK** to proceed with the creation of the ***Hole*** feature.

- The circular volume is removed from the solid model; the *CSG-Subtract* operation resulted in a single solid.

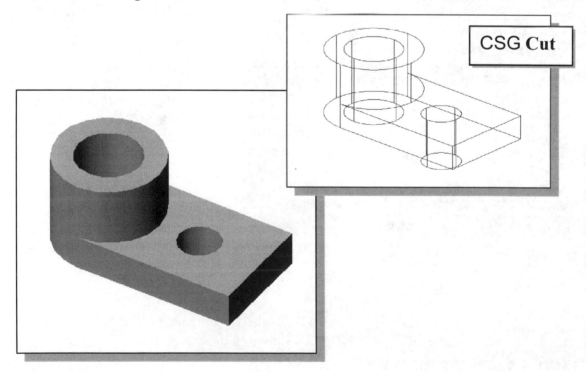

CSG **Cut**

Creating a Rectangular Cut Feature

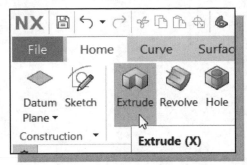

1. In the **Feature Toolbar** (toolbar next to the *Sketch* toolbar in the *Ribbon Bar*), select the **Extrude** icon as shown.

- We will next create a rectangular cut as the last solid feature of the **Locator**.

2. Pick the right vertical face of the base feature, with the work coordinate system aligned as shown.

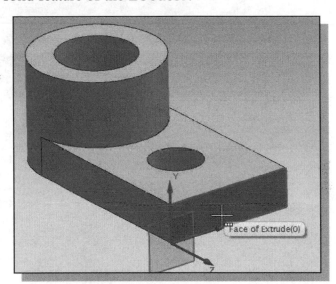

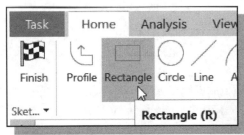

3. Select the **Rectangle** command by clicking once with the **left-mouse-button** on the icon in the *Sketch Curve* toolbar.

4. Create a rectangle that is aligned to the top and bottom edges of the base feature as shown.

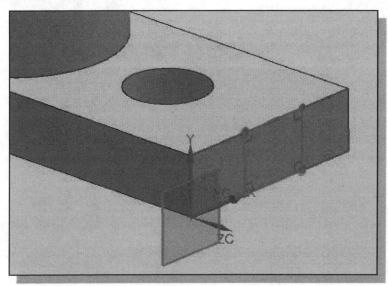

5. On your own, create and modify the two dimensions, **15 mm** and **20mm**, as shown in the figure.

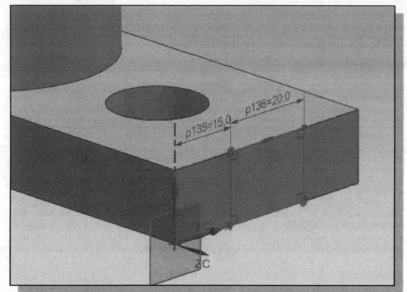

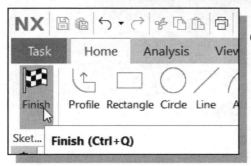

6. Select **Finish Sketch** by clicking once with the left-mouse-button (**MB1**) on the icon.

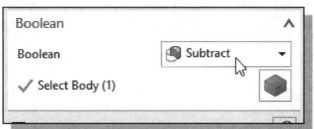

7. In the *Extrude* dialog box, set the *Boolean operation* option to **Subtract**.

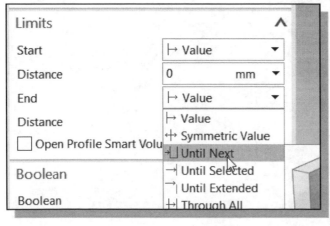

8. In the *Extrude* dialog box, set the *End option* to **Until Next** as shown.

9. In the *Extrude* pop-up window, set the operation option to **Subtract.** Set the extrusion direction toward the center of the solid model.

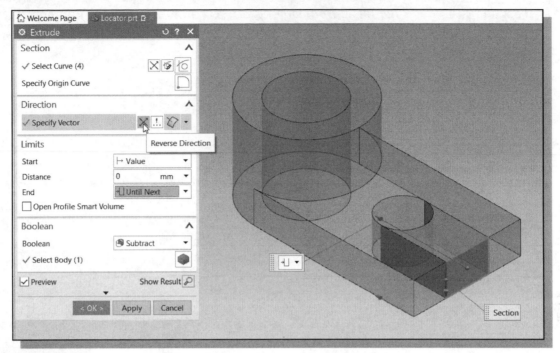

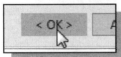

10. Click on the **OK** button to create the *cut feature* and complete the design.

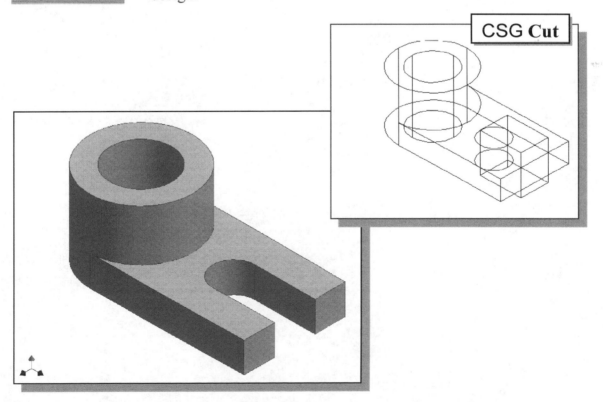

CSG Cut

Review Questions:

1. List and describe three basic *Boolean operations* commonly used in computer geometric modeling software.

2. What is a *primitive solid*?

3. What does *CSG* stand for?

4. Which *Boolean operation* keeps only the volume common to the two solid objects?

5. What is the main difference between a *CUT feature* and a *HOLE feature* in *NX*?

6. Using the CSG concepts, create *Binary Tree* sketches showing the steps you plan to use to create the two models shown on the next page:

Ex.1)

Ex.2)

Exercises:

1. **Latch Clip** (Dimensions are in inches. Thickness: **0.25** inches.)

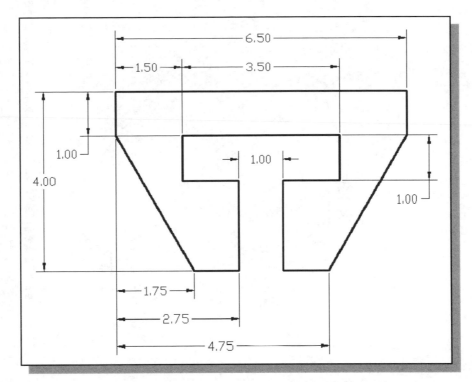

2. **Guide Plate** (Dimensions are in inches. Thickness: **0.25** inches. Boss height **0.125** inches.)

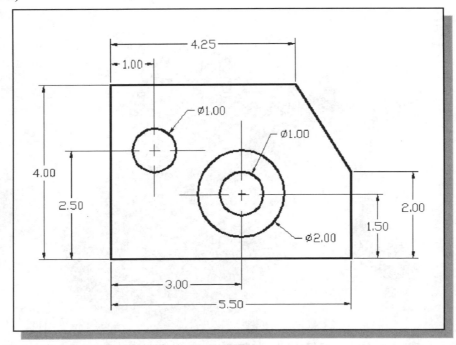

3. **Angle Slider** (Dimensions are in Millimeters.)

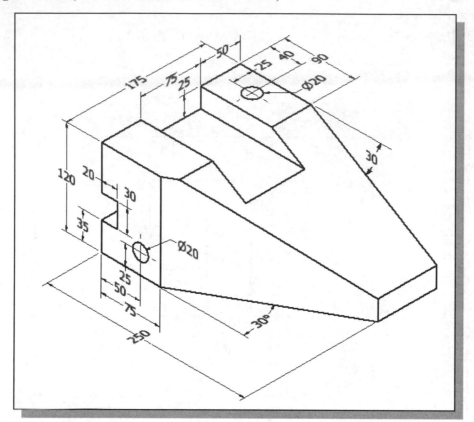

4. **Coupling Base** (Dimensions are in inches.)

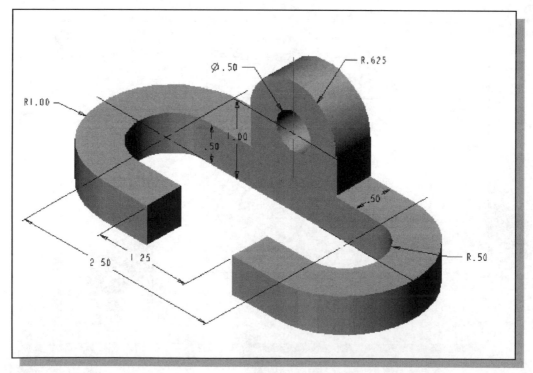

5. **Indexing Guide** (Dimensions are in inches.)

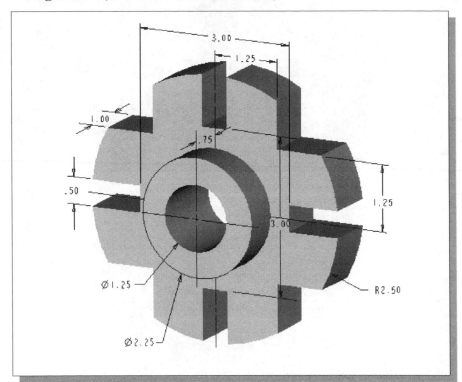

6. **L-Bracket** (Dimensions are in inches.)

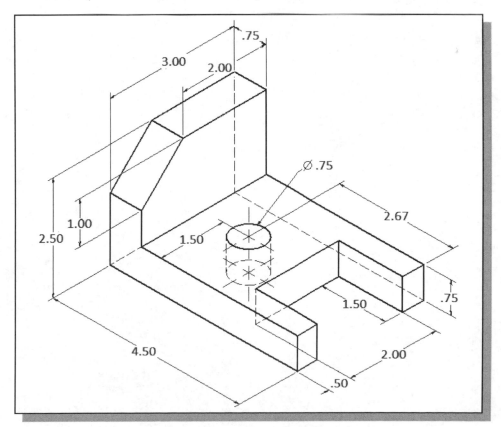

Notes:

Chapter 4
Model History Tree

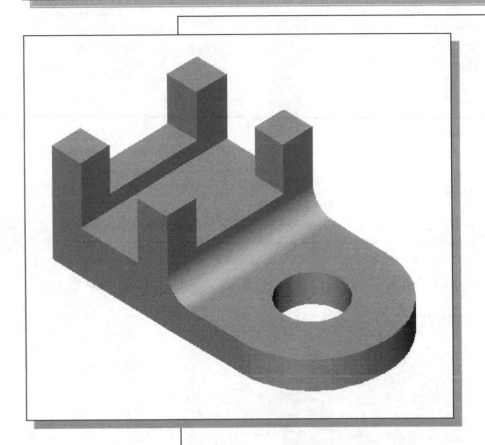

Learning Objectives

- ◆ **Using the BORN technique**
- ◆ **Use the Part Navigator**
- ◆ **Modify and Update Feature Dimensions**
- ◆ **Perform History-Based Part Modifications**
- ◆ **Change the Names of Created Features**
- ◆ **Implement Basic Design Changes**
- ◆ **Assigning and Calculating the associated Physical Properties**

The BORN Technique

In the previous chapters, we have concentrated on creating solid models relative to the *first feature* of the solid object. The first solid feature, the *base feature,* is the center of all features and is considered the key feature of the design. This is a common approach to create solid models, but this approach places much emphasis on the selection of the *base feature*. All subsequent features, therefore, are built by referencing the first feature, or **base feature**. In most cases, this approach is quite adequate and proper in creating solid models.

A more advanced technique of creating solid models is what is known as the "**Base Orphan Reference Node**" (**BORN**) technique. The basic concept of the BORN technique is to use a *Cartesian coordinate system* as the first feature prior to creating any solid features. With the *Cartesian coordinate system* established, we then have three mutually perpendicular datum planes (namely the *XY, YZ,* and *ZX planes*) available to use as sketching planes. The three datum planes can also be used as references for dimensions and geometric constructions. Using this technique, the first feature in the model isn't a solid feature and it is therefore called an "**orphan**," meaning that it has no history to be replayed. The technique of using reference geometry, such as the *Cartesian coordinate system,* in this fashion is therefore called the "Base Orphan Reference Node" (BORN) technique.

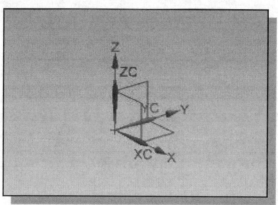

In *Siemens NX*, the BORN technique can be used by first creating a *Cartesian coordinate system* with three reference planes and three reference axes. All subsequent solid features can then use the coordinate system's reference geometry as sketching planes. The *base feature* is still important, but the *base feature* is no longer the <u>ONLY</u> choice for setting the orientation or selecting the sketching plane for subsequent solid features. In effect, the BORN technique is built-in when creating solid models in Siemens NX. Note that in Siemens NX, the default WCS is used to control the location of the solid features in the 3D Global space. This approach provides us with more options while we are creating parametric solid models. More importantly, this approach provides greater flexibility for part modifications and design changes.

Note that *Siemens NX* also comes with templates that contain pre-defined settings, which includes a set of reference coordinate systems. Note the BORN technique is implemented along with other pre-defined settings to reduce the amount of repetition needed in creating designs.

Model History Tree

In *SIEMENS NX*, the **design intents** are embedded into features in the **model history tree**. The structure of the model history tree resembles that of a **CSG binary tree**. A CSG binary tree contains only *Boolean relations*, while the **SIEMENS NX model history tree** contains all features, including *Boolean relations*. A history tree is a sequential record of the features used to create the part. This history tree contains the construction steps, plus the rules defining the design intent of each construction operation. In a history tree, each time a new modeling event is created previously defined features can be used to define information such as size, location, and orientation. It is therefore important to think about your modeling strategy before you start creating anything. It is important, but also difficult, to plan ahead for all possible design changes that might occur. This approach in modeling is a major difference of **FEATURE-BASED CAD SOFTWARE**, such as *NX*, from previous generation CAD systems.

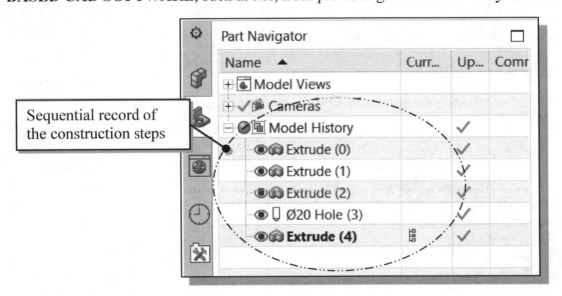

Feature-based parametric modeling is a cumulative process. Every time a new feature is added, a new result is created, and the feature is also added to the history tree. The database also includes parameters of features that were used to define them. All of this happens automatically as features are created and manipulated. At this point, it is important to understand that all of this information is retained, and modifications are done based on the same input information.

In *SIEMENS NX*, the model history tree gives information about modeling order and other information about the feature. Part modifications can be accomplished by accessing the features in the history tree. It is therefore important to understand and utilize the feature history tree to modify designs. *SIEMENS NX* remembers the history of a part, including all the rules that were used to create it, so that changes can be made to any operation that was performed to create the part. In *SIEMENS NX*, to modify a feature, we access the feature by selecting the feature in the model history tree displayed in the *Part Navigator area.*

The Saddle Bracket Design

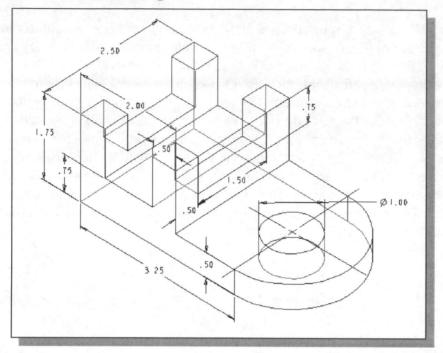

❖ Based on your knowledge of *SIEMENS NX* so far, how many features would you use to create the design? Which feature would you choose as the **BASE FEATURE**, the first solid feature, of the model? What is your choice in arranging the order of the features? Would you organize the features differently if additional fillets were to be added in the design? Take a few minutes to consider these questions and do preliminary planning by sketching on a piece of paper. You are also encouraged to create the model on your own prior to following through the tutorial.

Starting Siemens NX

1. Select the **NX** option on the *Start* menu or select the **NX** icon on the desktop to start *NX*.

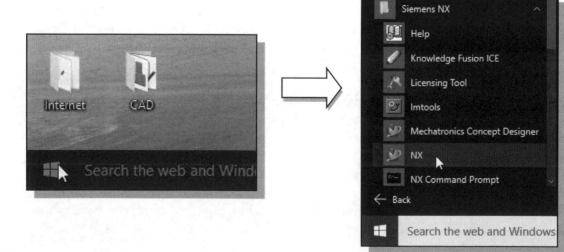

Modeling Strategy

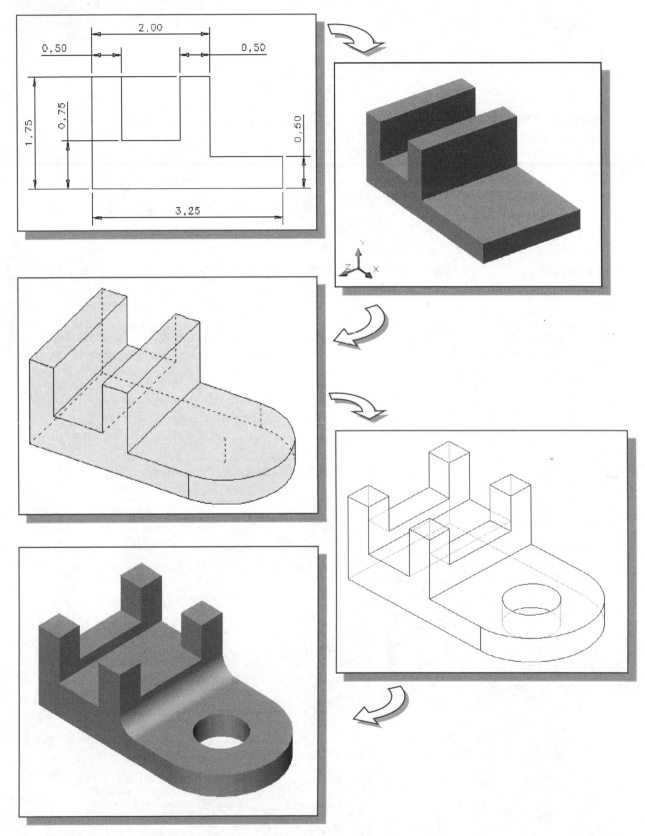

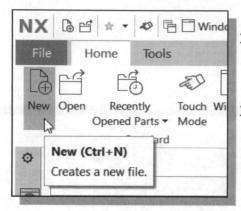

2. Select the **New** icon with a single click of the left-mouse-button (MB1) in the *Standard toolbar area*.

3. In the *New Part File* window, choose the Inches units as shown.

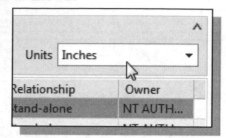

4. Select the **Blank** template file as shown.

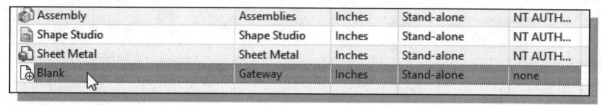

Assembly	Assemblies	Inches	Stand-alone	NT AUTH...
Shape Studio	Shape Studio	Inches	Stand-alone	NT AUTH...
Sheet Metal	Sheet Metal	Inches	Stand-alone	NT AUTH...
Blank	Gateway	Inches	Stand-alone	none

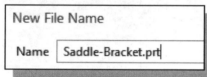

5. In the *New File Name* area, enter **Saddle-Bracket.prt** as the *File Name*.

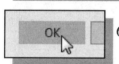

6. Click **OK** to proceed with the *New File* command.

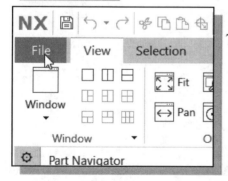

7. Select the **File** tab with a single click of the left-mouse-button (MB1) in the *Ribbon Bar area*.

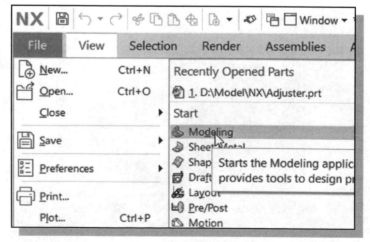

8. Pick **Modeling** in the pull-down list as shown in the figure.

Apply the BORN Technique

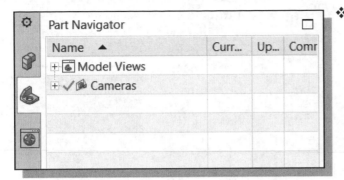

❖ In the *Part Navigator area*, notice a new part name appeared with seven work features established. Note that two main items are listed: **Model Views**, and **Cameras** (which contain pre-defined viewing angles of the model). There is no **history** information regarding the part, as we have not created anything yet.

1. In the *Ribbon bar area*, click on the triangle next to the **Datum Plane** icon to display the detail list and click on **Datum CSYS** as shown.

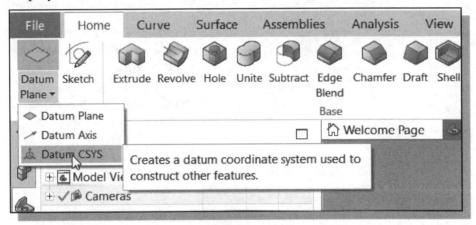

• In the *Graphics area*, notice a new **coordinate system** is placed on top of the *World Coordinate System* (**WCS**). Note that in the *Datum CSYS* dialog box, additional options are available for the placement of the new coordinate system.

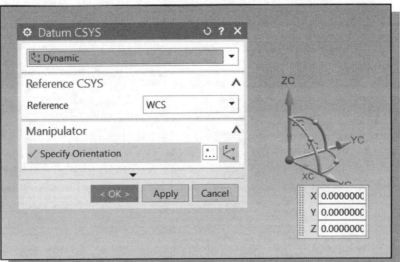

2. We will align the new **coordinate system** to the *World Coordinate System;* click **OK** to accept the default settings and create the new coordinate system.

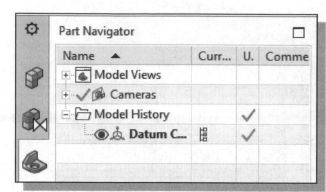

- In the *Part Navigator* area, the new coordinate system is added to the model history tree as the first **Model History** item.

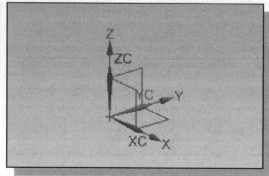

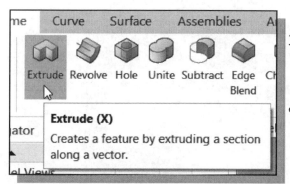

3. In the *Feature Toolbars* (toolbars aligned to the right edge of the main window), select the **Extrude** icon as shown.

- The *Extrude Feature Options Dialogue box*, which contains applicable construction options, is displayed as shown in the below figure.

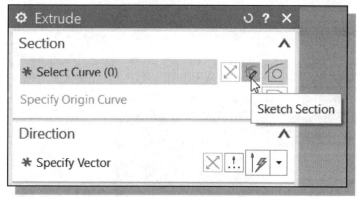

- Note the **Sketch Section** option is activated and NX expects us to align the sketch plane to begin creating a new *2D sketch*.

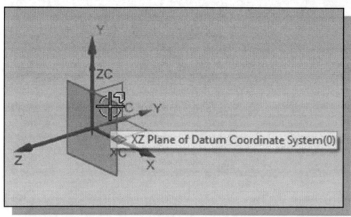

4. Select the **XZ Plane** of the Coordinate System to align the sketch plane of the *Base Feature*.

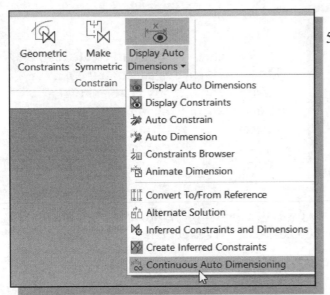

5. In the *Ribbon Bar area*, turn Off the **Continuous Auto Dimensioning** option by clicking on the triangle icon of *Display Sketch Constraints* as shown.

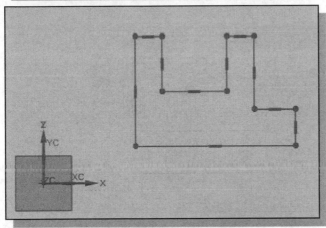

6. Create a 2D sketch with line-segments roughly close to the sketch below.

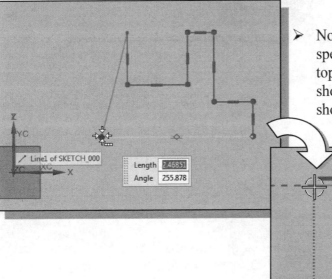

➢ Note that to assure the alignment to a specific location, pause the cursor on top of the alignment point and NX will show alignment to that location as shown in the figures.

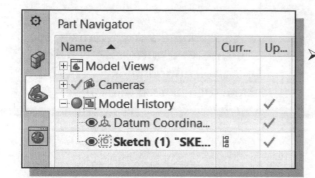

> Notice that as we begin to create features, **Model History** is added in the *part navigator*: Currently there are two new items: **Datum Coordinate System**, and **Sketch**.

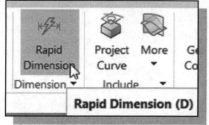

7. On your own, use the **Rapid Dimension** command to create and adjust the 2D sketch as shown in the below figure.

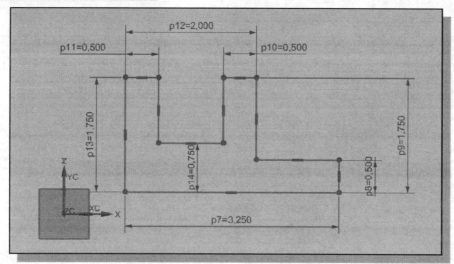

8. Click **Finish Sketch** to exit the NX sketcher mode and return to the *Feature option dialog window*.

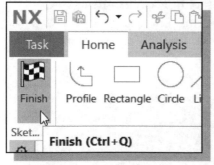

9. In the *Extrude dialog box*, set the first *Limits option* to **Symmetric Value** as shown.

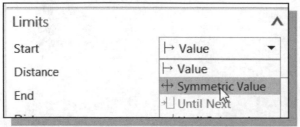

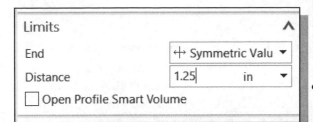

Limits ∧

End   ⊹ Symmetric Valu ▼

Distance   1.25   in   ▼

☐ Open Profile Smart Volume

10. Enter **1.25** inches as the extrusion distance as shown.

- Note the *Symmetric Value* option will automatically create the extrusion in both directions.

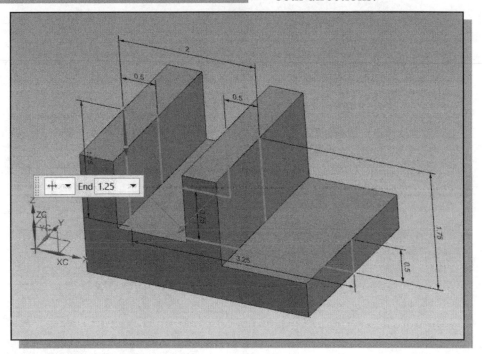

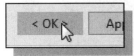

11. Click **OK** to accept the settings and create the base feature as shown.

The NX Part Navigator

- In the *NX* screen layout, the **Part Navigator** is located to the right of the graphics window. The *Part Navigator area* provides a visual structure of the features, constraints, and attributes that are used to create the part and/or assembly. The *Part Navigator* also provides right-click menu access for tasks associated specifically with the part or feature, and it is the primary focus for executing many of the *NX* commands.

1. Move the cursor on top of the datum coordinate system or the extruded solid and notice the corresponding feature is highlighted in the NX **Part Navigator**.

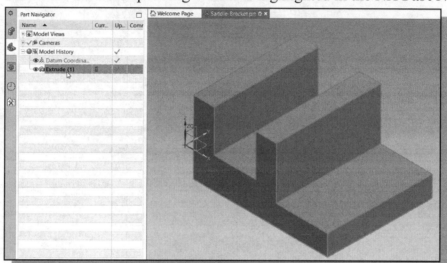

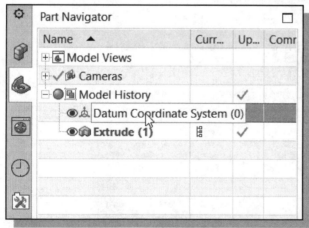

- In the *Part Navigator*, the *Model History* item contains the detailed information of the constructed model. Notice that we have created only one solid body and the solid body contains one coordinate system and one extruded feature.

2. Select the datum coordinate system in the *Part Navigator* and notice the selected item is highlighted in the graphics area.

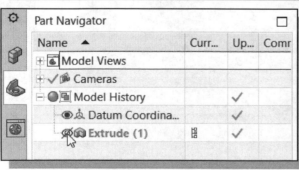

3. Click on the **eye icon** in front of the *Extrude* item to hide the corresponding item on the screen.

4. Redisplay the extruded feature by clicking on the **eye icon** again.

Create the Second Solid Feature

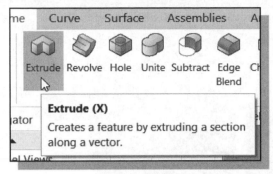

1. In the *Ribbon bar* area, select the **Extrude** icon as shown.

2. Select the **bottom horizontal face of the solid model** and set up the orientation of the sketch plane as shown.

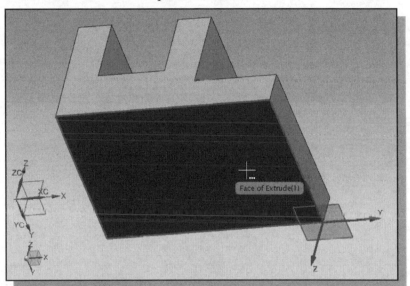

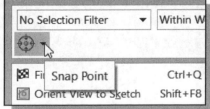

3. Select the **Circle** command by clicking once with the left-mouse-button on the icon in the *Sketch* toolbar.

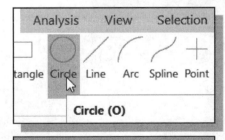

4. Right-click once to bring up the option list and click on the Snap point option and confirm the **Midpoint Snap** option is activated as shown.

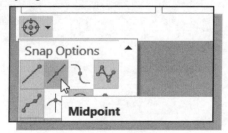

5. Move the cursor along the shorter edge of the base feature and pick the **midpoint** of the edge when the alignment symbol is displayed as shown. Note the coordinates of the point also confirm its location.

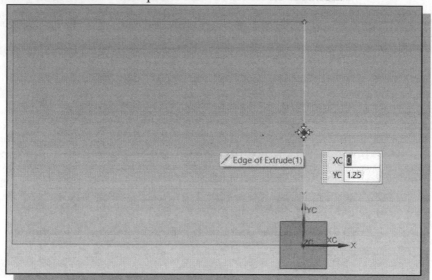

6. Select the **top corner** of the base feature to create a circle as shown below.

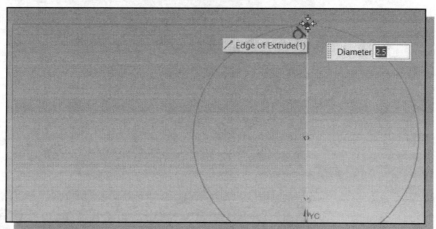

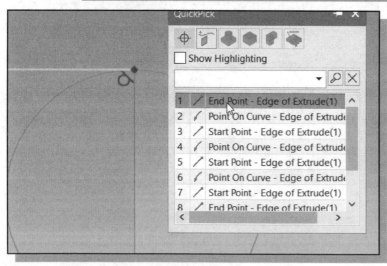

7. If necessary, use the **Select from list** option list to select the **Endpoint** of the top horizontal line as shown.

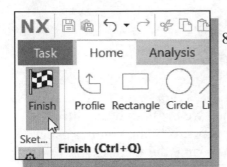

8. Click **Finish Sketch** to exit the NX sketcher mode and return to the *Feature option dialog window*.

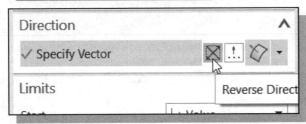

9. Click on the **Reverse Direction** icon to switch the extrusion direction.

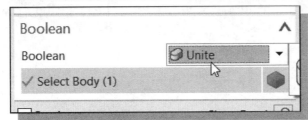

10. In the *Extrude* pop-up window, set the *Extrude* option to **Unite** as shown.

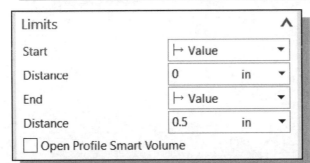

11. Enter **0.5** as the extrusion distance as shown.

12. Click **OK** to create the extrusion feature.

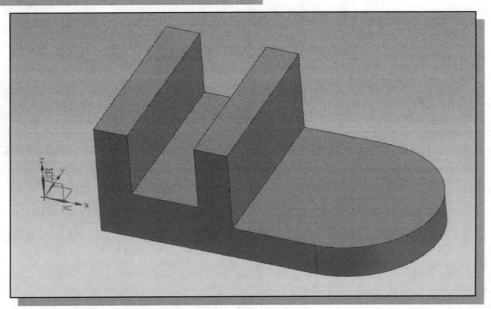

Use More Meaningful Feature Names

◆ Currently, our model contains two extruded features. The feature is highlighted in the display area when we select the feature in the *Part Navigator area*. Each time a new feature is created, the feature is also added in the *Part Navigator area*. By default, *NX* will use generic names for part features. However, when we begin to deal with parts with a large number of features, it will be much easier to identify the features using more meaningful names. Two methods can be used to rename the features: 1. **Clicking** twice (not double-clicking) on the name of the feature and 2. Using the **Option Menu** by clicking once with the right-mouse-button (MB3).

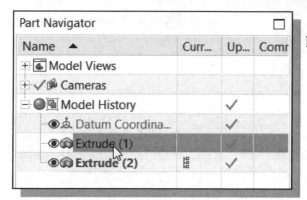

1. Select the first extruded feature in the *Model Part Navigator* area by left-clicking once on the name of the feature, *Extrude*. Notice the selected feature is also highlighted in the graphics window.

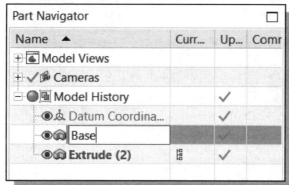

2. Left-mouse-click again, on the feature name, to enter the *Edit* mode as shown.

3. Enter **Base** as the new name for the first extruded feature.

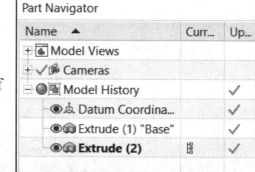

❖ Notice the new name is now shown as part of the extruded feature name.

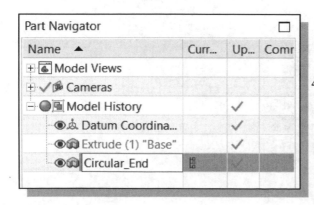

4. On your own, enter **Circular_End** as the second extruded feature name.

Adjust the Width of the Base Feature

❖ One of the main advantages of parametric modeling is the ease of performing part modifications at any time in the design process. Part modifications can be done through accessing the features in the history tree. *NX* remembers the history of a part, including all the rules that were used to create it, so that changes can be made to any operation that was performed to create the part. For our *Saddle Bracket* design, we will reduce the extrusion distance from 2.5 to 2.0 inches and the size of the base feature from 3.25 inches to 3.0 inches.

1. Inside the *Part Navigator* area, **right-mouse-click** on the first extruded feature to bring up the option menu and select the **Show Dimensions** option as shown.

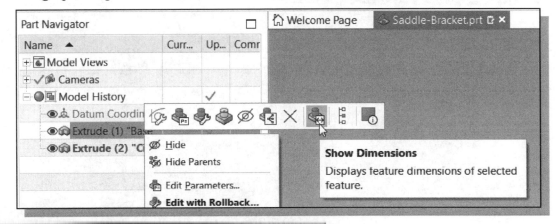

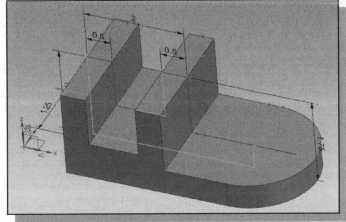

❖ The Show Dimensions option is used to display all of the dimensions related to the selected feature in the graphics area.

2. Right-mouse-click on the **Extrusion distance (1.25)** to bring up the option menu.

3. Select **Edit Value** in the *option menu* as shown.

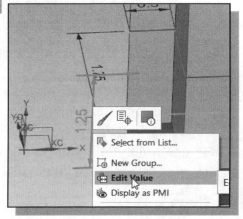

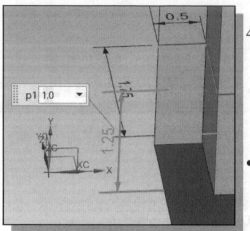

4. In the *dimension edit box*, enter **1.0** as the new extrusion distance. (Hint: hit the **Enter key** once after entering the new value.)

• Note that the new value is entered, but the model has not been updated.

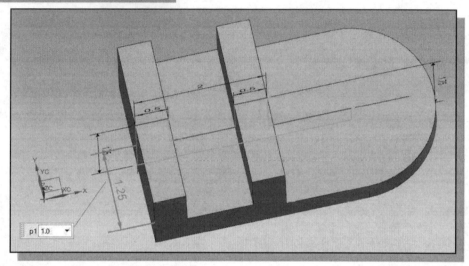

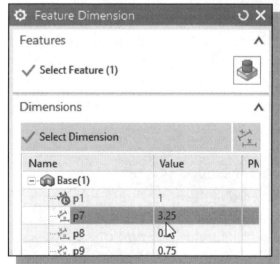

5. In the *Feature Dimensions dialog box*, select the width dimension **3.25** as shown.

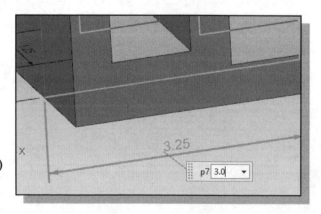

6. In the *dimension edit box*, enter **3.0** as the new width. (Hint: hit the **Enter key** once after entering the new value.)

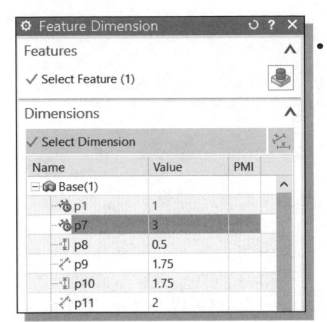

• Notice the **small clock icon** next to the modified dimension names indicating the model has not been updated.

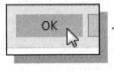

7. Click **OK** to accept the modification and update the model.

8. Hit the **[F5]** key once to execute the **Refresh** command to hide the displayed dimensions.

➢ Note that *NX* updates the model by re-linking all elements used to create the model. Also note that during the updating process, any problems caused by the changes and/or modifications will also be displayed in the part navigator area.

Add a Placed Feature

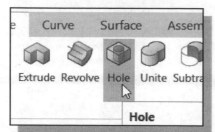

1. In the *Ribbon Bar* area, select the **Hole** command by releasing the left-mouse-button on the icon.

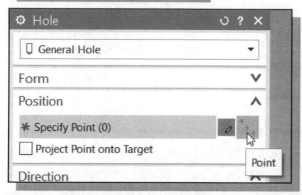

2. In the *Holes* window, notice the default hole type is set to **General Hole** as shown.

❖ Not that by default, the position of the hole is specified using the **Point** command.

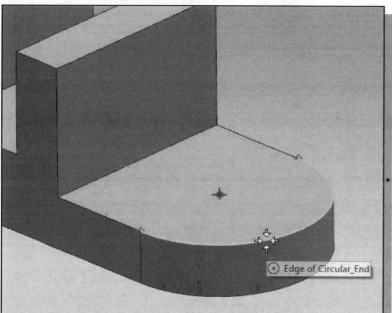

➢ We will align the center of the hole to the center of the circular feature on the right.

3. Move the cursor on top of the Top Arc and click once with the right-mouse-button to bring up the option list and pick the **Select from list** as shown.

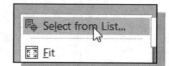

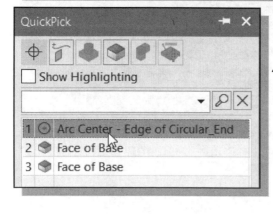

4. In the *Quick Pick dialog box*, select **the Arc Center** to align the center of the hole feature as shown.

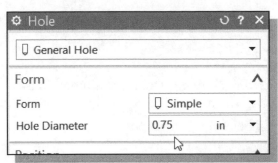

5. Enter **0.75 inch** as the diameter of the hole.

6. In the *Form and dimensions* options, set the **Depth Limit** to **Through Body** as shown.

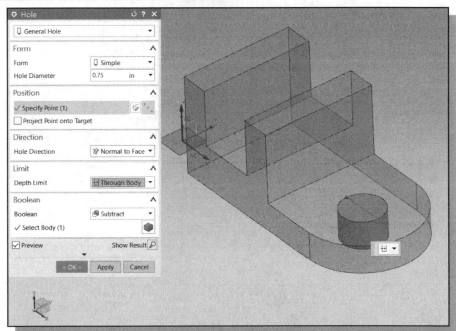

7. Click **OK** to accept the settings and create the feature.

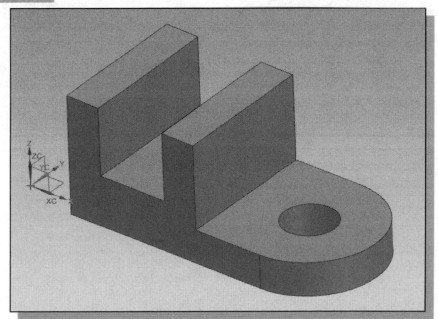

Create a Rectangular Cut Feature

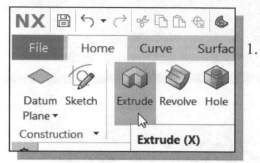

1. In the *Feature Toolbars* (toolbars aligned to the right edge of the main window), select the **Extrude** icon as shown.

2. Select the **right vertical face** of the base feature, in the orientation shown, of the sketch plane.

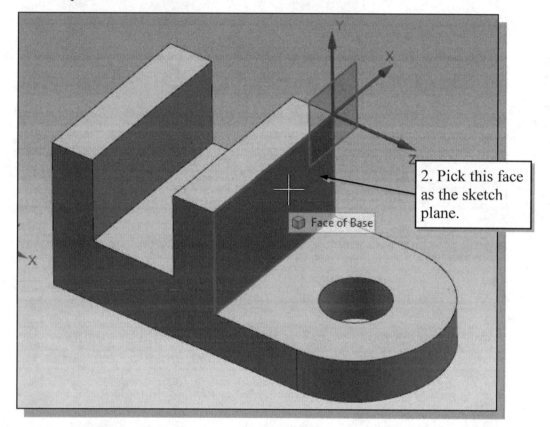

2. Pick this face as the sketch plane.

3. Select the **Rectangle** command by clicking once with the **left-mouse-button (MB1)** on the icon in the *Sketch Curve* toolbar.

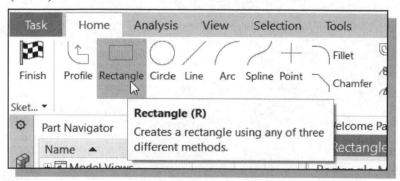

4. Create a rectangle of arbitrary size aligned to the upper edge of the solid model as shown. (Note the dimensions shown on your screen may be different.)

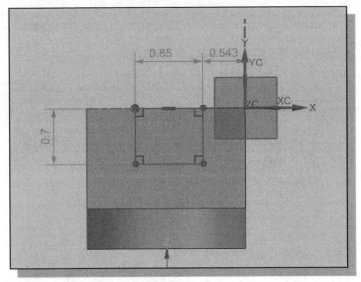

5. Inside the graphics window, click once with the **middle-mouse-button** (MB2) to end the Rectangle command.

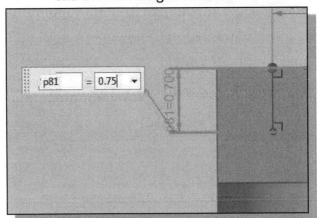

6. **Double left-click** on one of the dimensions to enter the **Edit Dimension** mode as shown

7. On your own, adjust the three dimensions of the sketch as shown in the below figure.

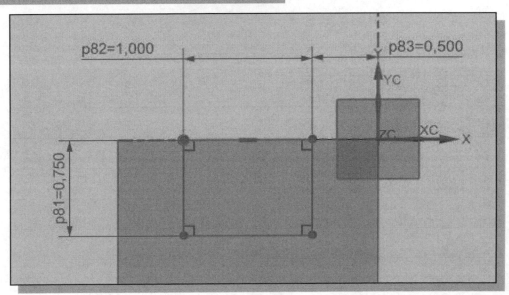

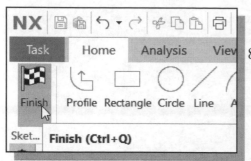

8. Select **Finish Sketch** by clicking once with the left-mouse-button (**MB1**) on the icon.

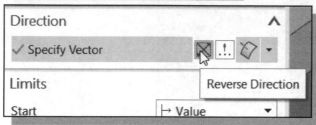

9. In the *Extrude* pop-up window, click on the **Reverse Direction** icon to switch the extrusion direction.

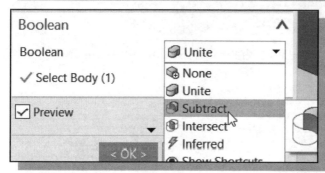

10. In the *Extrude* pop-up window, set the *Extrude* option to **Subtract** as shown.

❖ Notice the solid feature is highlighted when the *Subtract option* is set.

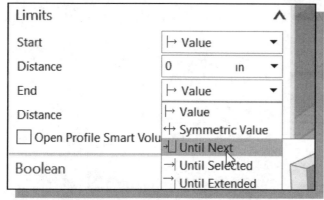

11. Select **Until Next** as the End extrusion distance option as shown.

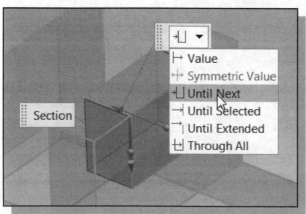

➢ Note the *Until Next* option can also be set in the graphics area as shown.

12. On your own, confirm the *Extrude settings* are as shown in the below figure.

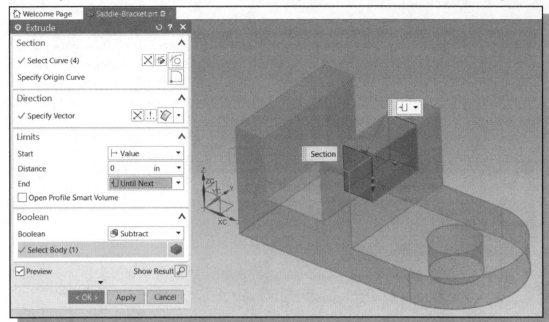

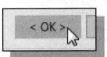 13. Click **OK** to create the extrusion feature.

➢ On your own, rename the rectangular cut feature to **Rect_Cut** and examine the created solid features by going through the *Part Navigator*.

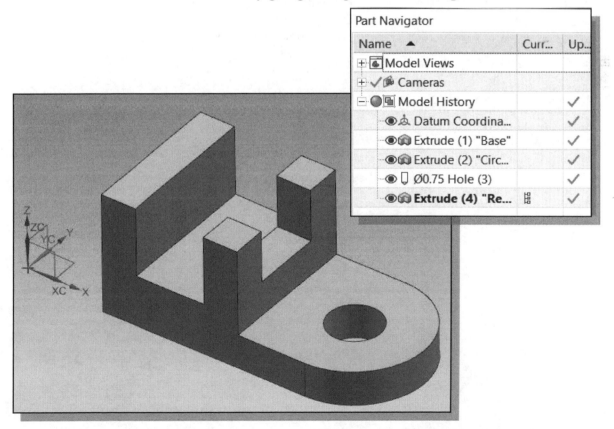

History-based Part Modifications

- *NX* uses the *history-based part modification* approach, which enables us to make modifications to the appropriate features and re-link the rest of the history tree without having to reconstruct the model from scratch. We can think of it as going back in time and modifying some aspects of the modeling steps used to create the part. We can modify any feature that we have created. As an example, we will adjust the depth of the rectangular cutout.

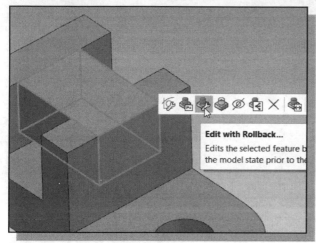

1. Inside the *Graphics* window, left-mouse-click once on the ***Rect_Cut*** feature and select the **Edit with Rollback** option.

- Note the same option is also selectable as an icon below the option menu.

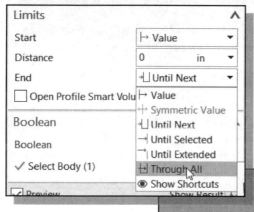

2. In the *Extrude* edit box, set the *End Limit* to the **Through All** option.

3. Click on the **OK** button to accept the settings.

A Design Change

❖ Engineering designs usually go through many revisions and changes. *NX* provides an assortment of tools to handle design changes quickly and effectively. We will demonstrate some of the tools available by changing the ***Base*** feature of the design.

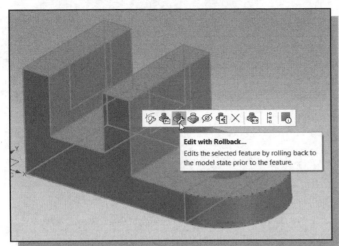

1. Inside the *Graphics* window, left-click on the ***Base*** feature and select **Edit with Rollback** option.

❖ We have entered the **Edit with Rollback** option. NX provides quite a few options to allow quick modification of the created features.

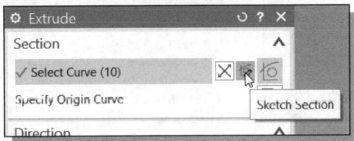

2. In the *Extrude* window, select the ***Sketch Section*** option by left-clicking once on the icon.

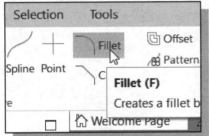

3. Select the **Fillet** command by clicking once with the **left-mouse-button (MB1)** on the icon in the *Sketch Curve* toolbar.

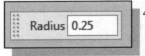

4. Enter **0.25** as the new radius of the fillet.

5. Create the **fillet** by selecting the corner as shown.

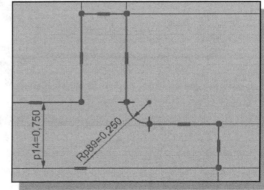

6. Inside the *graphics window*, click once with the middle-mouse-button (**MB2**) to end the Fillet command.

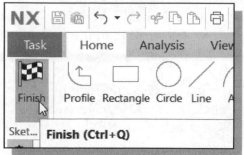

7. Select **Finish Sketch** by clicking once with the left-mouse-button (**MB1**) on the icon.

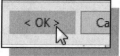

8. Click **OK** to accept the changes to the extrusion feature.

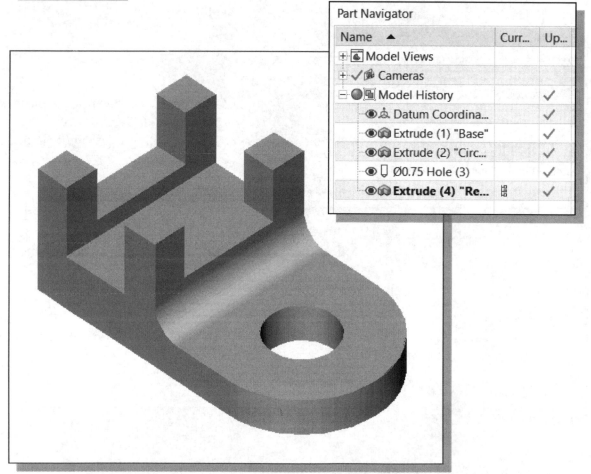

❖ In a typical design process, the initial design will undergo many analyses, tests, and reviews. The *history-based part modification* approach is an extremely powerful tool that enables us to quickly update the design. At the same time, it is quite clear that PLANNING AHEAD is also important in doing feature-based parametric modeling.

Assign and Calculate the Associated Physical Properties

➢ In parametric modeling, specific material properties can also be assigned to the virtual designs; this allows us to examine the detailed physical information of the model, such as mass, weight, center of gravity and moments of inertia. This information can be used to create bills of materials, prepare for the production and perform further analyses of the designs.

1. In the *NX ribbon toolbar*, select **Tools → Assign Materials** as shown.

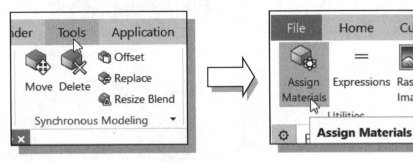

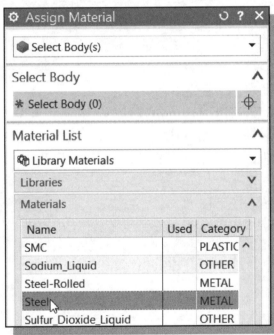

2. Scroll down the materials list and select **Steel** as shown in the figure.

3. Right-click on **Steel** and select **Information** to view the associated properties of the selected material.

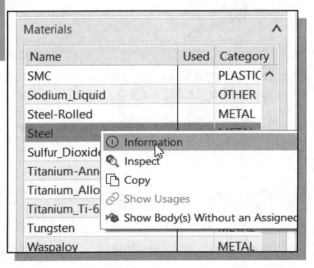

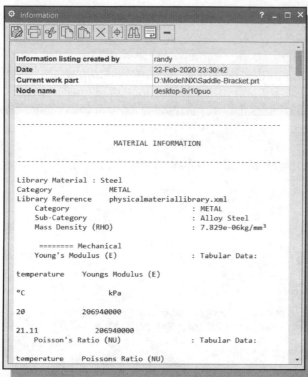

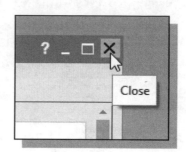

4. On your own, review the associated information. Click on the **Close** icon to close the *Information window*.

5. Inside the graphics window, **select the model** by clicking on it with the left-mouse-button.

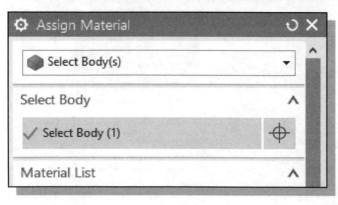

6. Click **OK** to exit the **Assign Material** option.

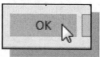

➢ Note Steel is now the material assigned to the model.

7. In the **Menu List**, select **Analysis → Measure Bodies...** as shown.

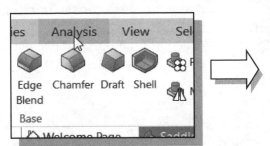
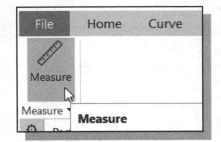

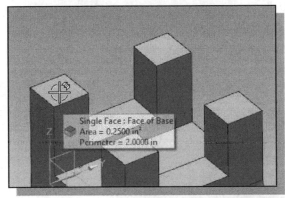

8. Move the cursor on one of the top surfaces of the model and note the area and perimeter of the surface is displayed next to the model.

9. Click on the surface to select and note the properties are shown.

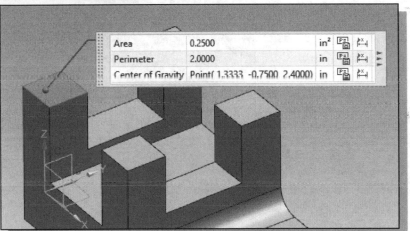

10. **De-select** the selected object by using the selection list and click on the **Remove button** as shown.

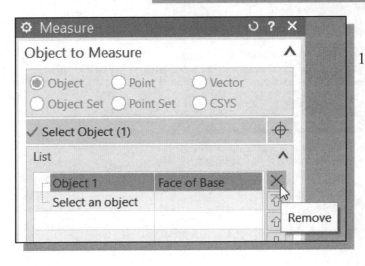

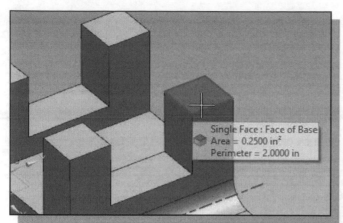

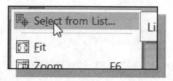

11. Pause the cursor on the top right surface of the model and click once with the right-mouse button and pick **Select from list** as shown.

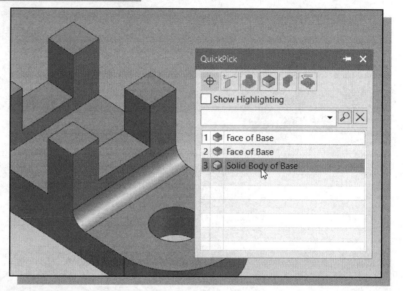

12. Select **Solid Body of Model** to see the related property of the model.

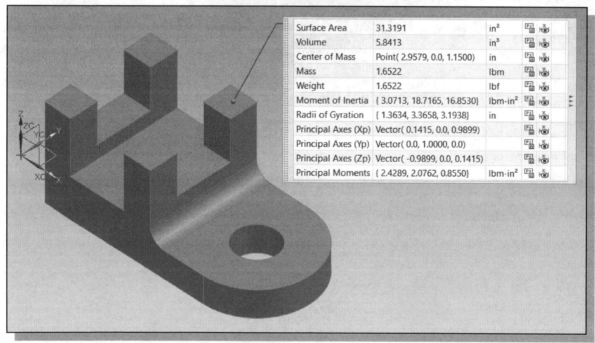

Surface Area	31.3191	in²		
Volume	5.8413	in³		
Center of Mass	Point(2.9579, 0.0, 1.1500)	in		
Mass	1.6522	lbm		
Weight	1.6522	lbf		
Moment of Inertia	{ 3.0713, 18.7165, 16.8530}	lbm-in²		
Radii of Gyration	{ 1.3634, 3.3658, 3.1938}	in		
Principal Axes (Xp)	Vector(0.1415, 0.0, 0.9899)			
Principal Axes (Yp)	Vector(0.0, 1.0000, 0.0)			
Principal Axes (Zp)	Vector(-0.9899, 0.0, 0.1415)			
Principal Moments	{ 2.4289, 2.0762, 0.8550}	lbm-in²		

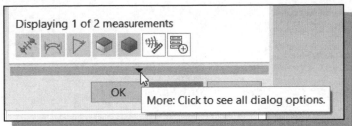

13. In the *Measure dialog box,* click on the **down arrow bar** to display additional options as shown.

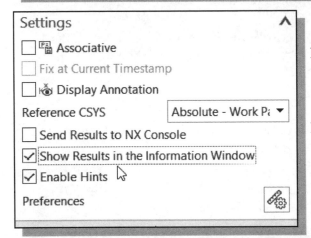

14. In the *Settings option* area, activate the **Show Results in the Information window** option as shown.

15. Click **Apply** to display the Information window.

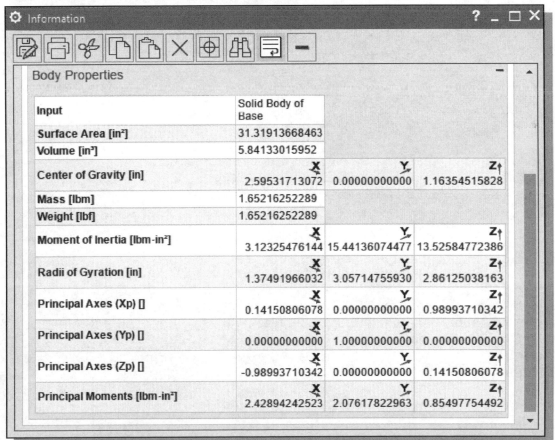

Body Properties

Input	Solid Body of Base		
Surface Area [in²]	31.31913668463		
Volume [in³]	5.84133015952		
	X	**Y**	**Z**
Center of Gravity [in]	2.59531713072	0.00000000000	1.16354515828
Mass [lbm]	1.65216252289		
Weight [lbf]	1.65216252289		
	X	**Y**	**Z**
Moment of Inertia [lbm-in²]	3.12325476144	15.44136074477	13.52584772386
	X	**Y**	**Z**
Radii of Gyration [in]	1.37491966032	3.05714755930	2.86125038163
	X	**Y**	**Z**
Principal Axes (Xp) []	0.14150806078	0.00000000000	0.98993710342
	X	**Y**	**Z**
Principal Axes (Yp) []	0.00000000000	1.00000000000	0.00000000000
	X	**Y**	**Z**
Principal Axes (Zp) []	-0.98993710342	0.00000000000	0.14150806078
	X	**Y**	**Z**
Principal Moments [lbm-in²]	2.42894242523	2.07617822963	0.85497754492

➢ The detailed physical properties of the model, using STEEL as the associated material, are displayed inside the information window.

Review Questions:

1. What are stored in the *NX History Tree*?

2. When extruding, what is the difference between *Value* and *Through All*?

3. Describe the *history-based part modification* approach.

4. Describe two methods available in *NX* to perform the *Edit with Rollback* option of solid features.

5. What are listed in the Part Navigator area?

6. Describe two methods available in *NX* to *modify the dimension values* of parametric sketches.

7. Create *History Tree sketches* showing the steps you plan to use to create the two models shown on the next page:

Ex.1)

Ex.2)

Exercises: (Dimensions are in inches.)

1. **C-Clip** (Dimensions are in inches. Plate thickness: **0.25 inches**.)

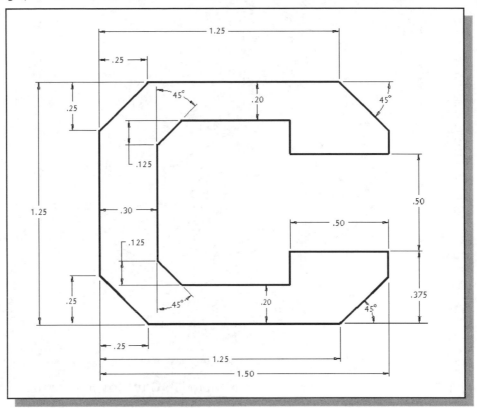

2. **Tube Mount** (Dimensions are in inches.)

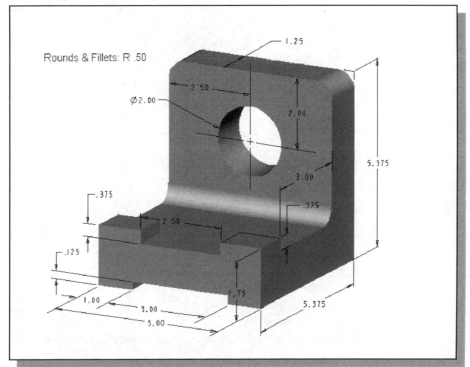

3. **Hanger Jaw** (Dimensions are in inches. Volume =?)

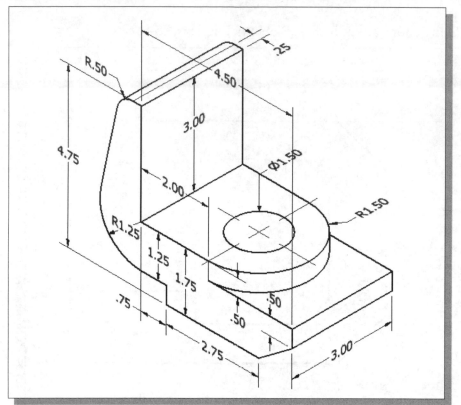

4. **Transfer Fork** (Dimensions are in inches. Material: **Cast Iron.** Volume =?)

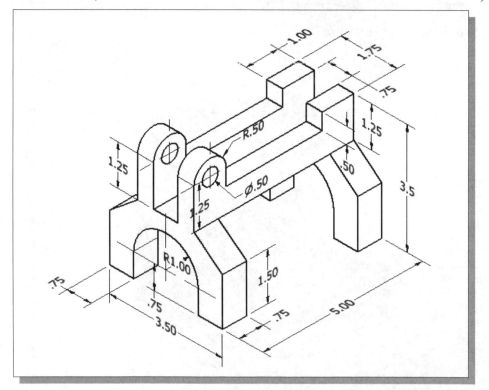

5. **Guide Slider** (Material: **Cast Iron**. Weight and Volume =?)

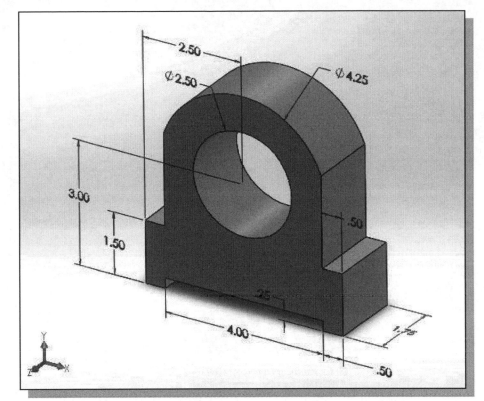

6. **Shaft Guide** (Material: **Aluminum-6061**. Mass and Volume =?)

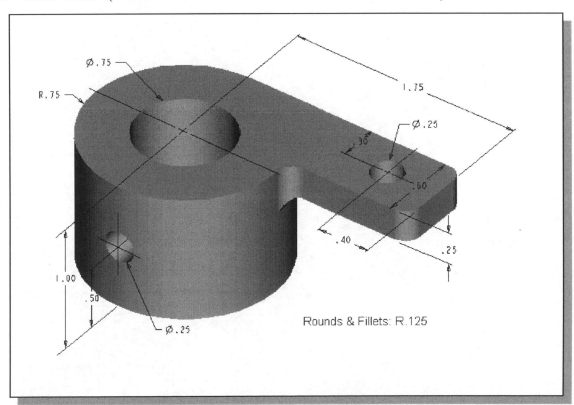

Notes:

Chapter 5
Parametric Constraints Fundamentals

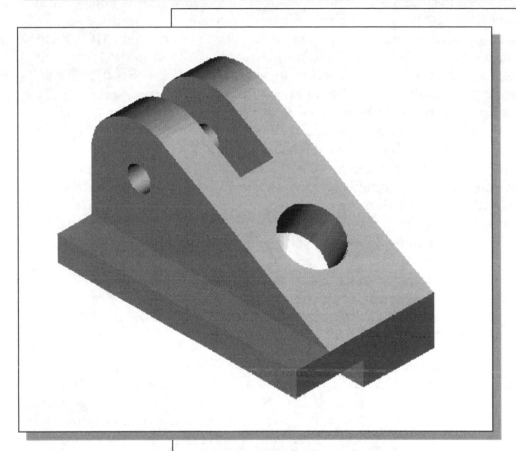

Learning Objectives

♦ **Create Parametric Relations**
♦ **Use Dimensional Variables**
♦ **Display, Add, and Delete Geometric Constraints**
♦ **Understand and Apply Different Geometric Constraints**
♦ **Display and Modify Parametric Relations**
♦ **Create Fully Constrained Sketches**

Constraints and Relations

A primary and essential difference between parametric modeling and previous generation computer modeling is that parametric modeling captures the *design intent*. In the previous lessons, we have seen that the design philosophy of *"shape before size"* is implemented through the use of *NX's* Sketcher and dimensioning commands. In performing geometric constructions, dimensional values are necessary to describe the **SIZE** and **LOCATION** of constructed geometric entities. Besides using dimensions to define the geometry, we can also apply geometric rules to control geometric entities. More importantly, *NX* can capture design intent through the use of **geometric constraints**, **dimensional constraints**, and **parametric relations**. In *NX*, there are two types of constraints: **geometric constraints** and **dimensional constraints**. For part modeling in *NX*, constraints are applied to *2D sketches*. **Geometric constraints** are **geometric restrictions** that can be applied to geometric entities; for example, *horizontal*, *parallel*, *perpendicular*, and *tangent* are commonly used *geometric constraints* in parametric modeling. **Dimensional constraints** are used to describe the SIZE and LOCATION of individual geometric shapes. One should also realize that depending upon the way the constraints are applied, the same results can be accomplished by applying different constraints to the geometric entities. In *NX*, **parametric relations** are user-defined mathematical equations composed of dimensional variables and/or *design variables*. In parametric modeling, features are made of geometric entities with both relations and constraints describing individual design intent. In this lesson, we will discuss the fundamentals of parametric relations and geometric constraints.

In *NX*, as we create 2D sketches, geometric constraints such as *horizontal* and *parallel* are automatically added to the sketched geometry. In most cases, additional constraints and dimensions are needed to fully describe the sketched geometry beyond the geometric constraints added by the system. Although we can use *NX* to build partially constrained or totally unconstrained solid models, the models may behave unpredictably as changes are made. In most cases, it is important to consider the design intent and to add proper constraints to geometric entities. In the following sections, a simple triangular model is used to illustrate the different tools that are available in *NX* to create/modify geometric and dimensional constraints.

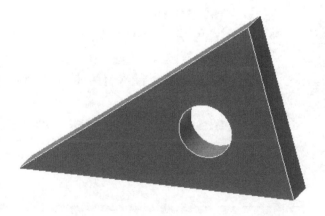

Starting Siemens NX

1. Select the **NX** option on the *Start* menu or select the **NX** icon on the desktop to start *NX*. The *NX* main window will appear on the screen.

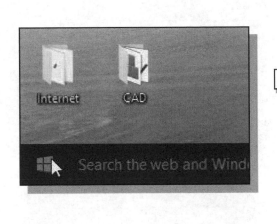 

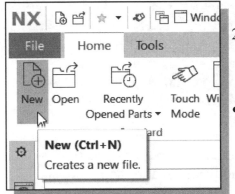

2. Select the **New** icon with a single click of the left-mouse-button (MB1) in the *Standard toolbar area*.

- In this chapter, we will use the Model Template that comes with the NX software. Using the template also allows us to use the BORN technique as a Datum Coordinate System is already established in the template.

3. Choose the **Millimeters** units as shown in the below figure.

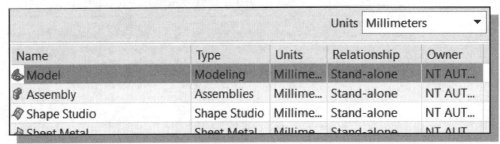

	Units	Millimeters ▼

Name	Type	Units	Relationship	Owner
Model	Modeling	Millime...	Stand-alone	NT AUT...
Assembly	Assemblies	Millime...	Stand-alone	NT AUT...
Shape Studio	Shape Studio	Millime...	Stand-alone	NT AUT...
Sheet Metal	Sheet Metal	Millime...	Stand-alone	NT AUT...

4. Select **Model** in the *Template list*. Note that the *Model template* will allow us to switch directly into the **Modeling task** as indicated in the *templates list*.

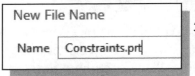

5. In the *New File Name* box, enter **Constraints.prt** as the *File Name*.

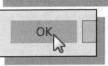

6. Click **OK** to proceed with the *New Part File* command.

7. In the resource bars area, click on the ***History*** option to view the parts that were created recently.

8. Click on the **Part navigator tab** to switch to the *Part Navigator* option.

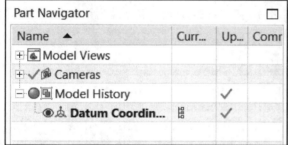

➢ In the *Part Navigator*, a *datum coordinate system* is listed under the *Model History* item. One of the pre-defined settings in the *Model template* is a Cartesian coordinate system, which enables the users to use the BORN technique in creating solid models.

9. Move the cursor on top of the coordinate system in the graphics area and notice the corresponding item is also highlighted in the model history tree as shown.

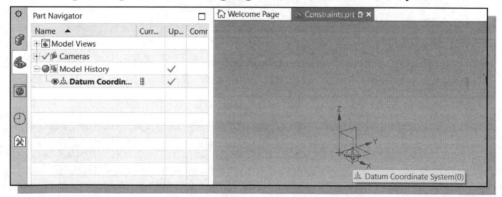

➢ Notice that with the Model template file, we entered the modeling mode directly with a pre-defined coordinate system. Using a template file also helps to eliminate repetitive steps and make our work much more efficient. Using template files also helps us maintain consistent design and drafting standards. Note that NX comes with several templates for both *English* and *Metric* units.

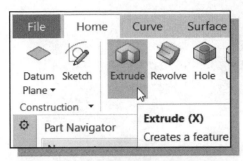

10. In the *Feature Toolbars* (toolbars aligned to the right edge of the main window), select the **Extrude** icon as shown.

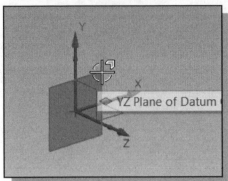

11. Select the **YZ Plane** of the established datum coordinate system as the sketch plane of the Base Feature.

12. On your own, turn **off** the **Display Sketch Constraints** option.

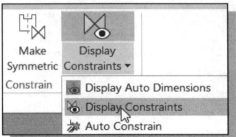

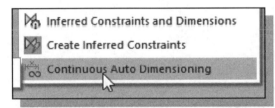

13. On your own, also turn **off** the **Continuous Auto Dimensioning** option.

14. Create a triangle of arbitrary size positioned near the upper right side of the screen as shown below. (Note that the base of the triangle is a horizontal line.)

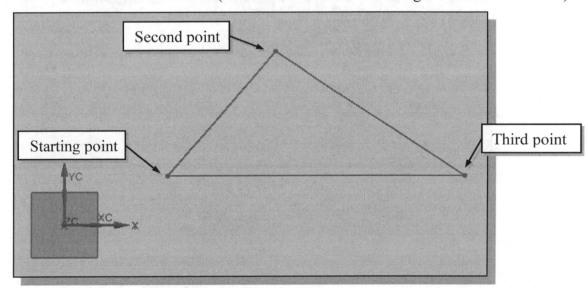

Display Existing Constraints

Two options are available to display the existing constraints that are applied to the 2D sketches.

➢ **Option One**

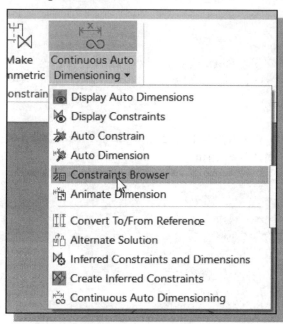

1. Switch **On** the **Relations Browser** in the *Constraints* toolbar. This command allows us to use the relations browser to **identify** and/or **remove** constraints that are already applied to the 2D sketches.

➢ Note the **Relations Browser** icon is activated as shown.

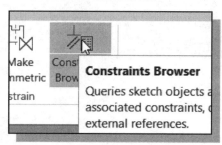

2. By default, the scope is set to **All in active sketch** option and the *Top-level Node objects* option is set to **Curves**.

3. NX displays the existing entities in the *objects* box. Currently several constraints are applied to the active sketch. A *Horizontal Constraint* is applied at the base and three *coincident constraints* are applied at the three corners. Expand the constraints list for line 3, and notice the three applied constraints, two coincident and one horizontal, are shown.

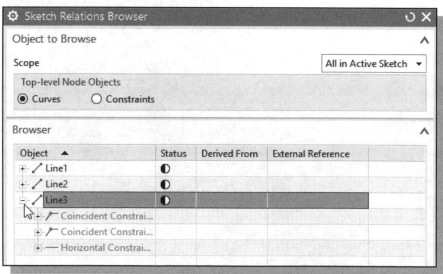

4. Click **Close** to exit the command.

➢ The **Relations Browser** command is generally used to display more detailed information on the existing constraints.

➢ **Option Two:**
    1. Select the **Display Sketch Constraints** command in the *Sketch Constraints* toolbar. This icon allows us to display constraints that are already applied to the 2D sketches.

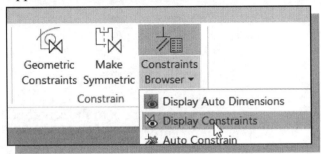

    ➢ In *NX*, constraints are applied as geometric entities are created. *NX* will attempt to add proper constraints to the geometric entities based on the way the entities were created. Constraints are displayed as symbols next to the entities as they are created. The current sketch consists of three line entities, three straight lines. The horizontal line has one constraint applied to it, a ***horizontal constraint***.

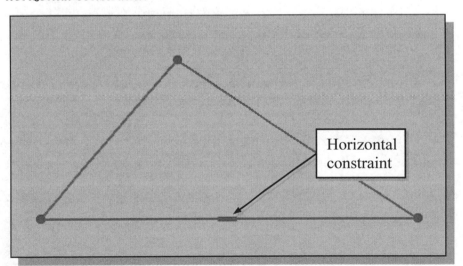

    ➢ The **Display Sketch Constraints** command can be used to quickly display the existing constraints. The constraints are displayed as symbols next to the entities. Note that this command can be made active to display constraints while the sketch is being made.

Apply Geometric Constraints Implicitly

• In *NX*, geometric constraints can be applied implicitly (as we create the geometry) or explicitly (apply the constraint individually).

NX displays the governing geometric rules as sketches are built. *NX* displays different visual clues, or symbols, to show alignments, perpendicularities, tangencies, etc. These constraints are used to capture the *design intent* by creating constraints where they are recognized. To prevent constraints from forming, hold down the [**Alt**] key while creating an individual sketch curve.

	Vertical	indicates a line is vertical
	Horizontal	indicates a line is horizontal
	Dashed line	indicates the alignment is to the center point or endpoint of an entity
	Parallel	indicates a line is parallel to other entities
	Perpendicular	indicates a line is perpendicular to other entities
	Coincident	indicates the cursor is at the endpoint of an entity
	Concentric	indicates the cursor is at the center of an entity
	Tangent	indicates the cursor is at tangency points to curves
	Midpoint	indicates the cursor is at the midpoint of an entity
	Point on Curve	indicates the cursor is on curves
	Equal Length	indicates the length of a line is equal to another line
	Equal Radius	indicates the radius of an arc is equal to another arc

Apply Geometric Constraints Explicitly

- In *NX*, geometric constraints can be applied explicitly, by first selecting the geometry and then the type of constraint desired.

    1. Select the **Geometry Constraints** command in the *Constraint* toolbar. This command allows us to apply the desired constraints individually.

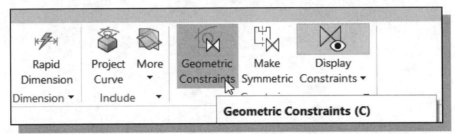

    2. Activate the **Perpendicular constraint** as shown in the figure.

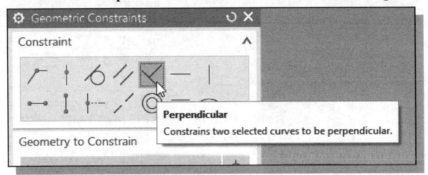

    3. Select the **inclined line** on the **left** and click once with the **middle mouse-button (MB2)** then select **the other inclined line** on the right.

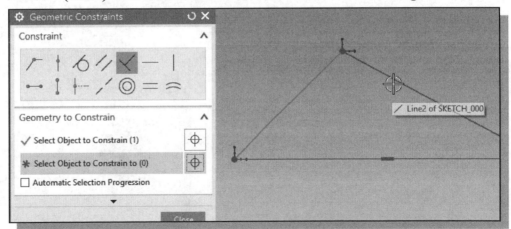

    4. The geometry is quickly adjusted so that the two selected lines are now perpendicular to each other, and notice *NX* indicates that **four** more constraints are needed to *fully constrain* the current sketch.

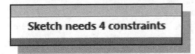

5. Click **Close** to exit the **Geometry Constraints** command.

- Five constraints are applied to the 2D sketch; four were done implicitly and the other explicitly.

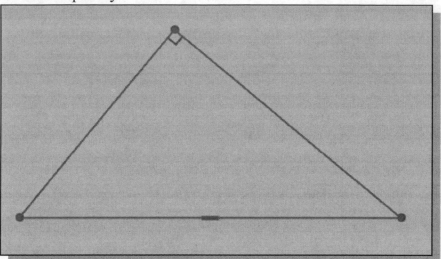

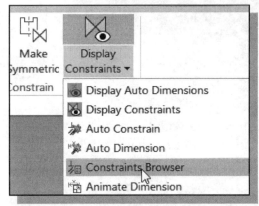

6. Turn **on** the **Relations Browser** in the *Sketch Constraints* toolbar.

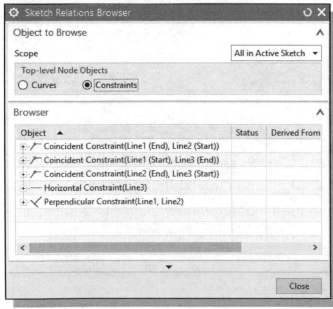

7. Select the **All in Active Sketch** option, the last option in the *List constraints for* group.

8. Set the *Top-level Node objects* option to **Constraints**.

- The five constraints are listed in the *Constraints* list as shown.

9. Click **Close** to exit the command.

Add Dimensional Constraints

- In *Parametric Modeling*, **geometric constraints** and **dimensional constraints** are interchangeable. For example, the *perpendicular* constraint we applied in the previous section can also be achieved by applying an angular dimension set to 90º. It is more important to consider the functionality of the design and embed the Design Intent through the use of proper geometric constraints and dimensional constraints.

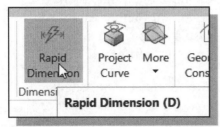

1. Select the **Rapid Dimension** command in the Sketch *Constraints* toolbar as shown.

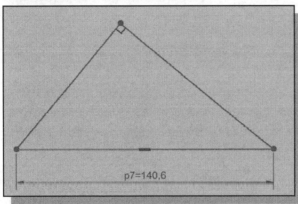

2. Click on the **horizontal line** to create a horizontal dimension as shown. Your number might look very different than what is shown here.

- ❖ Before continuing to the next step, consider this: What if we want to adjust the length of the line so that it is roughly 10 mm longer? Yes, we can simply change the dimension by adding 10 mm to the current displayed number. The more difficult question is: The line will be lengthened in which direction? To the left? To the right? Or maybe in both directions?

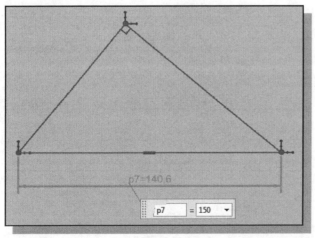

3. On your own, increase the displayed value by about **10 mm** and observe the adjustment done by *NX*.

- ❖ The adjustment will be made based on several factors, but the currently applied geometric constraints will definitely affect the results of the modification.

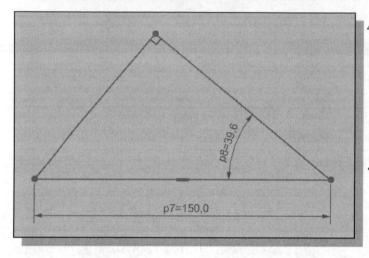

Sketch needs 3 constraints

❖ Also notice the message *"Sketch needs 3 constraints"* is displayed in the message area. Can you identify which constraints are needed to fully constrain the sketch?

4. On your own, create the angular dimension between the inclined line on the right and the horizontal line as shown. Again, the actual value displayed on your screen may look very different.

❖ Also notice the message *"Sketch needs 2 constraints"* is displayed in the message area. What dimensions are missing here?

❖ Use the **middle-mouse-button (MB2)** or click **OK** to exit the Rapid Dimension command.

➢ One option to identify the missing constraints is to perform a **drag and drop** with the left-mouse-button on the constructed geometry entities.

5. Click and drag with the **left-mouse-button (MB1)** on one of the inclined lines and noticed that the triangle can be moved to a new location. This is an indication that the triangle is probably fully defined, but its **LOCATION** relative to the 3D environment is not fully defined.

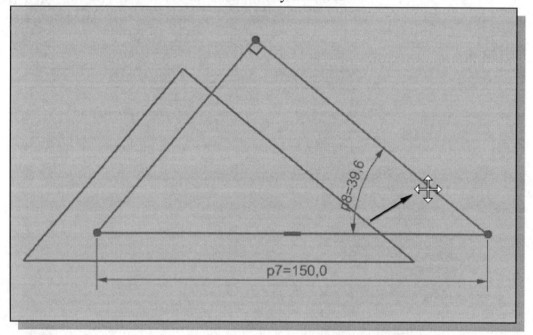

A Fully Constrained Sketch

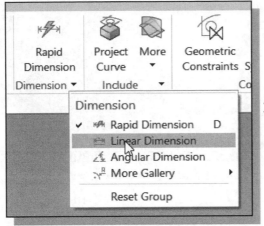

1. Click on the little triangle next to the **Rapid Dimension** command, in the Sketch *Constraints* toolbar, to expand the icon list and select **Linear Dimension** as shown.

2. Activate the *Linear Dimension* command by clicking on the icon as shown.

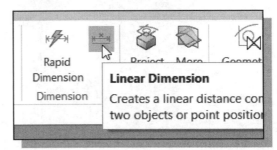

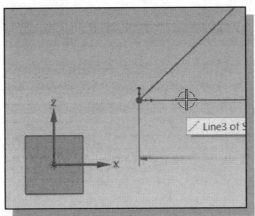

3. Select **Line3** by clicking near the left endpoint of the line as shown.

➢ Note the small arrows displayed at the three corners indicate the corners can still move in the 2D sketch plane.

4. Select the datum **origin** and place the perpendicular dimension below and to the left of the triangle as shown in the below figure.

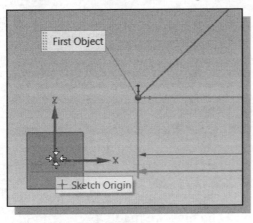

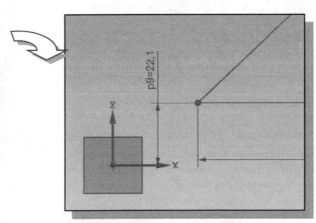

➢ Only one more constraint is needed to fully constrain the 2D sketch.

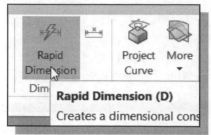

5. Select the **Rapid Dimension** command in the Constraints toolbar as shown.

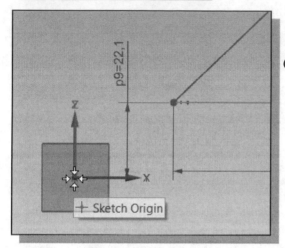

6. Select the **Sketch Origin** as the first entity to create a dimension as shown.

7. Select the endpoint of the horizontal line, **Line3**, as the second entity shown in the figure.

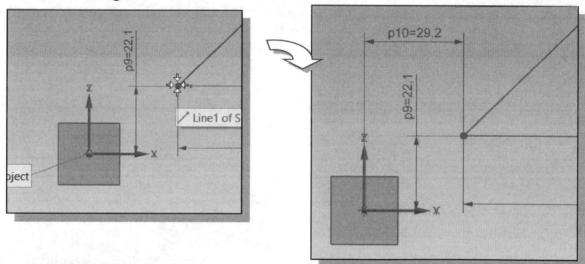

➤ Notice the message, near the bottom of the main window, indicating a fully constrained 2D sketch.

❖ A fully constrained 2D sketch is desirable as it means that we have full control of the 2D sketch and we know exactly what will happen as changes are made. For example, in the current design we have a horizontal dimension to control the length of the horizontal line. If the length of this line is modified to a greater value, *NX* can only lengthen the line toward the right side. This is due to the fact that the location of the left endpoint is governed by the two location dimensions we just added.

Over-Constraining 2D Sketches

- We can use *NX* to build partially constrained or totally unconstrained solid models. In most cases, these types of models may behave unpredictably as changes are made. On the other end, *NX* will also let us over-constrain a sketch, but with warnings. It is best not to have created over-constrained sketches; only create extra dimensions as references if necessary. These additional dimensions are generally called *reference dimensions*. *Reference dimensions* do not constrain the sketch; they only reflect the values of the dimensioned geometry.

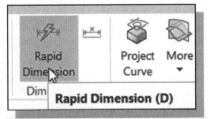

1. Select the **Rapid Dimension** command in the *Constraints* toolbar as shown.

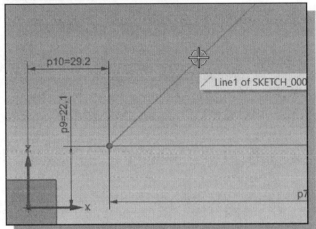

2. Select the **left inclined line** of the triangle as shown.

3. Place the dimension above the line; note the edit dimension option is automatically activated.

4. Notice a warning message indicating that we have over-constrained the 2D sketch is displayed in the message area. Click **OK** to close the dialog box.

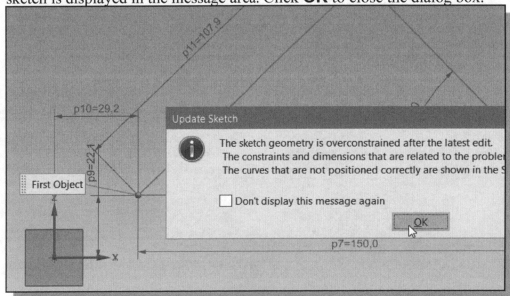

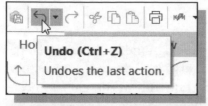

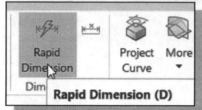

5. Click **Undo**, or use [**Ctrl+Z**], to remove the dimension we just added.

6. Select the **Rapid Dimension** command in the *Sketch Constraints* toolbar as shown.

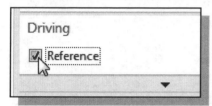

7. Activate the **Reference Dimension** option in the Driving option area as shown.

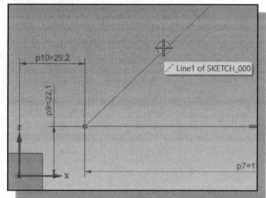

8. Select the **left inclined line** of the triangle as shown.

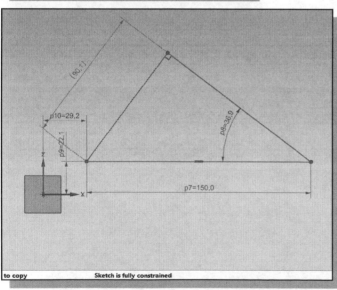

9. Place the dimension above the line; note the reference dimension is displayed as a single numeral value inside brackets and different color. Note that a reference dimension cannot be edited.

❖ As a general rule, it is best not to create over-constrained 2D sketches which can cause confusions.

Delete Existing Dimensions and Constraints

1. On your own, pre-select the **angle dimension** by clicking on the entities with the **left-mouse-button (MB1)**.

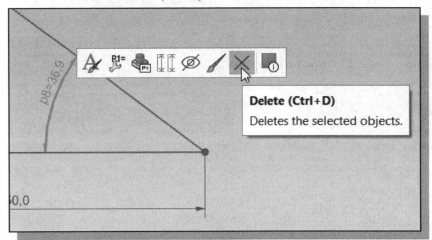

2. Click on the **Delete** icon or hit the **Delete** key once to remove the pre-selected dimensions.

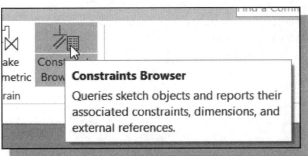

3. Select the **Relations Browser** command in the *Sketch Constraints* toolbar.

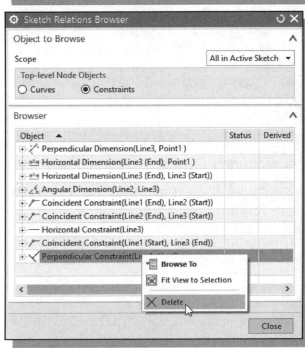

4. Select the **All in active sketch** in the *Scope* option, and **Constraints** in the *Top-level Node Objects* option.

5. Select the **Line1 Perpendicular to Line2** constraint in the *Browser list* as shown.

6. Right-click once and choose **Delete** to remove the selected constraint.

7. Click **Close** to exit the **Relations Browser** command.

8. Click and drag with the **left-mouse-button (MB1)** on one of the inclined lines and notice that we can adjust the upper portion of the triangle, but the horizontal line of the triangle remains fixed.

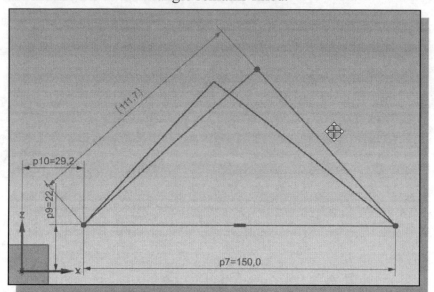

9. **Right-mouse-button (MB3)** click on the **right inclined line** to bring up the option menu. Notice the pop-up icon toolbar below the pop-up menu.

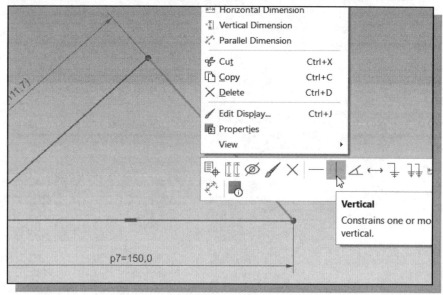

10. Select the **Vertical Constraints** command in the option menu or in the *pop-up* toolbar.

- Note that the *pop-up* toolbar, which provides options that are feasible to the current selected objects, can also be displayed by using the **left-mouse-button (MB1)**.

➤ Note that NX displays only options that are feasible to the current selected objects. This is known as the context sensitive approach, which makes the selection much easier.

❖ Currently two geometric constraints and three dimensions are applied to the sketch. How many more constraints are needed to make the sketch fully constrained?

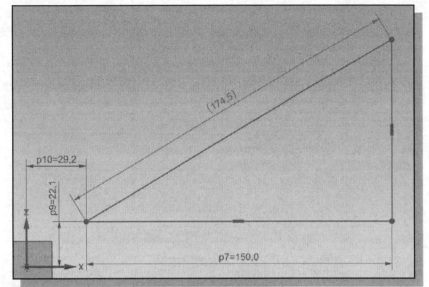

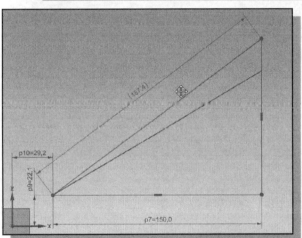

11. Click and drag with the **left-mouse-button (MB1)** on the inclined line and notice that we can still adjust the upper corner aligning to the right vertical line, but the lower two corners of the triangle remain fixed.

12. On your own, add a **vertical dimension** as shown. (Remember to un-check the *Reference* option.)

❖ Note that the sketch is now fully constrained. Consider what other constraints can be applied to the triangles while achieving the same results.

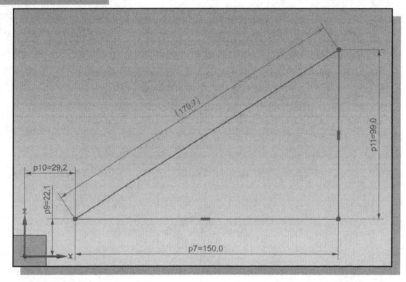

2D Sketches with Multiple Loops

❖ In *Parametric Modeling*, one option to combine multiple features into one feature is to create multiple loops within a 2D sketch. For example, if we need to create a hole in a triangular plate, one option is to create the triangular plate first and then create a hole feature. This approach will require two steps of solid features to complete. We can simplify the construction by creating two loops in one sketch. The outside loop (the triangle) will form the boundary of the material, and the inside loop (the circle) indicates a region of material that will be removed. The use of multiple loops can greatly simplify the content in the feature history tree.

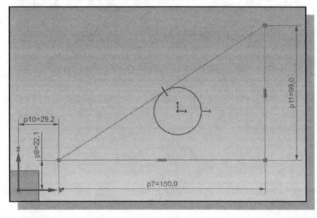

1. On your own, delete the *reference dimension* and create a **circle** of arbitrary size inside the triangle as shown in the figure.

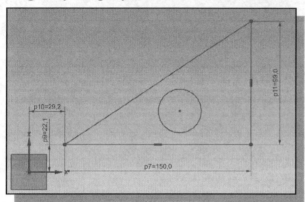

2. Select the **Geometric Constraints** command in the *Sketch Constraints* toolbar.

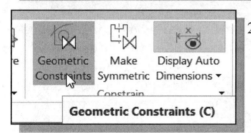

3. Click on the **Tangent** constraint icon in the *Sketch Constraints* toolbar. The sketched geometry is adjusted.

4. Pick the circle, near the top, and click once with the **middle-mouse-button (MB2)**.

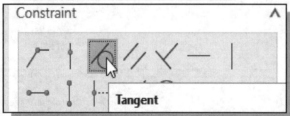

5. Pick the **inclined line**.

6. Click **Close** to exit the Geometric Constraints command.

•   How many more constraints or dimensions do you think will be necessary to fully constrain the circle? Which constraints or dimensions would you use to fully constrain the geometry?

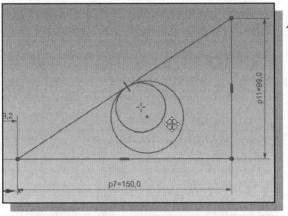

7. Move the cursor on top of the right side of the circle, and then drag the circle toward the right edge of the graphics window. Notice the size of the circle is adjusted while the system maintains the Tangent constraint.

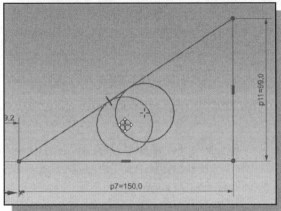

8. Drag the center of the circle in the lower left direction. Notice the Tangent constraint is always maintained by the system.

> ➢ On your own, experiment with adding additional constraints and/or dimensions to fully constrain the sketched geometry.

❖ One possibility is to have two **tangency** constraints and a **horizontal alignment** constraint as shown in the below figure (Circle center aligned to the mid-point of the vertical line.)

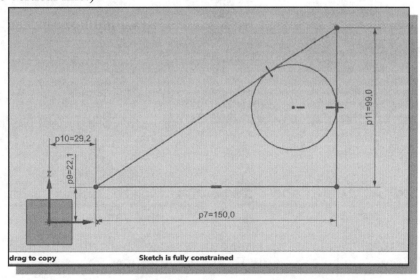

❖ The application of different constraints affects the geometry differently. The design intent is maintained in the CAD model's database and thus allows us to create very intelligent CAD models that can be modified and revised fairly easily.

9. On your own, modify the 2D sketch as shown below.

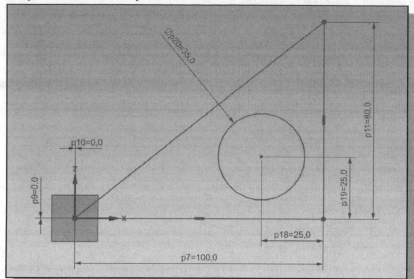

Inferred Constraint Settings

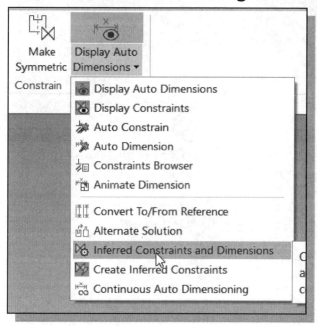

1. Display the **Inferred Constraints and Dimensions** option in the *Constraints* menu.

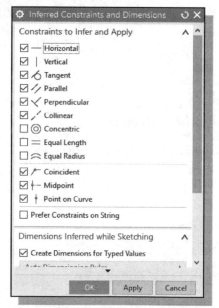

❖ In the **Inferred Constraint and Dimensions** dialog box, we can determine which constraints are recognizable and can be applied by the system.

2. On your own, adjust the settings and experiment with the effects of the different settings. Especially note the **Auto Dimensioning Rules**.

3. On your own, complete the **Extrude** command and create a 3D solid model with a plate thickness of **40mm**.

Parametric Relations

- In parametric modeling, dimensions are design parameters that are used to control the sizes and locations of geometric features. Dimensions are more than just values; they can also be used as feature control variables. This concept is illustrated by the following example.

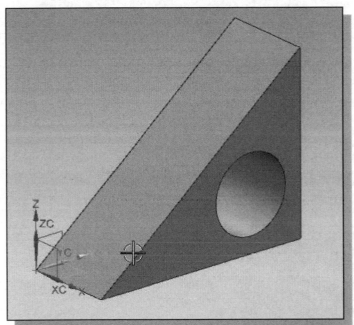

1. Double-click on the solid model to enter the **Edit with Rollback** option.

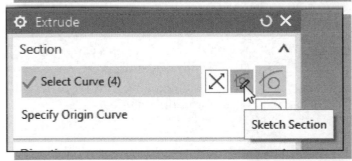

2. Click on the **Sketch Section** icon to enter the 2D sketcher mode.

- On your own, change the overall width of the triangle to **150** and the overall height of the rectangle to **100** and observe the location of the circle in relation to the edges of the triangle. Adjust the dimensions back to what are shown in the figure before continuing to the next page.

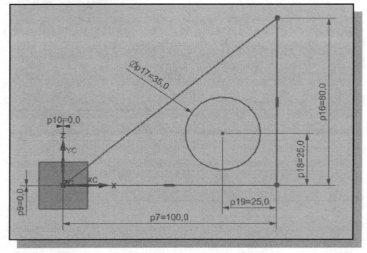

Dimensional Values and Dimensional Variables

Initially in *NX*, values are used to create different geometric entities. The text created by the Dimension command also reflects the actual location or size of the entity. **Dimensional constraints** are used to describe the SIZE and LOCATION of individual geometric shapes. Each dimension is also assigned a name that allows the dimension to be used as a control variable. The default format is "pxx," where the "xx" is a number that *NX* increments automatically each time a new dimension is added.

Let us look at our current design, which represents a triangular plate with a hole. The dimensional values describe the size and/or location of the plate and the hole. If a modification is required to change the width of the plate, the location of the hole will remain the same as described by the two location dimensional values. This is okay if that is the design intent. On the other hand, the *design intent* may require (1) keeping the location of the hole at a specific proportion to the length of the edges and (2) maintaining the size of the hole to be always one-third of the height of the plate. We will establish a set of parametric relations using the dimensional variables to capture the design intent described in statements (1) and (2) above.

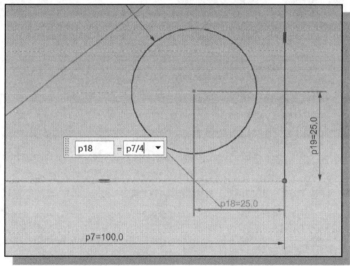

1. Select the horizontal location **dimension** of the circle by double-clicking on the dimension.

2. In the dimension edit box, enter **p7/4** as shown. (The **p7** variable name is the overall width of the triangle; note the variable name may be different for your triangle.)

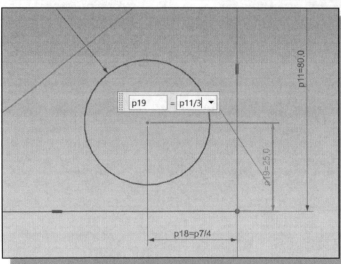

3. Double-click on the vertical location dimension of the circle and enter **p11/3** to make the dimension always equal to one third of the height of the overall height of the triangle. (Again, use the corresponding variable name for your equation.)

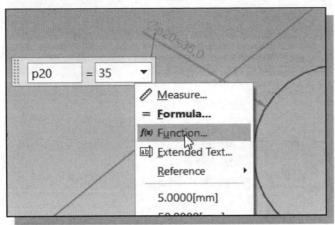

4. On your own, double click on the diameter dimension (**35**).

5. Click on the triangle to show the option list, and choose **Function** as shown.

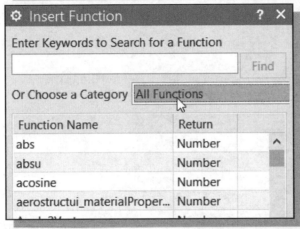

6. Select **All Functions** to view all of the functions that are available in *NX*. Note that the list includes basic trigonometry functions to complex gear calculation functions.

7. Click **Cancel** to exit the *Functions option*.

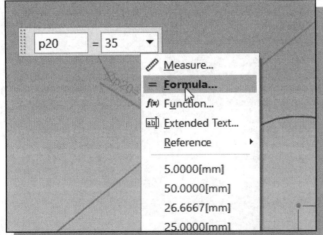

8. Choose **Formula** in the *Edit Dimension* box as shown.

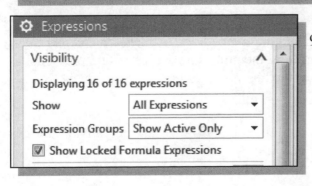

9. Select **All Expressions** in the *Listed Expressions* area to display all of the variables used in the model.

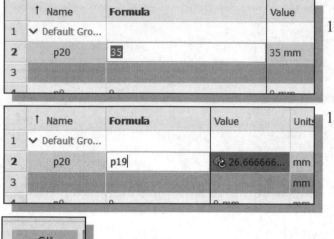

10. Click inside the **Formula Edit Box** as shown.

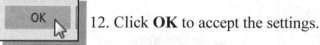

11. Select the vertical location dimension **p12** in the *Expressions dialog box* or type the name in the edit box as shown.

12. Click **OK** to accept the settings.

13. On your own, change the dimensions of the rectangle to **150 x 100** and observe the changes to the location and size of the circle.

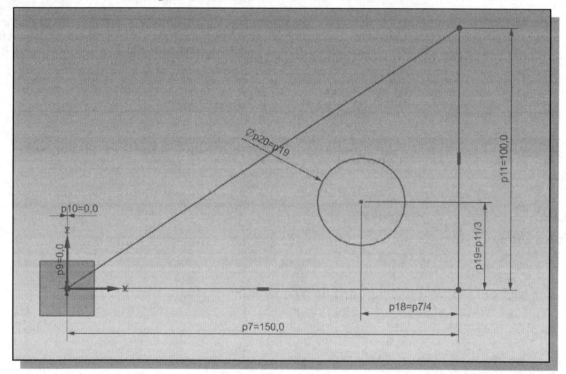

❖ *NX* automatically adjusts the dimensions of the design, and the parametric relations we entered are also applied and maintained. The dimensional constraints are used to control the size and location of the hole. The design intent, previously expressed by statements (1) and (2) at the beginning of this section, is now embedded into the model.

➢ On your own, experiment with establishing and modifying additional parametric relations in the model.

Review Questions:

1. Describe the basic concepts of the BORN technique.

2. What is the difference between *dimensional* constraints and *geometric* constraints?

3. How can we confirm a sketch is fully constrained?

4. How do we create derived dimensions?

5. Describe the procedure to Display/Edit user-defined equations.

6. How do we show and remove constraints?

7. List and describe three different geometric constraints available in *NX*.

8. Does *NX* allow us to build partially constrained or totally unconstrained solid models? What are the advantages and disadvantages of building these types of models?

9. How do we confirm the 2D sketch is fully constrained?

10. Create sketches showing the steps you plan to use to create the model shown on the next page:

Exercises: (Establish three parametric relations for each of the designs and experiment with the different modification techniques illustrated in this lesson.)

1. **Swivel Base** (Dimensions are in millimeters. Base thickness: **10 mm.** Boss: **5 mm.**)

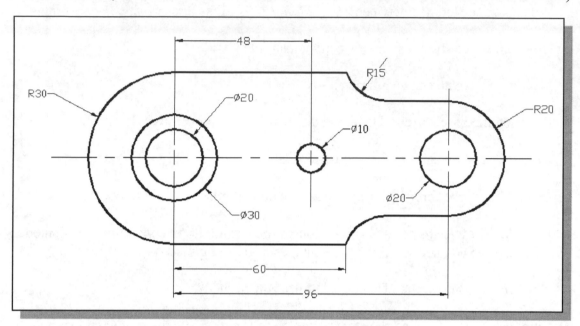

2. **Anchor Base** (Dimensions are in inches.)

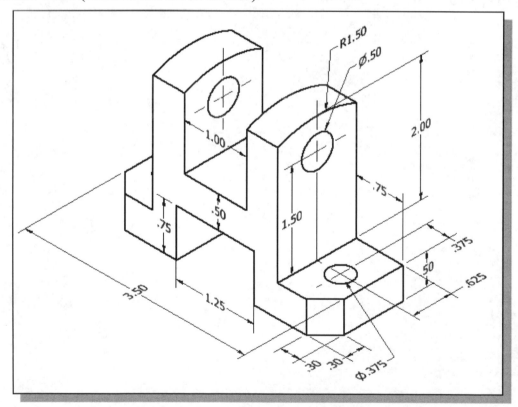

3. **Wedge Block** (Dimensions are in inches.)

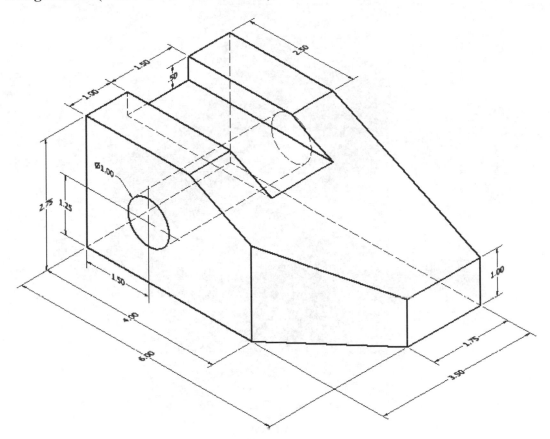

4. **Hinge Guide** (Dimensions are in inches.)

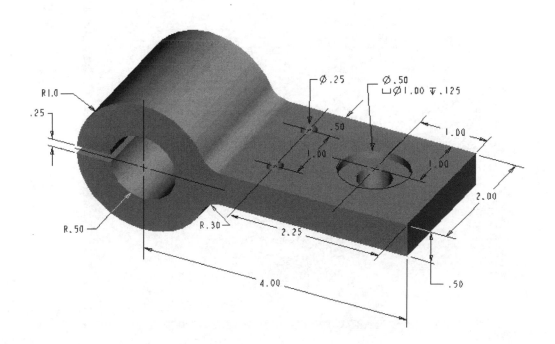

5. **Pivot Holder** (Dimensions are in inches.)

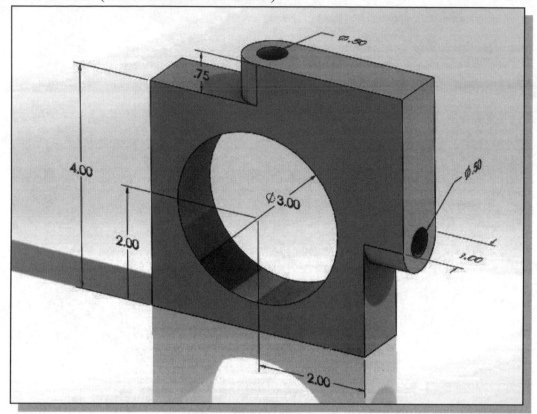

6. **Support Fixture** (Dimensions are in inches.)

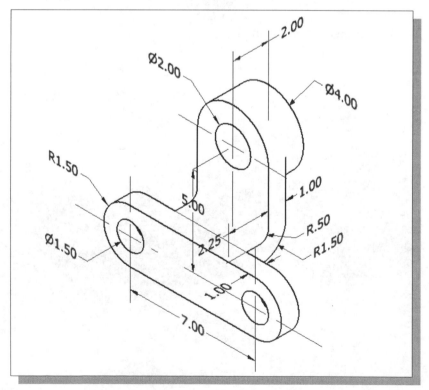

Chapter 6
Geometric Construction Tools

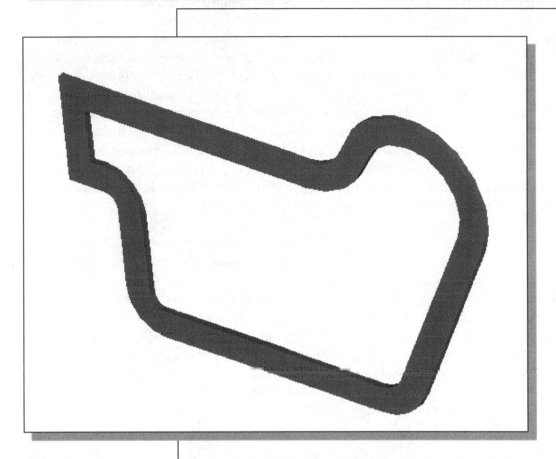

Learning Objectives

- ♦ **Applying Geometry Constraints**
- ♦ **Use the Trim/Extend Command**
- ♦ **Use the Offset Command**
- ♦ **Understand the Haystacking Approach**
- ♦ **Create Projected Geometry**
- ♦ **Edit with Click and Drag**

Introduction

The main characteristics of solid modeling are the accuracy and completeness of the geometric database of the three-dimensional objects. However, working in three-dimensional space using input and output devices that are largely two-dimensional in nature is potentially tedious and confusing. *NX* provides an assortment of two-dimensional construction tools to make the creation of wireframe geometry easier and more efficient. *NX* includes two types of wireframe geometry: ***curves*** and ***sections***. Curves are basic geometric entities such as lines, arcs, etc. Sections are a group of curves used to define a boundary. A *section* is a closed region and can contain other closed regions. Sections are commonly used to create extruded and revolved features. An *invalid section* consists of self-intersecting curves or open regions. In this lesson, the basic geometric construction tools, such as Trim and Extend, are used to create sections. The *NX's **Haystacking*** approach to creating sections is also introduced. Mastering the geometric construction tools along with the application of proper constraints and parametric relations is the true essence of *parametric modeling*.

In *NX*, **sections** are closed regions that are defined from sketches. Sections are used as cross sections to create solid features. For example, **Extrude**, **Revolve**, and **Sweep** operations all require the definition of at least a single section. The sketches used to define a section can contain additional geometry since the additional geometric entities are consumed when the feature is created. To create a **section** we can create single or multiple closed regions, or we can select existing solid edges to form closed regions. A section cannot contain self-intersecting geometry; regions selected in a single operation form a single section. As a general rule, we should dimension and constrain sections to prevent them from unpredictable size and shape changes. *NX* does allow us to create under-constrained or non-constrained sections; the dimensions and/or constraints can be added and/or edited later.

The Gasket Design

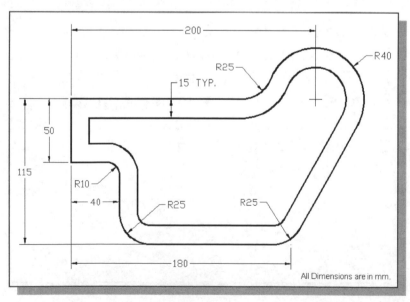

❖ Based on your knowledge of *NX* so far, how would you create this design? What is the more difficult geometry involved in the design? You are encouraged to create the design on your own prior to following through the tutorial.

Modeling Strategy

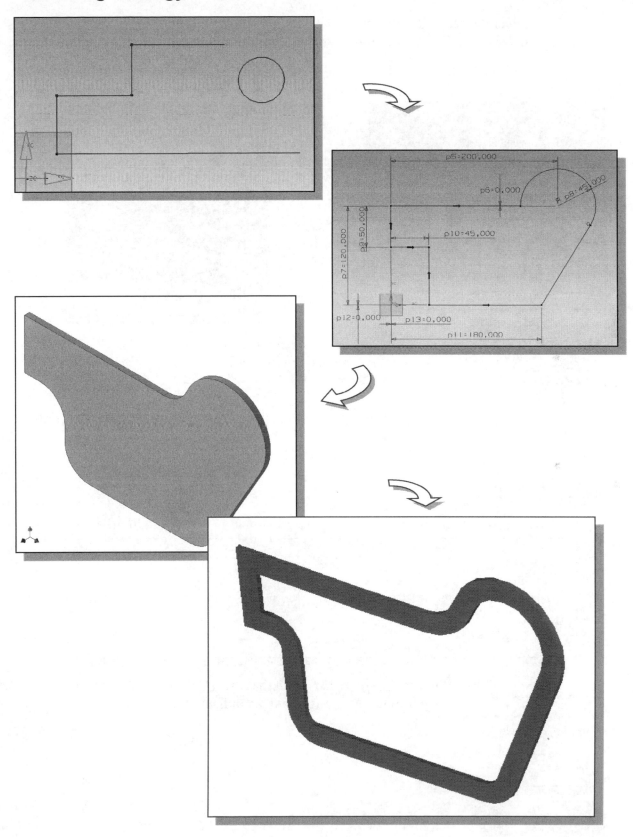

Starting NX

1. Select the **NX** option on the *Start* menu or select the **NX** icon on the desktop to start *NX*. The *NX* main window will appear on the screen.

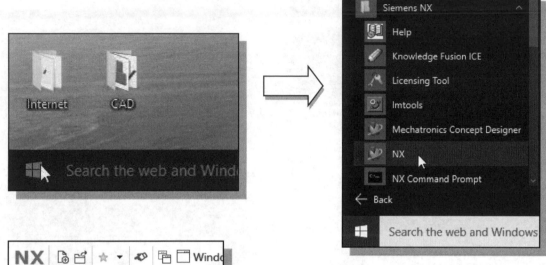

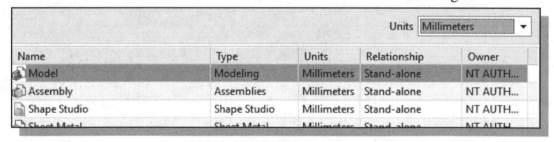

2. Select the **New** icon with a single click of the left-mouse-button (MB1) in the *Standard toolbar area*.

- In this chapter, we will use the **Model Template** that comes with the *NX* software. Using the template also allows us to use the *BORN* technique as a *Datum Coordinate System* is already established in the template.

3. Confirm the **Millimeters** units is set as shown in the below figure.

		Units	Millimeters	▼

Name	Type	Units	Relationship	Owner
Model	Modeling	Millimeters	Stand-alone	NT AUTH...
Assembly	Assemblies	Millimeters	Stand-alone	NT AUTH...
Shape Studio	Shape Studio	Millimeters	Stand-alone	NT AUTH...
Sheet Metal	Sheet Metal	Millimeters	Stand-alone	NT AUTH

4. Select **Model** in the *Template list*. Note that the *Model template* will allow us to switch directly into the **Modeling task** as indicated in the *templates list*.

5. In the *New File Name* box, enter **Gasket.prt** as the *File Name*.

6. Click **OK** to proceed with the *New Part File* command.

Create the Sketch of the Base Feature

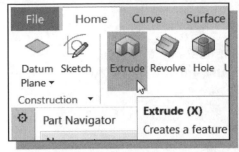

1. In the *Feature Toolbars*, select the **Extrude** icon as shown.

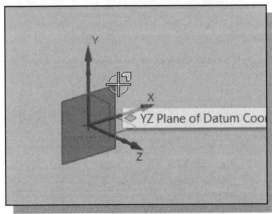

2. Select the **Y-Z Plane** of the established datum coordinate system as the sketch plane of the Base Feature.

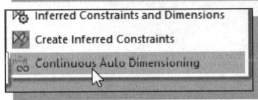

3. On your own, turn **off** the **Continuous Auto Dimensioning** option.

4. Create a sketch as shown in the figure below. Start the sketch from the top right corner. The line segments are all parallel and/or perpendicular to each other. We will intentionally make the line segments of arbitrary length, as it is quite common during the initial design stage that not all of the values are determined.

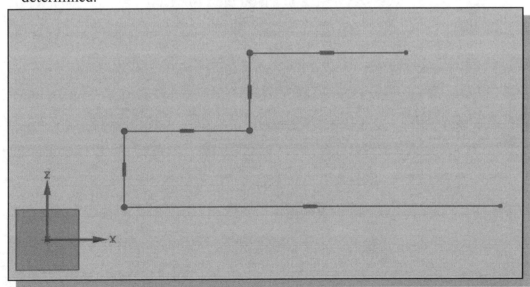

5. Inside the graphics window, middle-mouse-click (**MB2**) twice to end the Profile command.

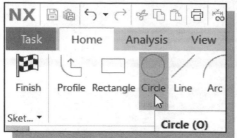

6. Select the **Circle** command by clicking once with the **left-mouse-button (MB1)** on the icon in the *Sketch* toolbar.

7. Pick a location that is above the bottom horizontal line as the center location of the circle.

8. Move the cursor toward the right and create a circle, by clicking with the left-mouse-button, of arbitrary size and not aligned to any of the existing geometry.

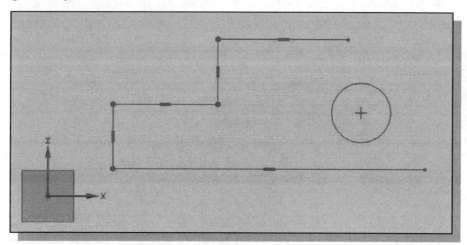

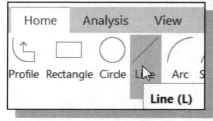

9. Click on the **Line** icon in the *2D Sketch Panel*.

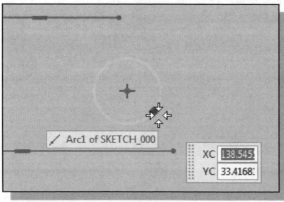

10. Move the cursor near the lower-right portion of the circle and pick a location on the circle as shown.

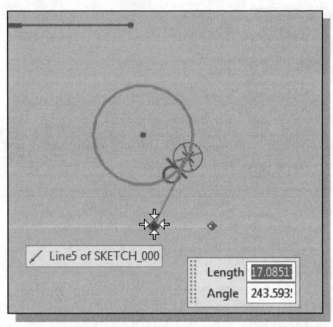

11. For the other end of the line, select a location that is on the lower horizontal line and about one-third from the right endpoint. Click when the **Tangent** constraint symbol is also added.

12. Inside the graphics window, middle-mouse-click (**MB2**) once to end the *Line* command.

Edit the Sketch by Dragging the Entities

❖ In *NX*, we can click and drag any under-constrained curve or point in the sketch to change the size or shape of the sketched section. As illustrated in the previous chapter, this option can be used to identify under-constrained entities. This *Editing by Drag and drop* method is also an effective visual approach that allows designers to quickly make changes.

1. Move the cursor on the **lower left vertical edge** of the sketch; click and drag the edge to a new location that is toward the right side of the sketch.

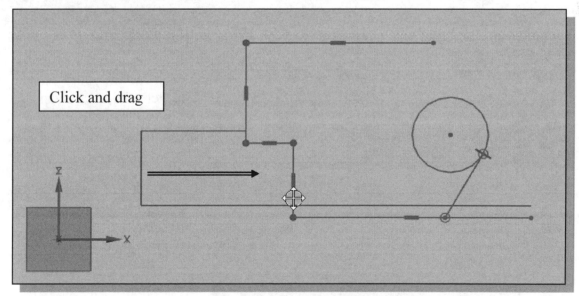

❖ Note that we can only drag the vertical edge horizontally; the connections to the two horizontal lines are maintained while we are moving the geometry.

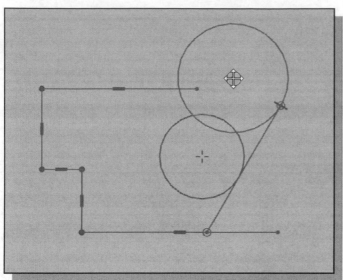

2. Click and drag the **center point** of the circle to a new location.

❖ Note that as we adjust the size and the location of the circle, the connection to the inclined line is maintained.

3. Click and drag the **inclined line** toward the right.

❖ Note that as we adjust the location of the inclined line the location of the circle and the size of the bottom horizontal edge are also adjusted.

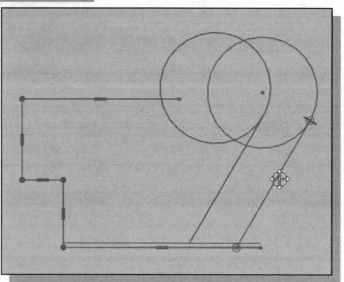

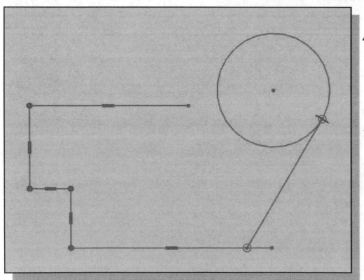

4. On your own, adjust the sketch so that the shape of the sketch appears roughly as shown.

❖ The *Editing by Drag and drop* method is an effective approach that allows designers to explore and experiment with different design concepts.

Add Additional Constraints

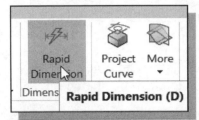

1. Select the **Rapid Dimension** command in the *Constraints* toolbar as shown.

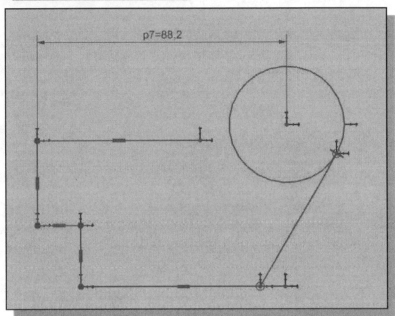

2. Add the **horizontal location dimension**, from the top left vertical edge to the center of the circle as shown. (Do not be overly concerned with the dimensional value; we are still working on creating a *rough sketch*.)

3. Click **Close** to end the *Dimension* command.

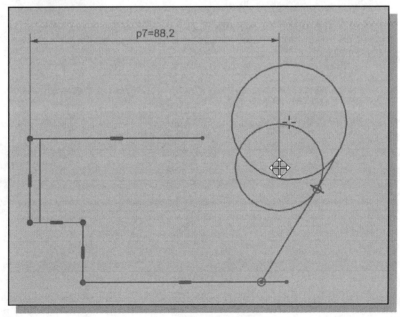

4. Click and drag the **circle** to a new location.

❖ Notice that the location dimension we added now restricts the horizontal movement of the center of the circle. The tangent relation to the inclined line is also maintained.

First Construction Method – Trim/Extend

➤ In the following sections, we will illustrate using the Trim and Extend commands to complete the desired 2D sketch.

❖ The **Trim** and **Extend** commands can be used to shorten/lengthen an object so that it ends precisely at a boundary. As a general rule, *NX* will try to clean up sketches by forming a closed region sketch.

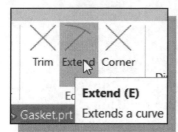

1. Choose **Extend** in the *Sketch Toolbar*.

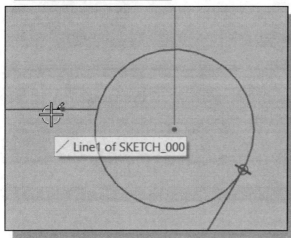

2. We will first extend the top horizontal line to the circle. Move the cursor near the right hand endpoint of the top horizontal line. *NX* will automatically display the possible result of the selection.

3. Inside the graphics window, middle-mouse-click (**MB2**) once to end the *Extend* command.

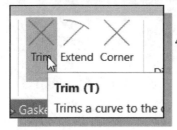

4. Choose **Quick Trim** in the *Sketch Toolbar*.

5. We will next trim the bottom horizontal line to the inclined line. Move the cursor near the right hand endpoint of the bottom horizontal line. *NX* will highlight the portion of the line that will be trimmed.

6. Left-mouse-click once on the line to perform the **Trim** operation.

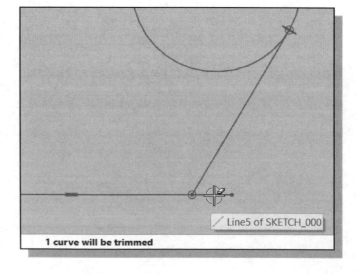

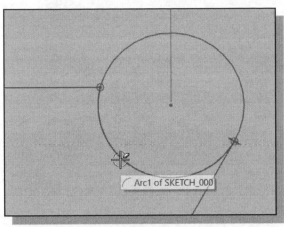

7. Left-mouse-click once on the lower portion of the **circle** to perform the *Trim* operation on it.

8. Inside the graphics window, middle-mouse-click (**MB2**) once to end the *Trim* command.

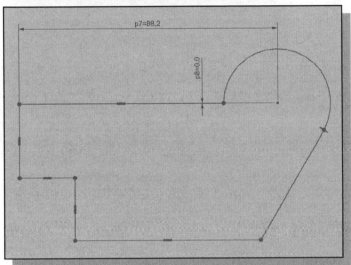

9. On your own, create the **vertical location dimension** of the center of the arc as shown below.

10. Adjust the dimension to **0.0** so that the horizontal line and the center of the circle are aligned horizontally.

11. On your own, create the additional **linear dimensions** as shown. Also confirm that **proper constraints** are applied to the horizontal and vertical lines.

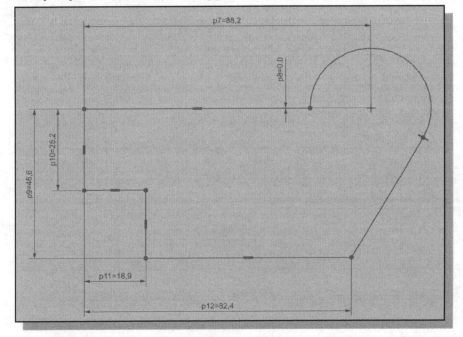

12. On your own, create the radius dimension for the arc as shown. Use the Radial method under the measurement list as shown.

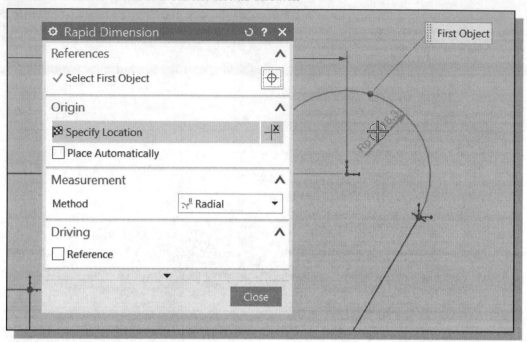

13. On your own, create the two locational dimensions to align the sketch to the origin of the CS as shown in the figure. Also adjust the dimensions to what is shown in the figure. The 2D sketch is now fully constrained.

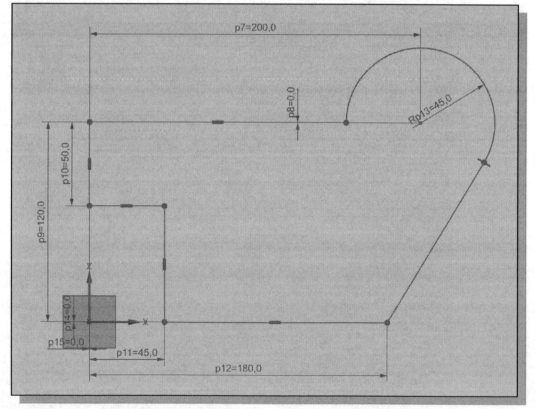

Create Fillets and Completing the Sketch

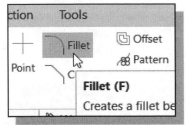

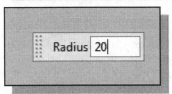

1. Select the **Fillet** command by clicking once with the **left-mouse-button (MB1)** on the icon in the *Sketch* toolbar.

2. Enter **20** as the new radius of the fillet.

3. On your own, create the four additional **fillets** as shown in the below figure. Note that the same radius value is used on all rounds and fillets; also note that all of the rounds and fillets are created with the proper constraints.

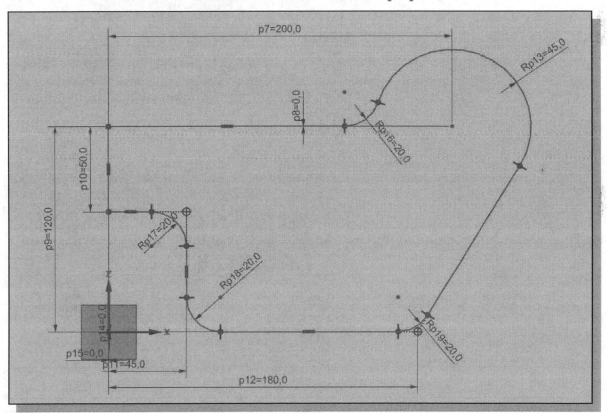

❖ On your own, try using the drag and drop option, and read the associated message, to confirm the 2D sketch is still fully constrained.

Complete the Extrusion Feature

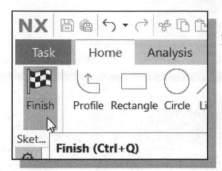

1. Click **Finish Sketch** to exit the NX sketcher mode and return to the *Feature option dialog window*.

2. Enter **5 mm** as the *End* extrusion distance as shown.

3. Click **OK** to create the extrusion feature.

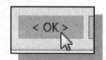

❖ Note that all the sketched geometric entities and dimensions are consumed and have disappeared from the screen when the feature is created.

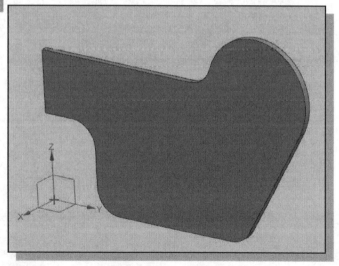

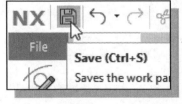

4. Click **Save** to save the current *Gasket* design.

❖ The above procedure illustrated the *Trim/Extend* approach to create the Gasket design; we will next look at the *Haystacking* approach to create the same design.

Second Construction Method – Haystacking Geometry

Haystacking is a grouping mechanism that allows us to select only the wireframe entities that we wish to include in the section. *Haystacking* is a tool that helps maintain design intent by reducing the amount of trimming necessary to build a section. Note that in *NX*, the *Haystacking* approach is done through the selection option.

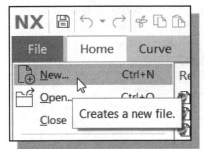

1. In the *File pull-down menu*, select **New** with a single click of the left-mouse-button (MB1).

2. Confirm the **Millimeters** units is set as shown in the below figure.

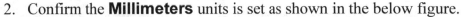
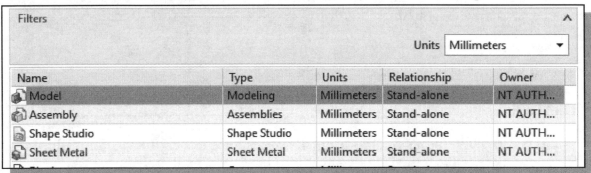

Name	Type	Units	Relationship	Owner
Model	Modeling	Millimeters	Stand-alone	NT AUTH...
Assembly	Assemblies	Millimeters	Stand-alone	NT AUTH...
Shape Studio	Shape Studio	Millimeters	Stand-alone	NT AUTH...
Sheet Metal	Sheet Metal	Millimeters	Stand-alone	NT AUTH...

3. Select **Model** in the *Template list*. Note that the *Model template* will allow us to switch directly into the *Modeling task* as indicated in the *templates list*.

4. In the *New Part File* window, enter **Gasket1.prt** as the *File Name*.

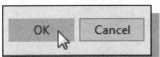

5. Click **OK** to proceed with the New Part File command.

6. In the *Feature Toolbars*, click the **Extrude** button as shown.

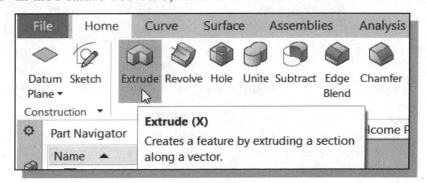

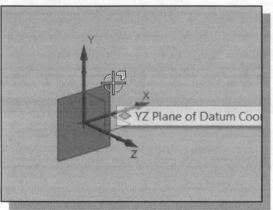

7. Select the **YZ Plane** of the established datum coordinate system as the sketch plane of the *Base Feature*.

8. Click **OK** to accept the selection of the *Sketch Plane*.

9. On your own, confirm the **Display Sketch Constraints**, **Display Sketch Auto Dimension**, **Create Inferred Constraints** and **Continuous Auto Dimensioning** options are activated.

10. We will first create **six line segments**, starting the sketch from the top right corner. Note that the first five lines are either horizontal or vertical.

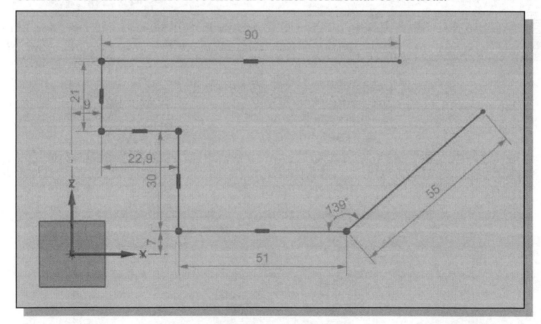

*   Note that with the **Create Inferred Constraints** and **Continuous Auto Dimensioning** options switched on, NX will automatically create both constraints and dimensions to maintain a fully constrained sketch.

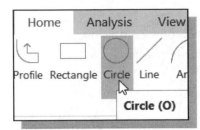

11. Next create a **circle** at the right-endpoint of the top horizontal line.

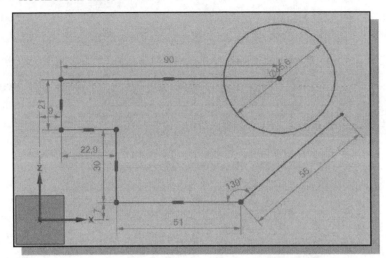

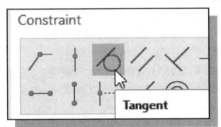

12. Apply a **Tangent** constraint to the inclined line and the circle.

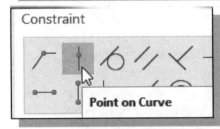

13. Apply a **Point On Curve** constraint to the end of the inclined line and the circle.

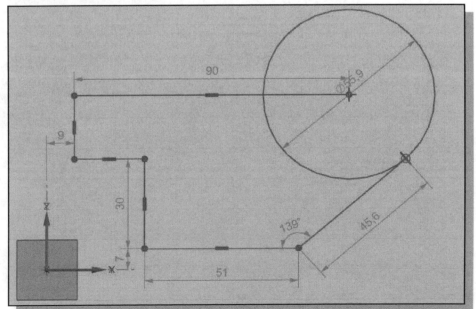

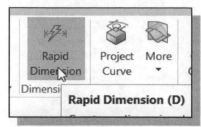

14. Select the **Rapid Dimension** command in the *Constraints* toolbar as shown.

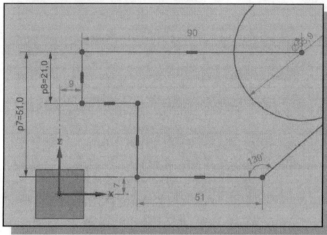

15. Create the **overall height** (p7) and the **height** (p8) dimensions and place them to the left side as shown.

- Note the newly created dimensions have the variable names (p7 and p8) associated with the displayed values.

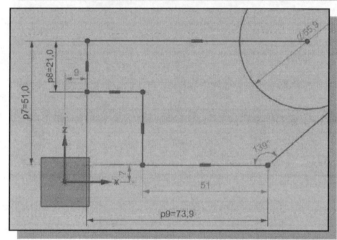

16. Create the **width dimension** (p9) and place it below the sketch as shown.

- Note that *NX* automatically removes a dimension every time a new dimension is created.

17. On your own, create and modify the dimensions as shown in the figure.

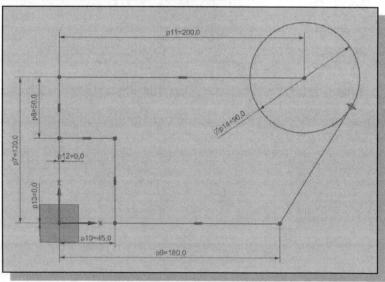

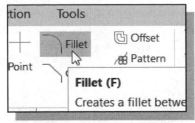

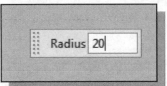

18. Select the **Fillet** command by clicking once with the **left-mouse-button (MB1)** on the icon in the *Sketch* toolbar.

19. Enter **20** as the new radius of the fillet.

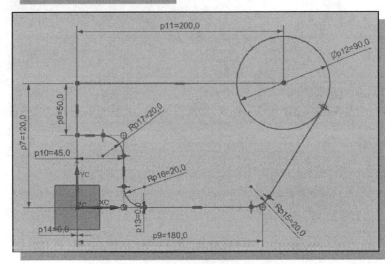

20. On your own, create the three **fillets** as shown in the figure. Note that all of the rounds and fillets are created with the proper constraints.

21. Complete the sketch by adding a **circle** tangent to both the large circle and the top horizontal line as shown.

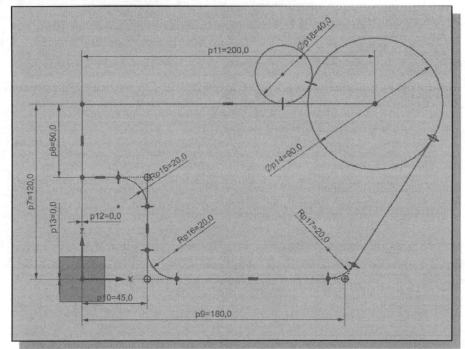

❖ On your own, examine the constraints and confirm the 2D sketch is fully constrained.

Use the NX Selection Intent Option

❖ To complete the extrusion, we will select only the segments we need by using the available selection options.

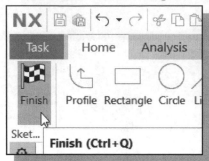

1. Click **Finish Sketch** to exit the NX sketcher mode and return to the *Feature option dialog window*.

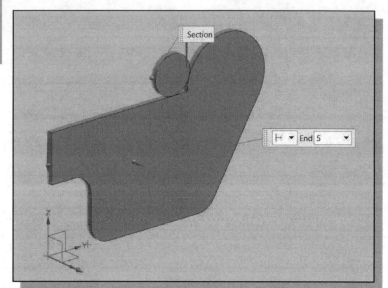

❖ Note the **NX Auto-Select** option has pre-selected some entities to form a section for extrusion. In this case, the pre-selected entities do not form the section we want.

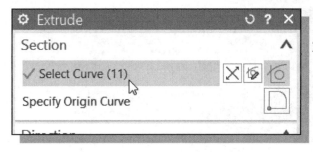

2. In the *Extrude* dialog box, click on **Select Curve** to edit the curve selection.

3. To de-select the selected curves, hold down the **[Shift]** key and click on the curve.

4. On your own, **de-select** all of the objects.

5. In the graphics window, right-click once to bring up the option menu. In the *Selection Filter* option area, choose **Tangent Curve** and **Stop at Intersection** options.

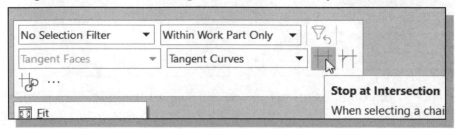

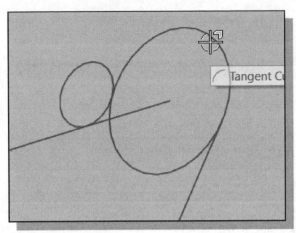

6. Select the top of the large *circle* and NX will show the selected portion as a **heavier line** and also show the two intersections to the adjacent entities as shown.

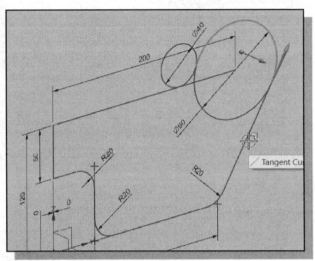

7. Select the adjacent **inclined line** and notice NX automatically selects the tangent curves.

8. On your own, select the adjacent entities to form the section as shown in the below figure.

9. Hit once with the middle button to end the selection.

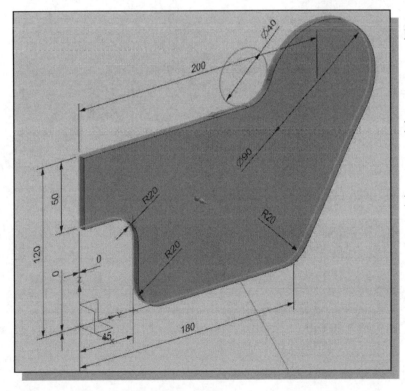

10. Choose all of the line segments necessary to form the profile as shown.

11. Enter **5 mm** as the extrusion distance as shown.

12. Click **OK** to create the extrusion feature.

❖ Note the hay-stacking approach is a very flexible and fast method to create 2D sections with less editing to the sketch.

Create an Associative Offset Cut Feature

➢ To complete the design, we will create a cutout feature by using the Offset Curve command.

1. In the *Feature* toolbar select the **Extrude** command by left-clicking once on the icon.

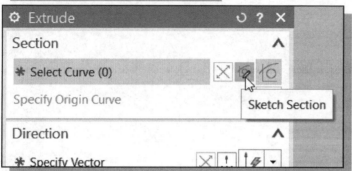

2. In the *Extrude dialog box*, pick on the **Sketch Section** option as shown.

3. Select the **front face**, near the upper left corner, of the 3D model in the orientation as shown.

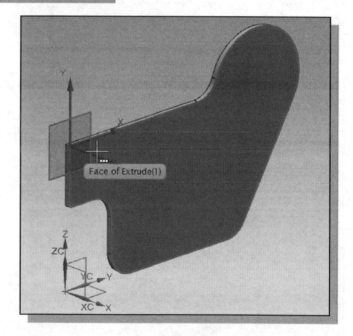

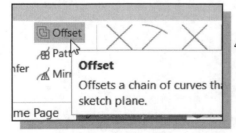

4. In the *Curve toolbar*, select **Offset** command as shown.

5. Select any surface edges of the model and **all curves** are selected and highlighted.

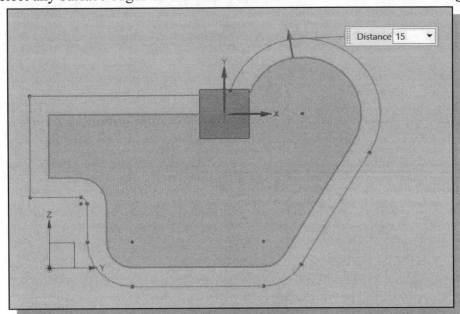

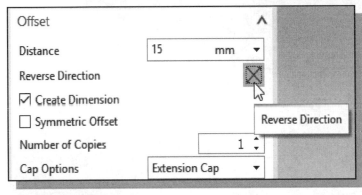

6. Enter **15 mm** in the offset distance input box. Also click on the **Reverse Direction** icon to set the offset toward the inside of the sketch.

7. Confirm the **Create Dimension** option is activated.

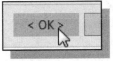

8. Click on the **OK** button to accept the settings and complete the design.

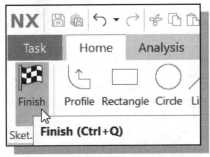

9. Click **Finish Sketch** to exit the NX sketcher mode and return to the *Feature option dialog box*.

10. Select the front surface of the model to set the extrusion direction.

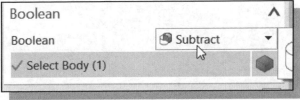

11. Set the *Boolean* option to **Subtract** as shown.

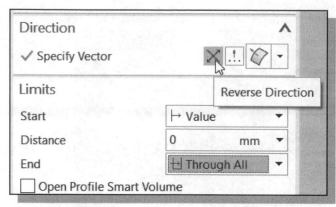

12. Click the **Reverse Direction** button to assure the extrusion will be cutting through the existing solid.

13. In the *Extrude* dialog box, set the termination *End Limit* to the **Through All** option.

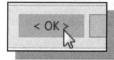

14. Click on the **OK** button to accept the settings and complete the design.

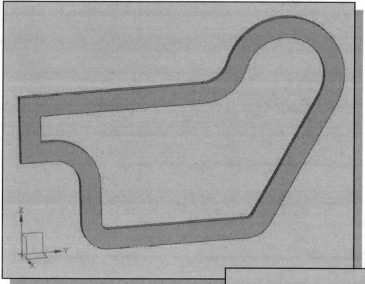

> The offset geometry is associated with the original geometry. On your own, adjust the overall height of the design to **150 millimeters** and confirm that the entire model, including the offset geometry, is adjusted accordingly.

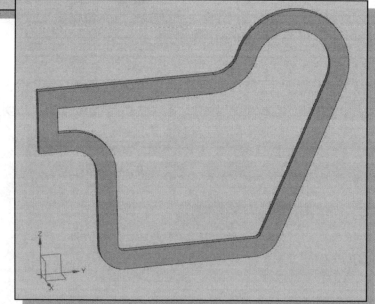

Review Questions:

1. What are the two types of wireframe geometry available in *NX*?

2. Can we create a section with extra 2D geometry entities?

3. How do we access the *NX's* **Edit Sketch** option?

4. How do we create a *Haystacking Section* in *NX*?

5. Can we build a section that consists of self-intersecting curves?

6. Describe the general procedure to create an associative Offset Curve.

7. Can we remove a portion of an existing 2D curve? How is this done?

8. What is the main advantage of using the *Haystacking section* approach?

9. Create sketches showing the steps you plan to use to create the model shown on the next page:

Exercises:

1. **V-slide Plate** (Dimensions are in inches. Plate Thickness: **0.25**)

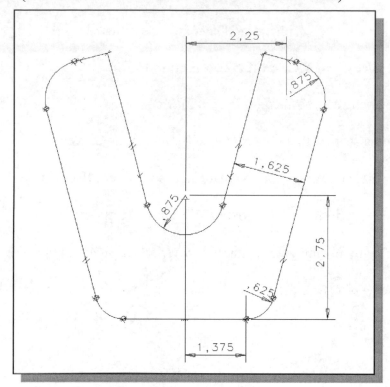

2. **Shaft Support** (Dimensions are in millimeters. Note the two R40 arcs at the base share the same center.)

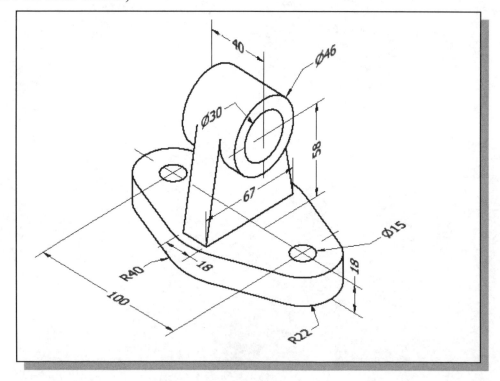

3. **Vent Cover** (Thickness: **0.125** inches.)

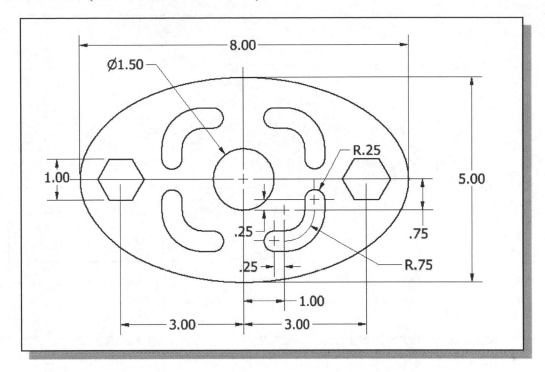

4. **Anchor Base** (Dimensions are in inches.)

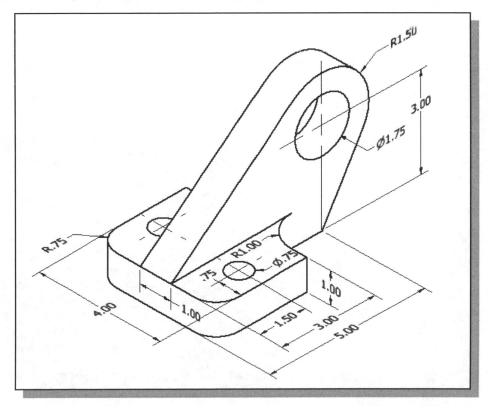

5. **Tube Spacer** (Dimensions are in inches.)

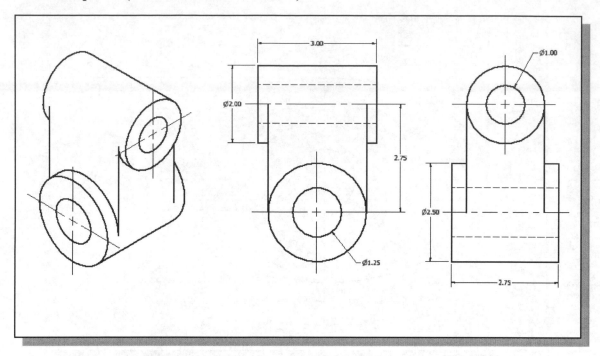

6. **Pivot Lock** (Dimensions are in inches. The circular features in the design are all aligned to the two centers at the base.)

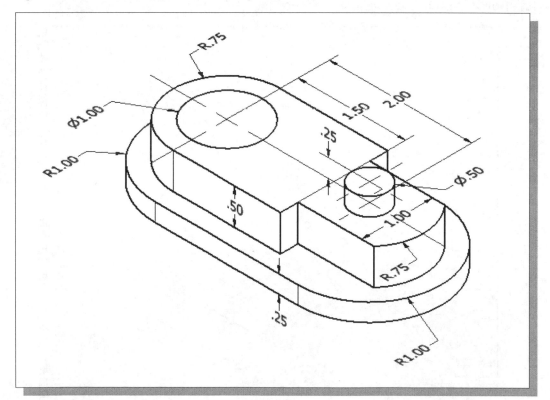

Chapter 7
Parent/Child Relationships

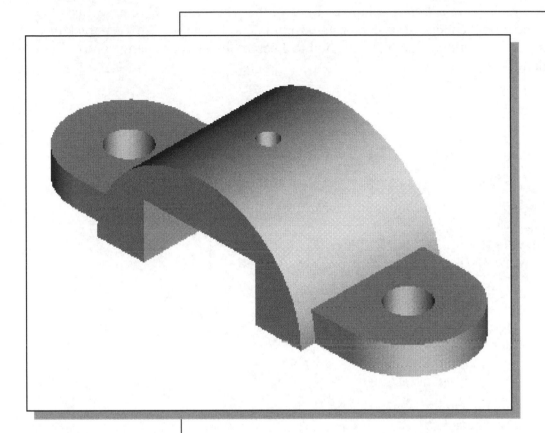

Learning Objectives

♦ **Understand the Importance of Parent/Child Relations in Features**

♦ **Use the Suppress Feature Option**

♦ **Resolve Undesired Feature Interactions**

♦ **Applying the BORN Technique to create Flexible Designs**

Introduction

The parent/child relationship is one of the most powerful aspects of ***parametric modeling***. In *NX*, each time a new modeling event is created, previously defined features can be used to define information such as size, location, and orientation. The referenced features become **PARENT** features to the new feature, and the new feature is called the **CHILD** feature. The parent/child relationships determine how a model reacts when other features in the model change, thus capturing design intent. It is crucial to keep track of these parent/child relations. Any modification to a parent feature can change one or more of its children.

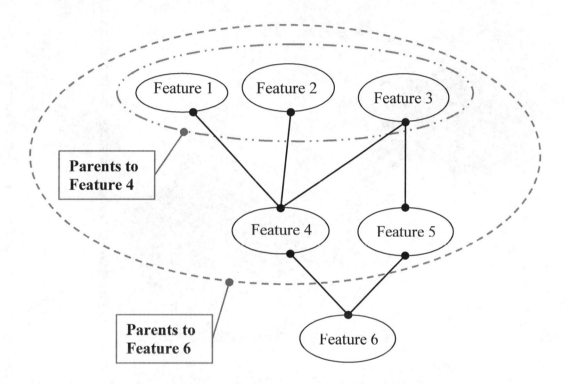

Parent/child relationships can be created *implicitly* or *explicitly*; implicit relationships are implied by the feature creation method and explicit relationships are entered manually by the user. In the previous chapters, we first select a sketching plane before creating a 2D profile. The selected surface becomes a parent of the new feature. If the sketching plane is moved, the child feature will move with it. As one might expect, parent/child relationships can become quite complicated when the features begin to accumulate. It is therefore important to think about modeling strategy before we start to create anything. The main consideration is to try to plan ahead for possible design changes that might occur which would be affected by the existing parent/child relationships. Parametric modeling software, such as NX, also allows us to adjust feature properties so that any feature conflicts can be quickly resolved. In this chapter, we will concentrate the discussion on parent/child relationships and also look at some of the tools that are available to handle design changes quickly and effectively.

The U-Bracket Design

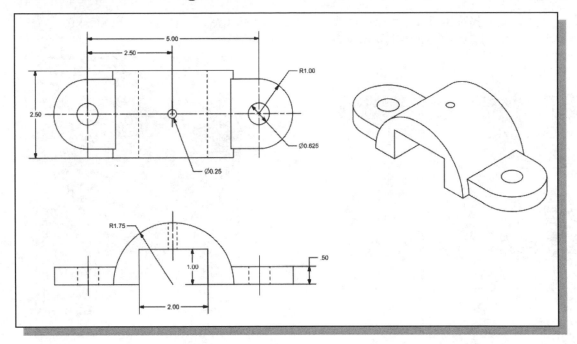

> Based on your knowledge of *NX* so far, how many features would you use to create the model? Which feature would you choose as the **base feature**? What is your choice for arranging the order of the features? Would you organize the features differently if the rectangular cut at the center is changed to a circular shape?

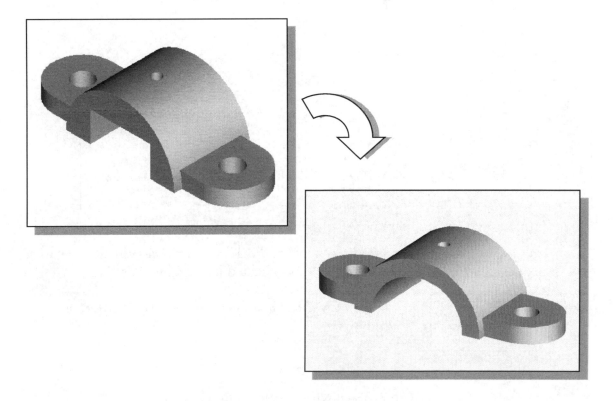

Create the Base Feature

1. Select the **NX** option on the *Start* menu or select the **NX** icon on the desktop to start *NX*. The *NX* main window will appear on the screen.

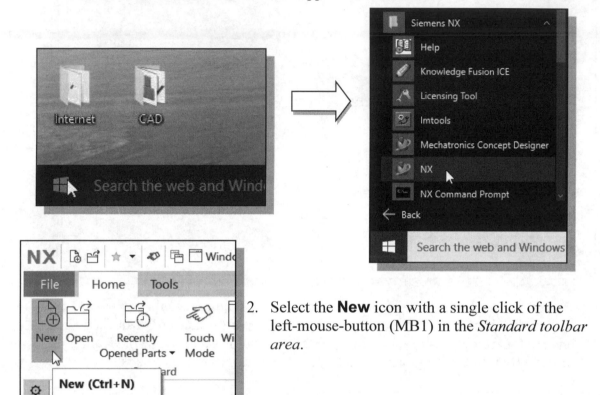

2. Select the **New** icon with a single click of the left-mouse-button (MB1) in the *Standard toolbar area*.

3. Select the **inches** units as shown in the below figure.

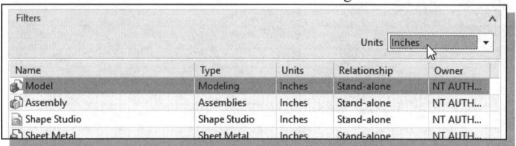

4. Select **Model** in the *Template list*. Note that the *Model template* will allow us to switch directly into the **Modeling task** as indicated in the *templates list*.

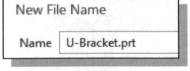

5. In the *New File Name* section, enter **U-Bracket** as the *File Name*.

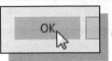

6. Click **OK** to proceed with the New File command.

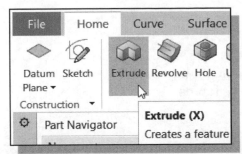

7. In the *Form Feature Toolbar,* select the **Extrude** icon as shown.

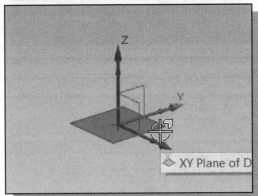

8. Select the **XY plane** of the displayed Work Coordinate System to align the *Sketch Plane.*

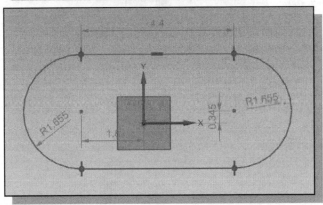

9. On your own, create the 2D sketch with two horizontal lines and two tangent arcs as shown. (Hint: Drag the **MB1** to switch to the Arc option while sketching with the *Profile* command.)

10. Click twice with the middle-mouse-button (**MB2**) to end the Profile command.

11. On your own, create and adjust the dimensions as shown in the below figure.

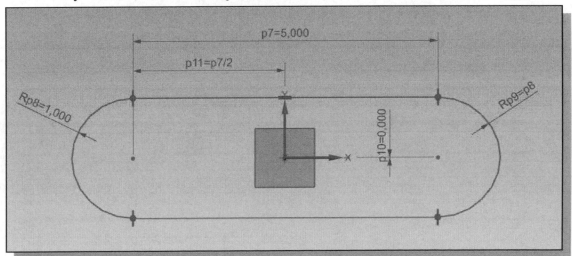

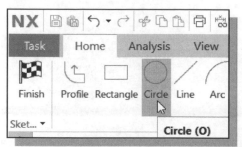

12. Select the **Circle** command by clicking once with the **left-mouse-button (MB1)** on the icon in the *Sketch Curve* toolbar.

13. Create the **two circles** aligned to the centers of the two larger arcs as shown.

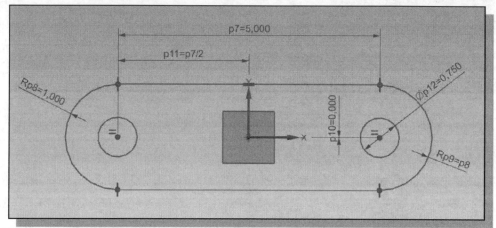

➢ How many more dimension/constraints are needed to make the sketch fully constrained?

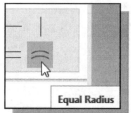

14. Create the necessary constraints and dimensions to make the sketch fully constrained as shown in the figure. (Hint: Try the **Equal Radius** constraint.)

Complete the Base Feature

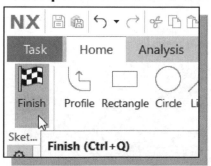

1. Click **Finish Sketch** to exit the NX sketcher mode and return to the *Feature option dialog window*.

2. Enter **0.5** in as the extrusion distance as shown.

3. Click **OK** to create the extrusion feature.

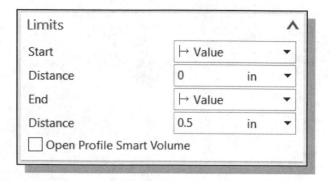

The Implied Parent/Child Relationships

- As we build solid features, NX records the individual steps and the parent/child relationships. Currently, several parent/child relationships have been established implicitly: *XY plane* was selected as the sketch plane; the *X* and *Y Axes* were used as references to position the 2D sketch in the 3D environment.

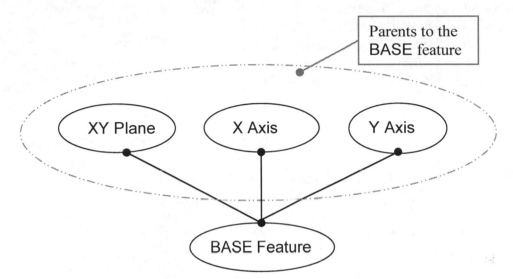

1. At the bottom of the *Part Navigator*, left-mouse-click (**MB1**) on the **Dependencies** option to show more details on the model history tree.

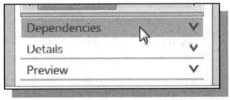

2. Select the **Extrude (1)** item as shown.

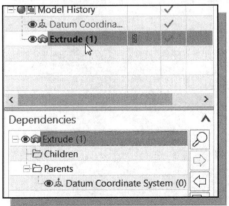

❖ Note the *datum coordinate system* is listed as the **parent feature** of the base feature.

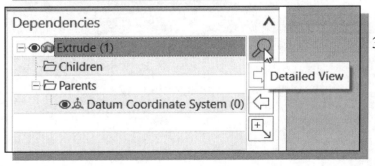

3. On your own, examine the different options in the *Object Dependency Browse* window.

Create the Second Solid Feature

- For the next solid feature, we will create the top section of the design. Note that the center of the base feature is aligned to the *origin* of the established datum coordinate system. This was done intentionally so that additional solid features can be created referencing the same location. For the second solid feature, the *XZ plane* will be used as the sketch plane.

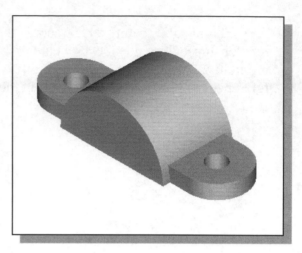

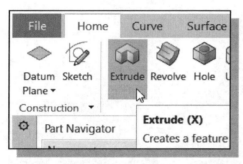

1. In the *Feature Toolbar*, select the **Extrude** command as shown.

2. Select the **XZ Plane** of the datum coordinate system as the sketch plane of the second solid feature.

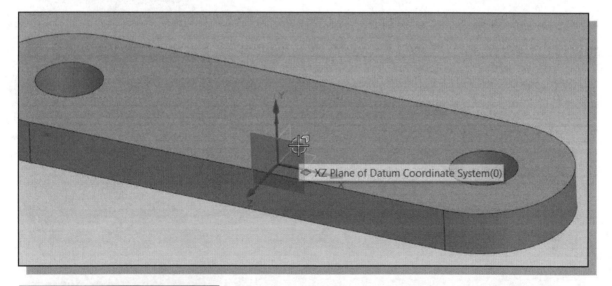

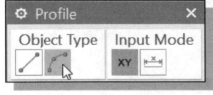

3. Activate the **Arc** option by clicking the icon in the option list near the upper-left corner of the screen.

4. Click on the **bottom edge** of the base feature as shown in the below figure. Select the 1st **point on curve** *option* in the *Quick pick window* if necessary.

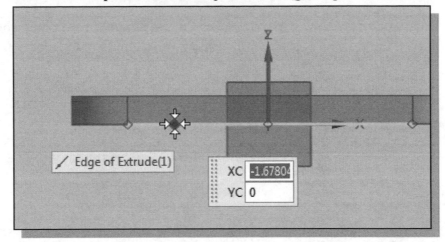

5. Move the cursor along the bottom edge of the base feature to the right and place the second endpoint of the arc also on the edge. (Select the 1st ***point on curve*** constraint if necessary.)

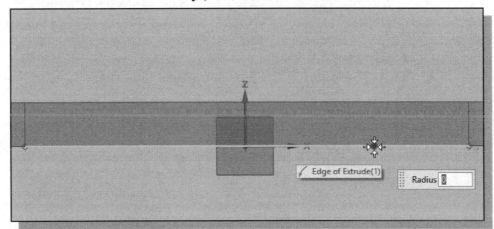

6. Click on any location of the arc, in the **positive Y direction**, to proceed with the profile command.

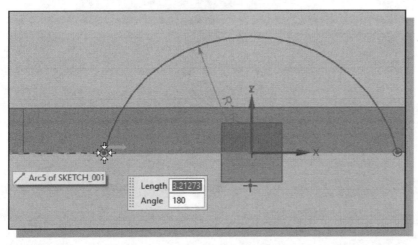

7. Complete the sketch by adding a **straight line segment** connecting to the starting point of the sketch.

Fully Constraining the Sketch

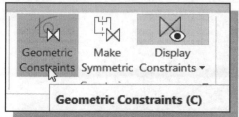

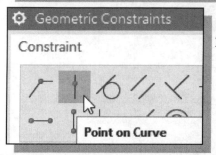

1. Select the **Geometric Constraints** command in the *Sketch Constraints* toolbar.

2. Choose the **Point on Curve** option as shown.

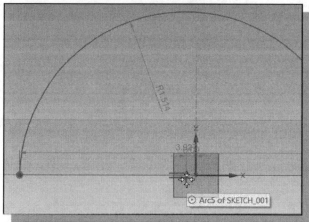

3. Select the **center** of the arc as shown. (Hint: Observe the constraint icon next to the cursor.)

4. Click once with the **middle-mouse-button** to accept the selection.

5. Move the cursor near the origin of the coordinate system and use the **Select from list** option to align to the YZ plane of Datum Coordinate System as shown.

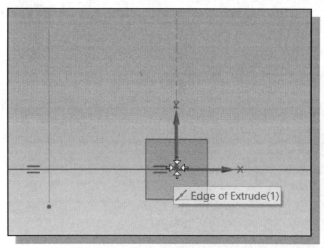

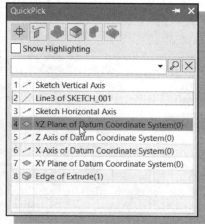

6. On your own, repeat the above step and align the arc center to the X axis of Datum Coordinate System

7. On your own, adjust the **radius dimension** and confirm the 2D sketch is fully constrained.

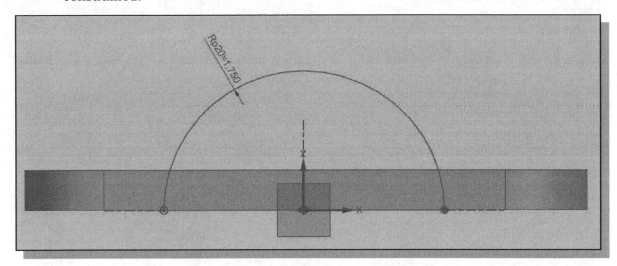

❖ In the above example, we used two constraints and one dimension to fully constrain the sketch. What other combinations can you think of that will produce the same result?

8. On your own, rotate the solid model and confirm the new sketch is located at the center of the part, aligned to the ZX plane of the datum coordinate system.

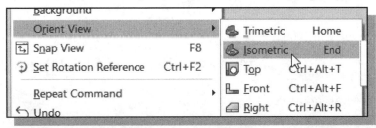

9. Hit the [End] key to reset the display to isometric as shown.

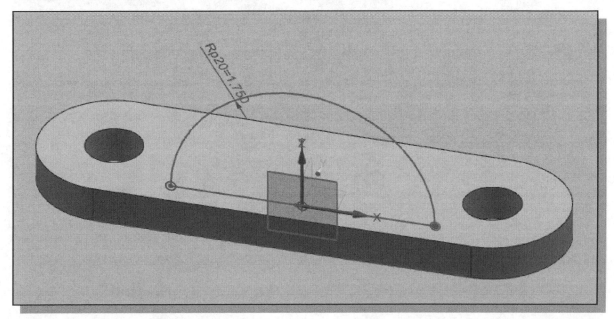

Complete the Extrude Feature

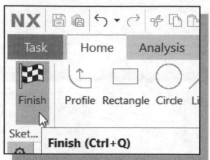

1. Click **Finish Sketch** to exit the NX sketcher mode and return to the *Feature option dialog window*.

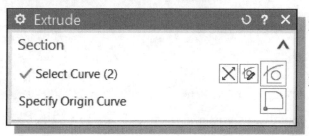

2. Note that two curves are currently selected: **one arc and one line**.

3. Click on the arc to set the extrusion direction.

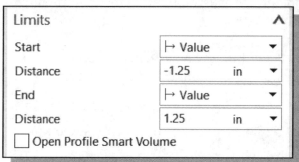

4. Enter **–1.25** in as the extrusion *Start* location, and **1.25** as the *End* location as shown. (The Limits option allows different extrusion distance in both directions.)

5. In the *Extrude* pop-up window, select **Unite** as the *Extrude* option as shown.

6. Click **OK** to create the extrusion feature.

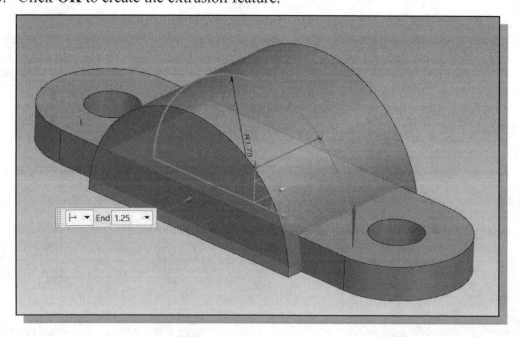

Creating a Subtract Feature

- A rectangular cut will be created as the next solid feature.

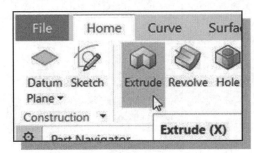

1. In the *Feature* toolbar select the **Extrude** command by left-clicking once on the icon.

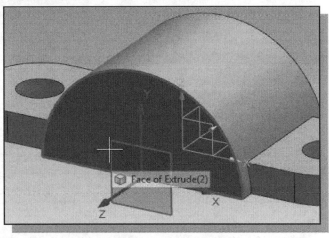

2. On your own, use the front vertical face as the sketch plane shown.

3. On your own, create a rectangle and apply the dimensions as shown below.

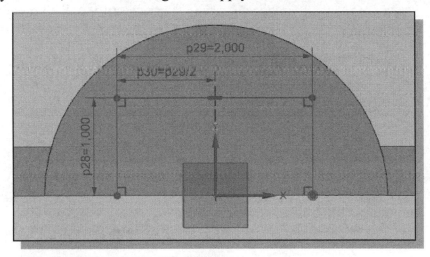

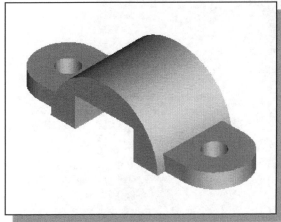

4. On your own, use the **Subtract – Through All** options in both directions to create a cutout that cuts through the entire 3D solid model as shown.

Another Subtract Feature

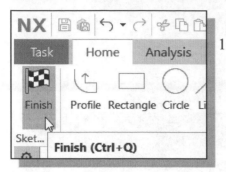

1. In the *Feature* toolbar select the **Extrude** command by left-clicking once on the icon.

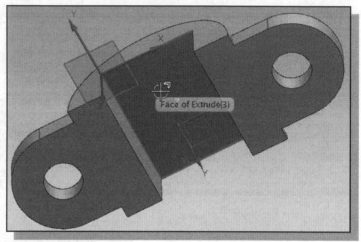

2. On your own, set up the horizontal face of the last cut feature as the *sketching plane*.

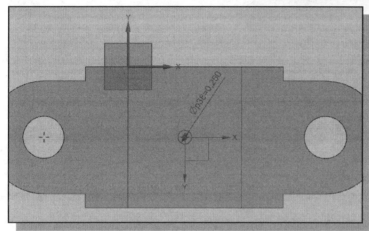

3. On your own, create a **circle** of size **0.25** in and the center point is aligned to the **origin** of the datum coordinate system.

4. On your own, use the **Subtract – Through All** options to create a cutout that cuts through the entire 3D solid model as shown.

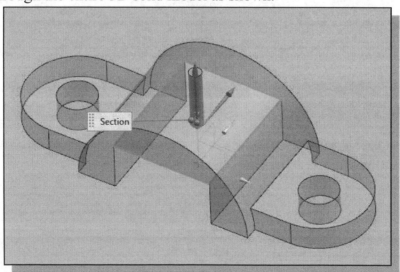

Examine the Parent/Child Relationships

Part Navigator

Name ▲	Curr...	Up...
⊞ 🖼 Model Views		
⊞ ✓ 📷 Cameras		
⊟ ●🖼 Model History		✓
👁⚓ Datum Coordina...		✓
👁🪹 Extrude (1) "Base"		✓
👁🪹 Extrude (2) "Mai...		✓
👁🪹 Extrude (3) "Rect...		✓
👁🪹 **Extrude (4) "Ce...**		✓

1. On your own, rename the feature names to **Base**, **Main-Body**, **Rect-Cut** and **CenterDrill** as shown in the figure.

❖ The *Part navigator* window now contains the four extruded features. All of the parent/child relationships were established implicitly as we created the solid features. As more features are created, it becomes much more difficult to make a sketch showing all the parent/child relationships involved in the model. On the other hand, it is not really necessary to have a detailed picture showing all the relationships among the features. In using a feature-based modeler, the main emphasis is to consider the interactions that exist between the **immediate features**. Treat each feature as a unit by itself, and be clear on the parent/child relationships for each feature. Thinking in terms of *features* is what distinguishes *feature-based modeling* and the previous generation solid modeling techniques. Let us take a look at the last feature we created, the **CenterDrill** feature. What are the parent/child relationships associated with this feature? (1) Since this is the last feature we created, it is not a parent feature to any other features. (2) Since we used one of the surfaces of the rectangular cutout as the sketching plane, the **Rect-Cut** feature is a parent feature to the **CenterDrill** feature. (3) We also used the **Datum Coordinate System** as references to align the location of the *CenterDrill*; therefore, the **Datum Origin** is also a parent to the last feature.

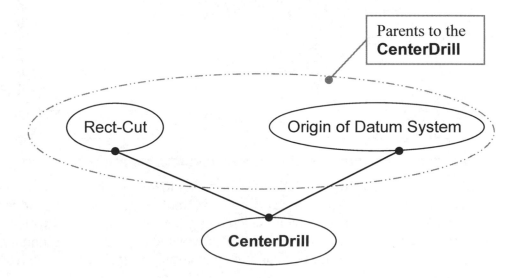

A Design Change

➢ Engineering designs usually go through many revisions and changes. For
   example, a design change may call for a circular cutout instead of the current
   rectangular cutout feature in our model. *NX* provides an assortment of tools to
   handle design changes quickly and effectively. In the following sections, we will
   demonstrate some of the more advanced tools available in *NX*, which allow us to
   perform the modification of changing the rectangular cutout (2.0 × 1.0 inch) to a
   circular cutout (radius: 1.25 inch).

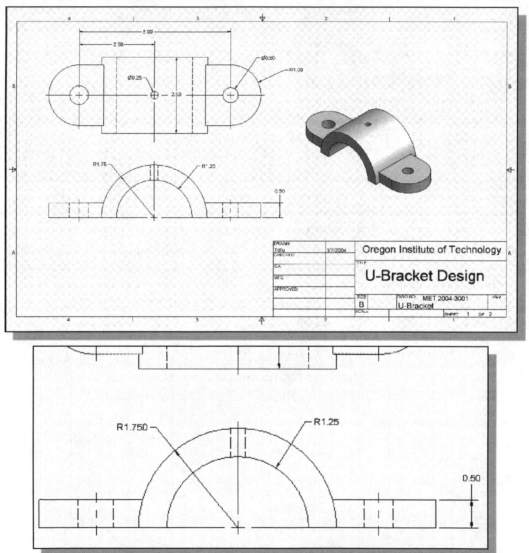

❖ Based on your knowledge of *NX* so far, how would you accomplish this
   modification? What other approaches can you think of that are also feasible? Of
   the approaches you came up with, which one is the easiest to do and which is the
   most flexible? If this design change were anticipated right at the beginning of the
   design process, what would be your choice in arranging the order of the features?
   You are encouraged to perform the modifications prior to following through the
   rest of the tutorial.

Feature Suppression

❖ With *NX*, we can take several different approaches to accomplish this modification. We could (1) create a new model, or (2) change the shape of the existing cut feature using the **Edit with Rollback** command, or (3) perform **feature suppression** on the rectangular cut feature and add a circular cut feature. The third approach offers the most flexibility and requires the least amount of editing to the existing geometry. **Feature suppression** is a method that enables us to disable a feature while retaining the complete feature information; the feature can be reactivated at any time. Prior to adding the new cut feature, we will first suppress the rectangular cut feature.

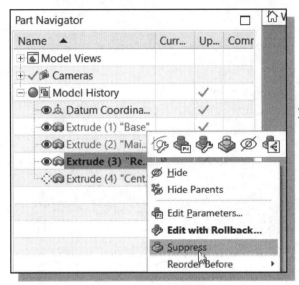

1. Open the *Part Navigator* window, click once with the right-mouse-button on top of *Rect-Cut* to bring up the option menu.

2. Pick **Suppress** in the pop-up menu.

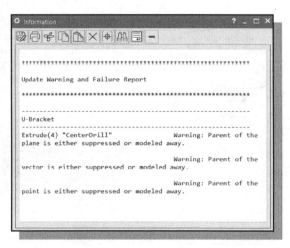

• NX indicates the *Centerdrill* feature cannot be properly placed as some of the references associated to the *Rect_Cut* feature cannot be resolved.

3. Click **Close** to close the warning dialog box.

4. In the *Part Navigator* window, **Suppress** the *CenterDrill* feature by using the option list with the right-mouse-button.

❖ The embedded feature relations are maintained by the system, a child feature cannot exist without its parent(s), and any modification to the parent *(Rect-Cut)* may affect the child *(CenterDrill)*.

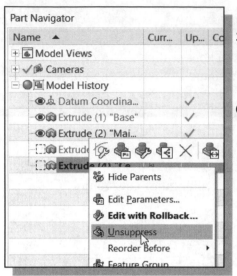

5. Open the *Part Navigator* window, and click once with the right-mouse-button on top of ***CenterDrill*** to bring up the option menu.

6. Pick **Unsuppress** in the pop-up menu.

➤ In the *Graphics area* and the *Part Navigator* window, both the ***Rect-Cut*** feature and the ***CenterDrill*** feature have been re-activated. The child feature cannot exist without its parent(s); the parent *(Rect-Cut)* must be activated to enable the child *(CenterDrill)*.

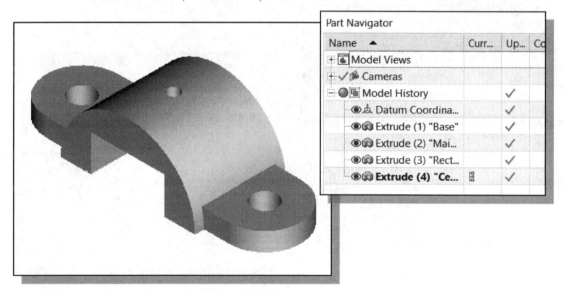

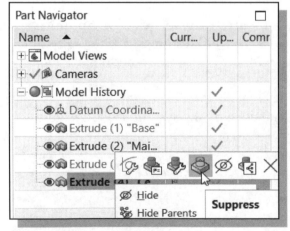

❖ Note that we can also click on the fourth icon in the icon toolbar with the **right-mouse-button (MB1)** to **suppress** features.

A Different Approach to the Center-Drill Feature

❖ The main advantage of using the BORN technique is to provide greater flexibility for part modifications and design changes. In this case, the *CenterDrill* feature can be placed on the *XY datum plane* and therefore not be linked directly to the *Rect-Cut* feature.

1. Open the *Part Navigator* window, click once with the right-mouse-button (**MB3**) on top of **CenterDrill** to bring up the option menu and pick **Delete** as shown.

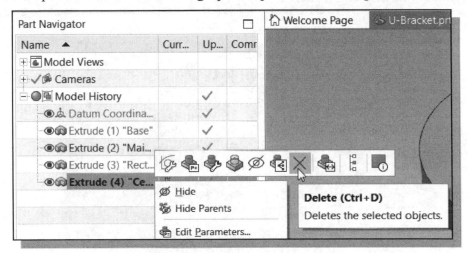

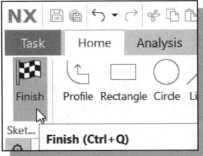

2. In the *Feature Toolbar*, select the **Extrude** icon as shown.

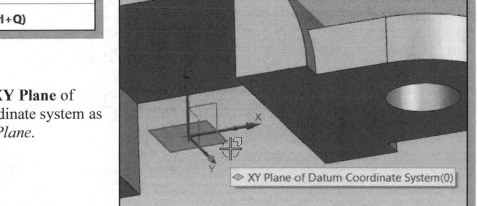

3. Select the **XY Plane** of datum coordinate system as the *Sketch Plane*.

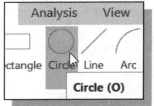

4. Select the **Circle** command by clicking once with the **left-mouse-button (MB1)** on the icon in the *Sketch Curve* toolbar.

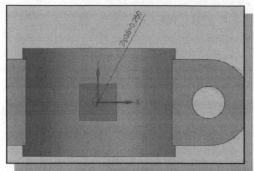

5. On your own, create a **circle** of size **0.25** in and the center point is aligned to the **origin** of the datum coordinate system.

6. On your own, use the **Subtract – Through All** options to create a cutout that cuts through the entire 3D solid model as shown.

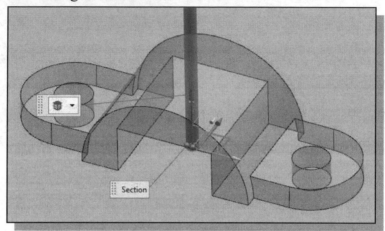

Examine the Parent/Child Relationships

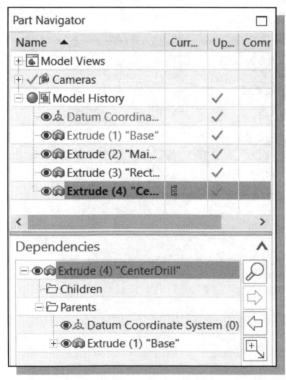

1. Open the *Part Navigator* and left-mouse-click (**MB1**) on the newly created **Extrude** feature.

2. Expand the **Dependency Browse** option to show details on the selected feature.

3. Note the two **Parent features** are highlighted in red in the *Model History* window.

❖ Note the two **parent features** of the selected feature include (1) The XY datum plane where the sketch plane was aligned to, and (2) The base feature, where material was removed. The Rect_Cut feature is no longer a parent feature to the CenterDrill feature.

Suppress the Rect-Cut Feature

❖ Now the *CenterDrill* feature is no longer a child of the *Rect-Cut* feature, any changes to the *Rect-Cut* feature will not affect the *CenterDrill* feature anymore.

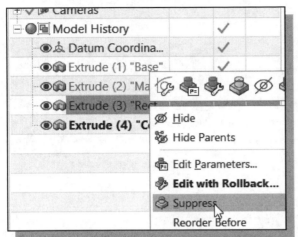

1. Open the *Part Navigator* and suppress the *Rect-Cut* feature by clicking once with the right-mouse-button (MB3) on the feature as shown.

❖ The *Rect-Cut* feature is now disabled without affecting the *New Center_Drill* feature.

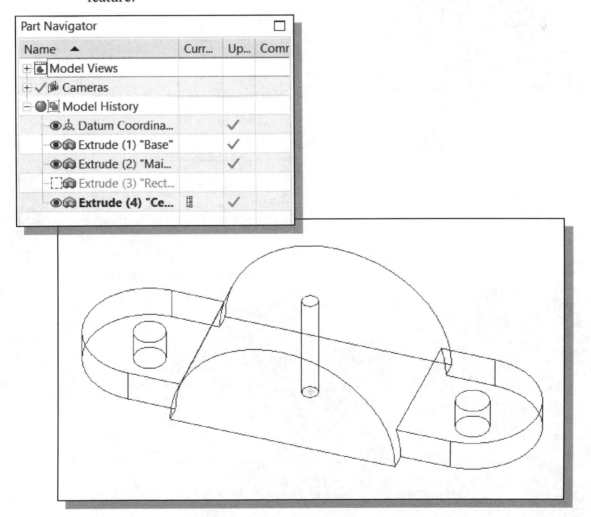

Create a Circular Cut Feature

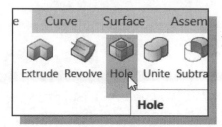

1. In the *Feature Operation* toolbar, select the **Hole** command by releasing the left-mouse-button on the icon.

2. Select the Point option in the Position option as shown.

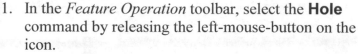
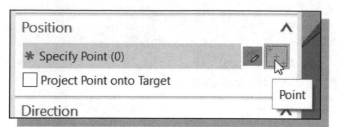

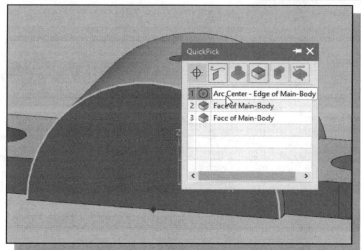

3. Use the quick pick option and select the front **Arc center** of the *main_body* feature as shown in the figure.

4. Set the *Hole Direction* to **Along Vector** as shown.

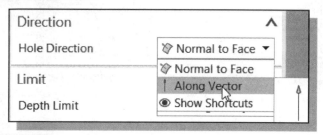

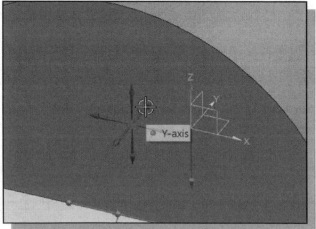

5. Select the **Y axis** to set the vector direction as shown.

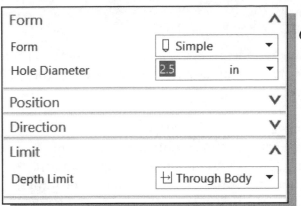

6. Set the *Diameter* to **2.5** inch. And *Depth limit* to **Through Body** as shown.

7. Confirm the *Boolean* option is set to **Subtract** as shown.

8. Click **OK** to create the feature.

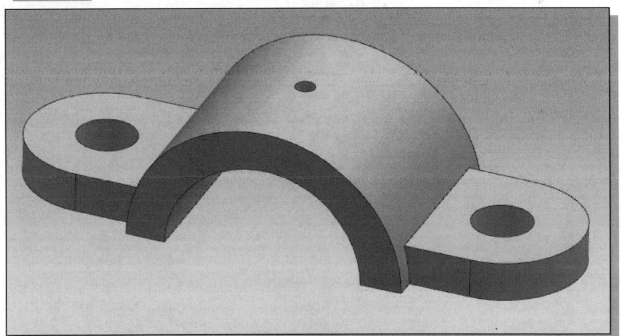

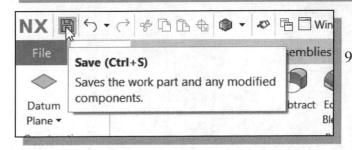

9. Save the **U-Bracket** model; this model will be used again in the next chapter.

A Flexible Design Approach

In a typical design process, the initial design will undergo many analyses, testing, reviews and revisions. *NX* allows the users to quickly make changes and explore different options of the initial design throughout the design process.

The model we constructed in this chapter contains two distinct design options. The *feature-based parametric modeling* approach enables us to quickly explore design alternatives and we can include different design ideas into the same model. With parametric modeling, designers can concentrate on improving the design and the design process to be much quicker and more effortless. The key to successfully using parametric modeling as a design tool lies in understanding and properly controlling the interactions of features, especially the parent/child relations.

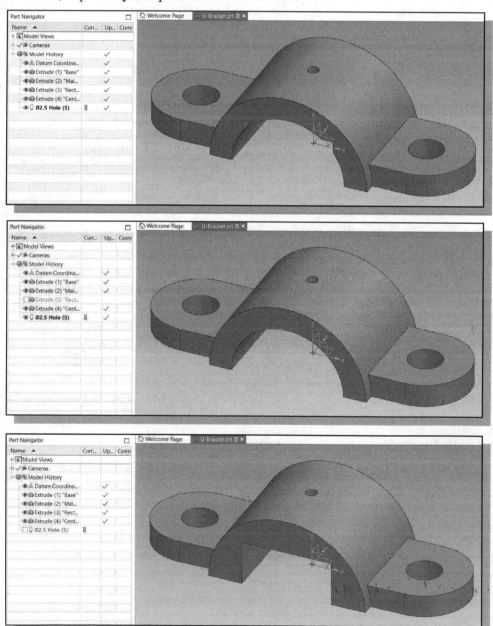

Review Questions:

1. Why is it important to consider the parent/child relationships in between features?

2. Describe the procedure to **suppress** a *feature*.

3. What is the basic concept of the BORN technique?

4. What happen to a feature when it is suppressed?

5. How do you unsuppress a suppressed feature in a model?

6. Describe the advantages of parametric modeling in creating alternate designs over the traditional CAD systems.

7. Create sketches showing the steps you plan to use to create the models shown on the next page:

Exercises: (Dimensions are in inches)

1. **Swivel Yoke** (Material: **Cast Iron**)

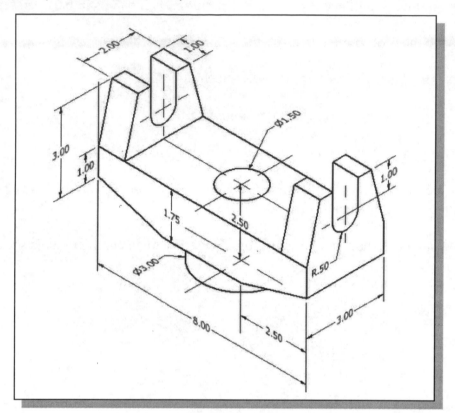

2. **Angle Bracket** (Material: **Carbon Steel**)

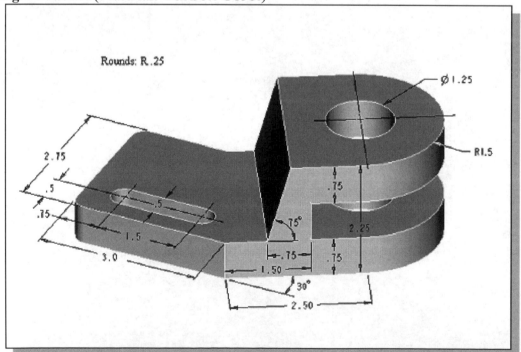

3. **Connecting Rod** (Material: **Carbon Steel**)

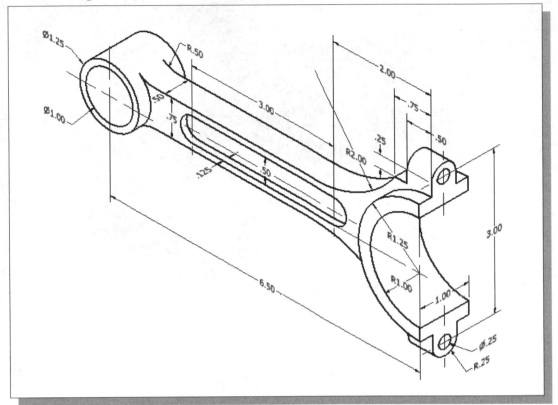

4. **Tube Hanger** (Material: **Aluminum 6061**)

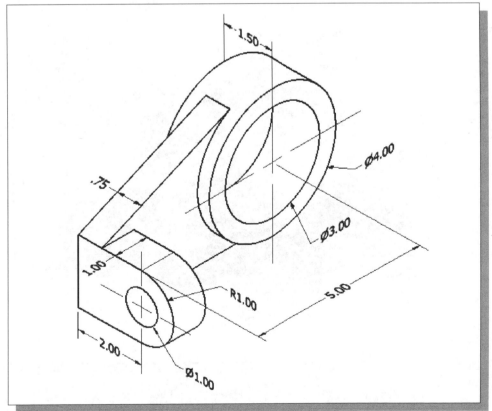

5. **Angle Latch** (Dimensions are in millimeters. Material: **Brass**)

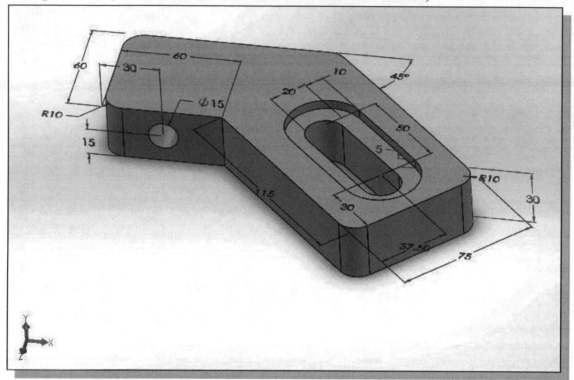

6. **Inclined Lift** (Material: **Mild Steel**)

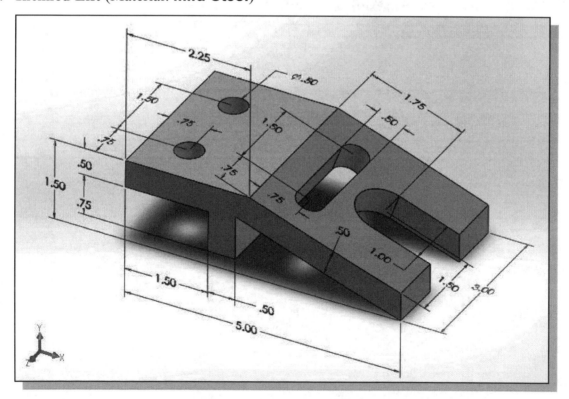

Chapter 8
Part Drawings and Associative Functionality

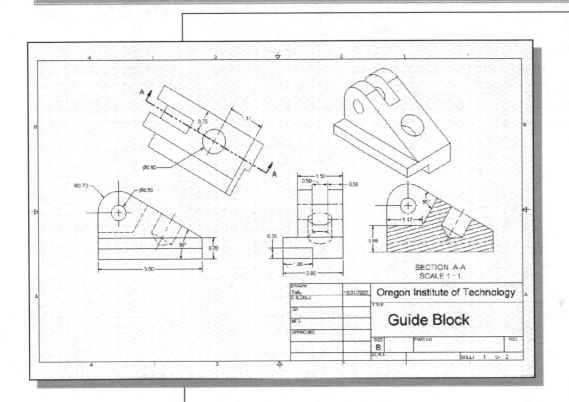

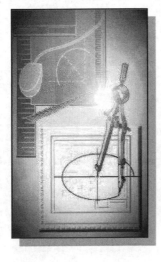

Learning Objectives

- **Create Drawing Layouts from Solid Models**
- **Understand Associative Functionality**
- **Using the default Borders and Title Block in the Layout Mode**
- **Arrange and Manage 2D Views in Drafting mode**
- **Display and Hide Feature Dimensions**
- **Create Reference Dimensions**

Drawings from Parts and Associative Functionality

With the software/hardware improvements in solid modeling, the importance of two-dimensional drawings is decreasing. Drafting is considered one of the downstream applications of using solid models. In many production facilities, solid models are used to generate machine tool paths for *computer numerical control* (CNC) machines. Solid models are also used in *rapid prototyping* to create 3D physical models out of plastic resins, powdered metal, etc. Ideally, the solid model database should be used directly to generate the final product. However, the majority of applications in most production facilities still require the use of two-dimensional drawings. Using the solid model as the starting point for a design, solid modeling tools can easily create all the necessary two-dimensional views. In this sense, solid modeling tools are making the process of creating two-dimensional drawings more efficient and effective.

NX provides associative functionality in the different *NX* modes. This functionality allows us to change the design at any level, and the system reflects it at all levels automatically. For example, a solid model can be modified in the *Part Modeling Mode* and the system automatically reflects that change in the *Drafting mode*. And we can also modify a feature dimension in the *Drafting mode*, and the system automatically updates the solid model in all modes.

In this lesson, the general procedure for creating multi-view drawings is illustrated. The *U_Bracket* design from the last chapter is used to demonstrate the associative functionality between the model and drawing views.

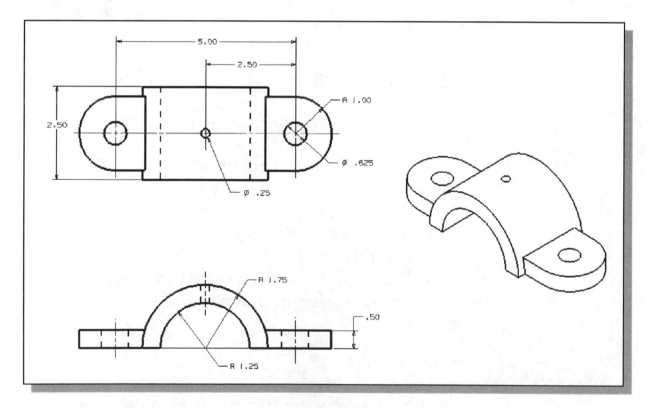

Start NX

1. Select the **NX** option on the *Start* menu or select the **NX** icon on the desktop to start *NX*. The *NX* main window will appear on the screen.

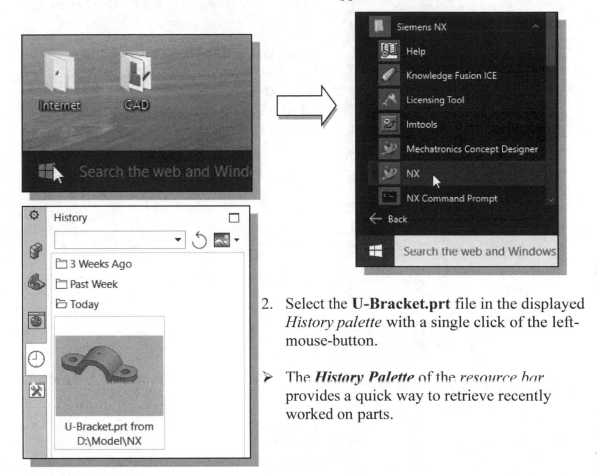

2. Select the **U-Bracket.prt** file in the displayed *History palette* with a single click of the left-mouse-button.

➢ The *History Palette* of the *resource bar* provides a quick way to retrieve recently worked on parts.

Drafting mode – 2D Paper Space

➢ *NX* allows us to generate 2D engineering drawings from solid models so that we can plot the drawings to any exact scale on paper. An engineering drawing is a tool that can be used to communicate engineering ideas/designs to manufacturing, purchasing, service, and other departments. Until now we have been working in ***model space*** to create our design in ***full size***. We can arrange our design on a two-dimensional sheet of paper so that the plotted hardcopy is exactly what we want. This two-dimensional sheet of paper is generally known as ***paper space***. We can place borders and title blocks in the *paper space*. In general, each company uses a set of standards for drawing content, based on the type of product and also on established internal processes. The appearance of an engineering drawing varies depending on when, where, and for what purpose it is produced. However, the general procedure for creating an engineering drawing from a solid model is fairly well defined. In *NX*, creation of 2D engineering drawings from solid models consists of four basic steps: drawing sheet formatting, creating/positioning views, annotations, and printing or plotting.

NX Drafting Mode

1. Click on the **File** tab in the *Ribbon* bar area to display the available options.

2. Select **Drafting** from the option list.

❖ In the *Sheet* dialog box, *NX* displays different settings for a drawing sheet that will be used.

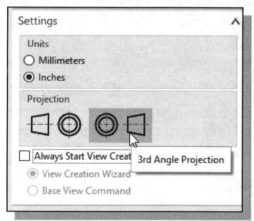

3. Select **A-8.5X11** from the **standard size** list.

4. Confirm the **Scale** is set to **1:1**, which is *Full Scale*.

➢ Note that **Sheet 1** is the default drawing sheet name that is displayed in the *Drawing Sheet Name* area.

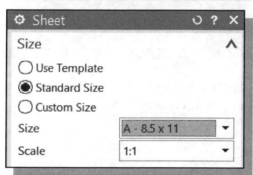

5. In the *settings area*, confirm the *Units* are set to **Inches** and the *Projection* type is set to **Third Angle of projection** as shown.

6. **Uncheck** the *Always Start View Creation Command* option as shown.

7. Click **OK** to accept the settings and proceed with the creation of the drawing sheet.

➢ Note that a new graphics window appears on the screen. The dashed rectangle indicates the size of the current active drawing sheet.

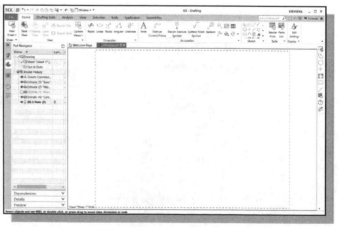

Add a Base View

❖ In *NX Drafting mode*, the first drawing view we create is called a **base view**. A *base view* is the primary view in a drawing; other views can be derived from this view. When creating a *base view*, *NX* allows us to specify one of the principal views to be shown. By default, *NX* will treat the *world XZ plane* as the front view of the solid model.

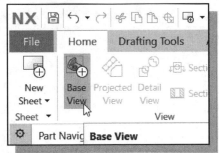

1. Click on the **Base View** icon in the *Drawing Layout toolbar* to create a new base view.

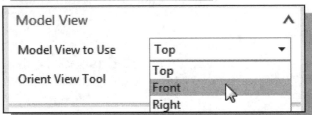

2. In the *Model View* option area, set the view option to **Front** as shown.

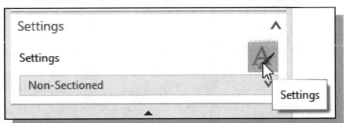

3. In the *Settings* option area, click the **View Style Settings** icon as shown.

4. In the Settings dialog box, click the **Hidden Lines** tab as shown.

5. Choose **Dashed, 0.35mm,** as the *line type* to be used for hidden lines in the *Base view*.

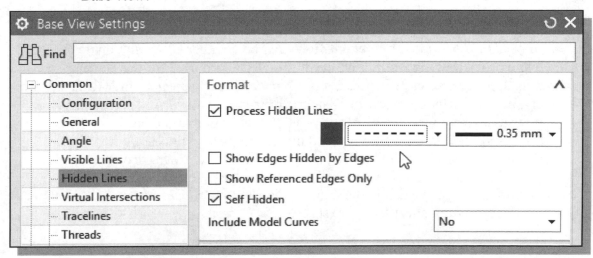

6. Click on the **Visible Lines** tab and select **0.70mm** as the line type for the *Base View*.

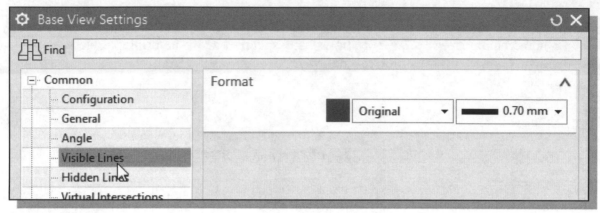

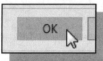

7. Click **OK** to accept the settings and exit the *View Style* command.

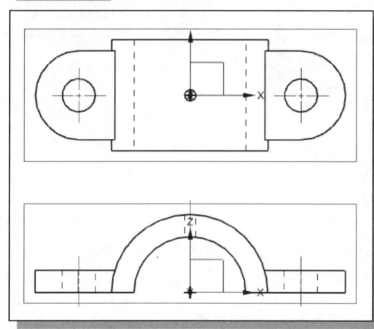

8. Place the **Base View**, the *Front view* of the *U-Bracket* model, near the lower half of the drawing sheet as shown.

9. Move the cursor above the *Front View* and place the **Top view**, and close the projected view dialog box. Note the projection is done automatically, relative to the established *Base View*.

10. Click **Close** to end the *Base View* command.

❖ In *NX Drafting mode*, **projected views** can be created with a first-angle or third-angle projection, depending on the drafting standard used for the drawing. We must have a **base view** before a *projected view* can be created. In *NX*, Base View can be a 2D principal view or isometric or trimetric view. And *projected views* are done using orthographic projections. Note that orthographic projections are always aligned to the base view and inherit the base view's scale and display settings. We will also add an isometric view in the following sections. Before adding the isometric view, we will first make two adjustments to the current drawing sheet: (1) Change the drafting display to *Monochrome* and (2) Change to a *B-Size* drawing sheet for more space.

Drawing Display Option

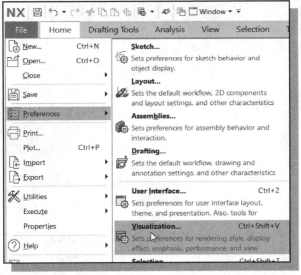

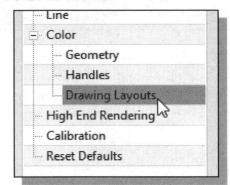

1. Select **Visualization** in the *Preferences menu list* as shown.

2. Select **Drawing Layouts** under **Color** as shown.

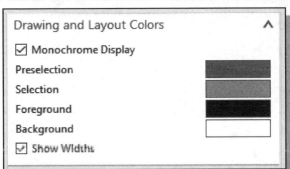

3. Confirm the **Monochrome Display** and the **Show Widths** options are switched on for the drawings part settings as shown.

4. Click **OK** to accept the settings and exit the *Visualization Preferences* window.

Change the Size of the Drawing Sheet

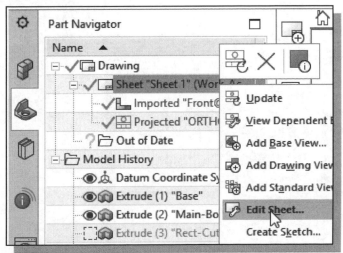

1. In the *Part Navigator* window, right-mouse-click (**MB3**) on **Sheet "Sheet1"** to bring up the option list.

2. Select **Edit Sheet** in the option menu.

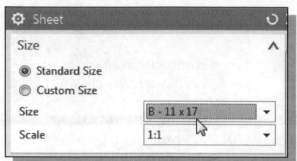

3. Inside the *Edit Sheet* dialog box, set the *Sheet Size* to **B-11X17** as shown in the figure.

4. Click on the **OK** button to accept the settings and proceed with updating the drawing views.

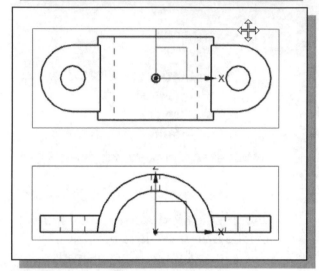

5. **Click and drag** the view borders of the views to reposition them.

   ➤ Note the dashed lines indicating alignments in between views.

6. Click on the **Base View** in the *Drawing Layout toolbar* to create a new base view.

7. Select **Isometric** in the view list, and place an isometric view to the right side of the drawing sheet as shown.

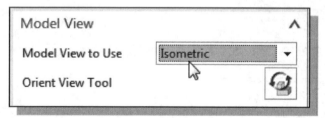

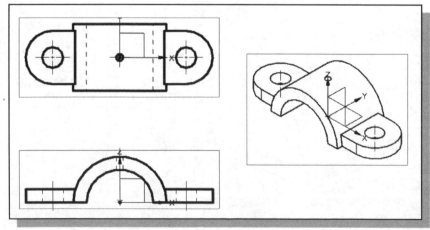

Turn Off the Datum Coordinate System

➢ The display of the datum planes and/or WCS can be turned off in the drafting mode or in the Modeling mode.

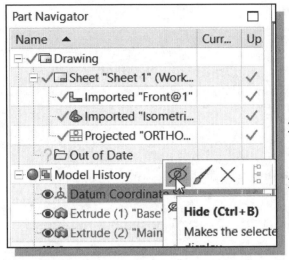

1. In the *Part Navigator* window, right-mouse-click **(MB3)** on **Datum Coordinate System** to bring up the option list.

2. Click **Hide** to turn off the *Datum Coordinate System* as shown.

3. The **Hide** command can be used to temporarily turn off the display of objects.

4. Right-mouse-click on the small cross hair in the *Front View*, and select **Hide** to turn off the display.

5. Repeat the above step for the other two views so that your drawing appeared as shown.

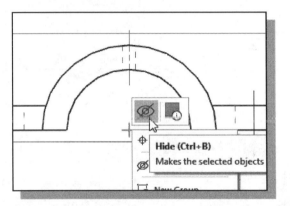

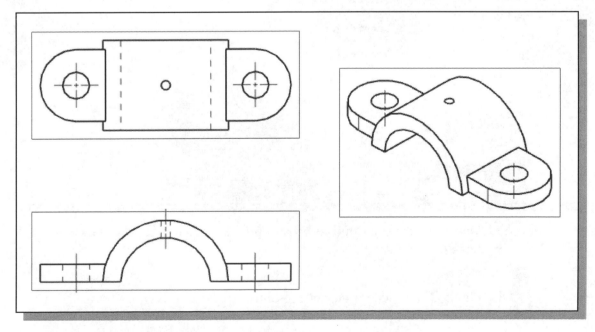

Text Orientation Setup

- In *NX*, the default text orientation is set to be aligned to the dimension line, which is based on the ISO standard. For our drawing, we will use the ANSI drafting and set up text orientation to horizontal.

1. Select **Drafting** in the **Preferences** menu list as shown.

2. Select **Dimension → Text → Orientation and Location** in the *Drafting Preferences* dialog box as shown.

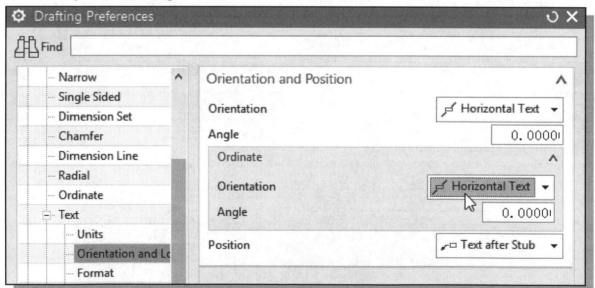

3. Set the *Orientations* for the dimensions and position to **Horizontal Text** as shown in the figure.

4. Click **OK** to accept the settings and exit the command.

Display Feature Dimensions

- Feature dimensions (parameters) can be displayed in 2D views in *NX*. We have the options of selecting which feature parameters to be displayed in which views. Note that in NX 9, this command is not available in the *Ribbon Bar area*. This command is only accessible through the **Command Finder** option.

    1. Enter **Feature Parameters** in the **Command Finder** box as shown.

    2. Select **Feature Parameters** in the **Command Finder** dialog box as shown.

    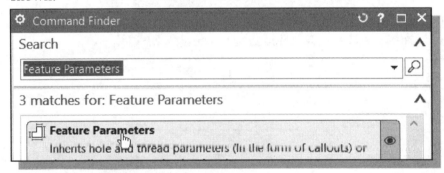

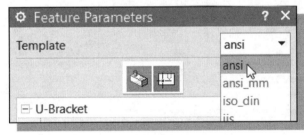

3. In the *Feature Parameters* window, set the **ANSI** standard as the default settings.

- ❖ Two steps are required: (1) select the feature used and (2) select the view to place the dimensions.

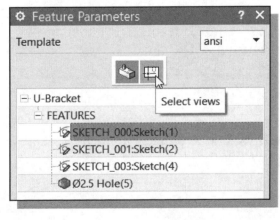

4. Expand the *FEATURES* list and choose the first sketch, which is related to the **Base** feature (*Sketch(1)*), as shown in the figure.

5. Click the **Select Views** icon to proceed to the next step.

6. Uncheck **Use Sketch Dimension Style**.

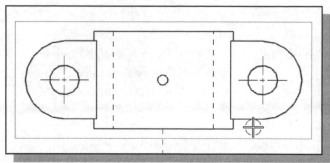

7. Inside the graphics window, select the **Top View** by clicking once with the **left-mouse-button** (**MB1**) as shown.

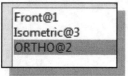

❖ Note that the corresponding view name is highlighted as soon as we click on a view in the graphics window.

8. Click **Apply** to accept the settings.

❖ The feature dimensions that were used to create the **Base Feature** have now been retrieved and placed in the top view.

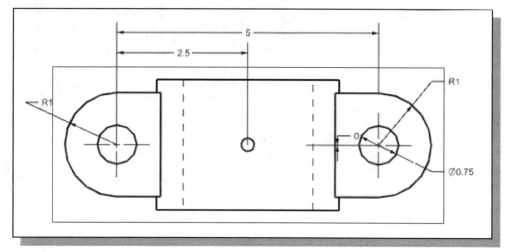

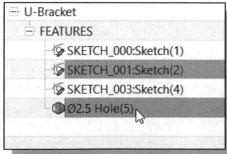

9. Hold down the [**Ctrl**] key and select both the sketches related to the **MainBody (Sketch 2)** and **Circular_Cut (5)** features, in the *Features list* as shown.

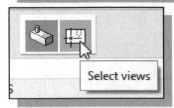

10. Click the **Select Views** icon to proceed to the next step.

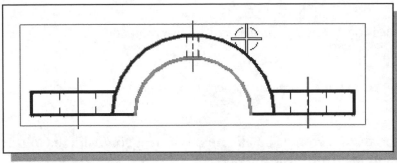

11. Inside the *graphics window*, select the **Front View** as shown.

12. Click **Apply** to accept the settings.

❖ Note that more than one feature can be selected when placing feature dimensions with the *Feature Parameters* command.

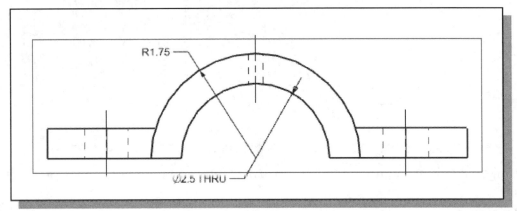

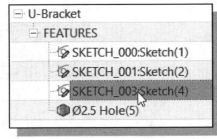

13. On your own, repeat the above process and retrieve the feature dimensions of the *CenterDrill (4)* feature and place them in the **Top View** as shown in the below figure.

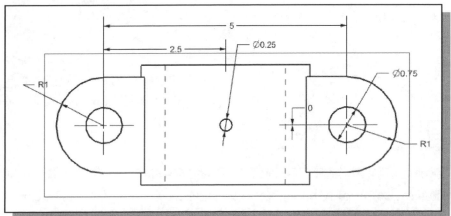

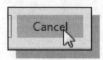 14. Click **Cancel** to end the *Feature Parameters* command.

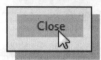 15. Click **Close** to exit the *Command Finder* command.

16. On your own, reposition the dimensions by **clicking and dragging** with the left-mouse-button on the individual dimension.

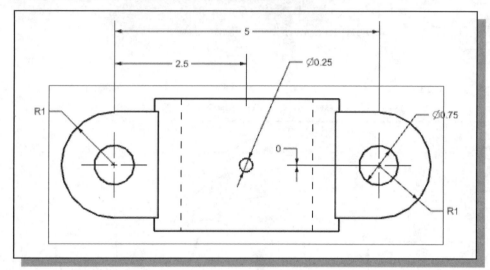

❖   Rearrange the dimensions so that your screen appears roughly as shown in the below figure.

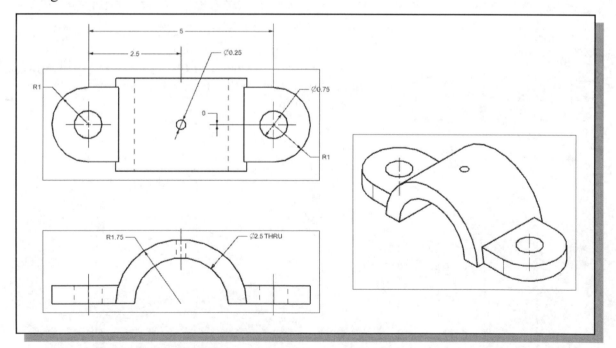

➢   Note that several commonly used drafting standards, such as *ISO, DIN*, can also be loaded through the **Drafting Standard** command with the *Command Finder*.

Adjust the Display of Tangency Edges

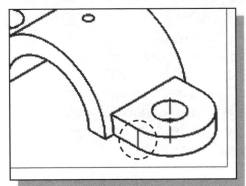

❖ In *NX*, by default, the tangency edges are displayed as visible lines as shown in the figure. We can turn off the display through the *View Style* command.

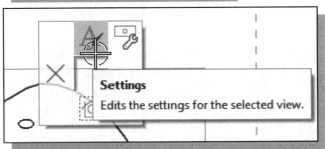

1. Hold down the **right-mouse-button** on the view border of the *Isometric view* to bring up the option list.

2. Select the **Settings** icon as shown.

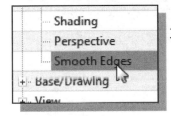

3. Under the Common list, left-mouse-click (MB1) on **Smooth Edges** as shown.

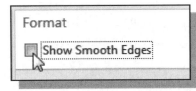

4. Turn **off** the **Smooth Edges** option as shown.

5. Click **OK** to accept the settings.

➢ Note the tangency edges are turned off as shown in the below figure.

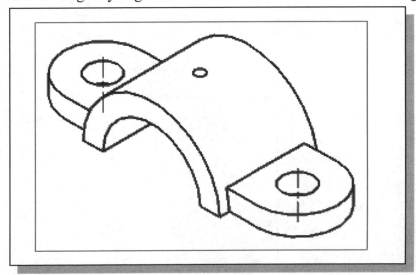

Edit the Display of Arrows on Diameter Dimensions

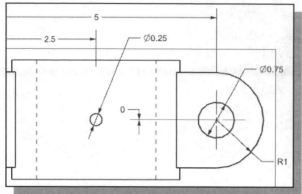

❖ In *NX*, by default, the diameter dimensions are displayed with two arrows as shown in the figure. We can adjust the settings on the dimension so that the displayed dimensions are more in line with the *ANSI* drafting standard.

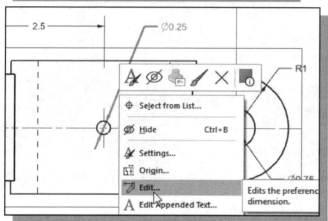

1. Hold down the **right-mouse-button** on the diameter dimension of the small drill hole in the *top view* to bring up the option list.

2. Select the **Edit** icon as shown.

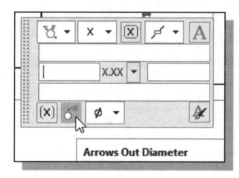

3. Left-mouse-click (MB1) on **Arrow Out diameter** to switch the arrow to outside.

4. Click **Close** to accept the settings.

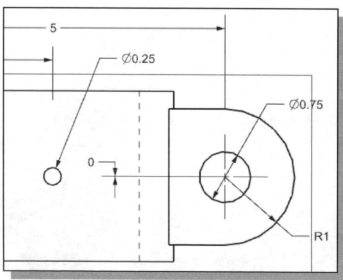

Hide Feature Dimensions

❖ In NX, two options are available to remove any of the displayed Feature dimensions: (1) The **Delete** command or (2) The **Hide** command. Using the *Delete* command, removed feature dimensions can be redisplayed by using *the Feature Parameters* command, which is a more intuitive approach. The *Hide* command is more of a general *Hide* command, and the *Show* command is required to re-display any of the hidden items.

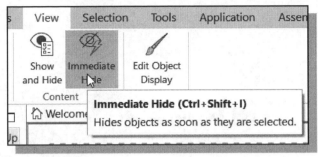

1. Select the **Immediate Hide** command in the *View toolbar* as shown.

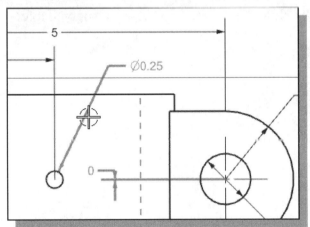

2. Select the **two dimensions** in the top view as shown.

3. Click **Close** to accept the selections.

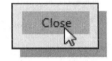

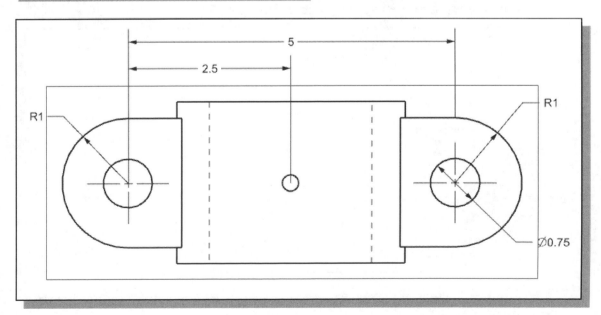

Unhide the Hidden Dimensions

❖ The *Show* command can be used to redisplay any of the objects that were turned off using the *Hide* command.

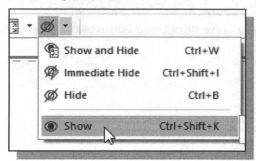

1. Use the quick key [**Ctrl+Shift+K**] to activate the **Show** command.

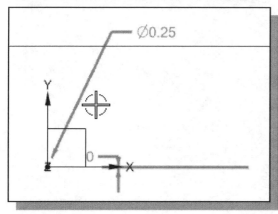

2. Select the **two dimensions** associated with the *top view* as shown.

❖ Note the datum features are also available to be redisplayed.

3. Click **OK** to accept the selections.

❖ The dimensions that were removed from the display are redisplayed on the screen as shown in the below figure.

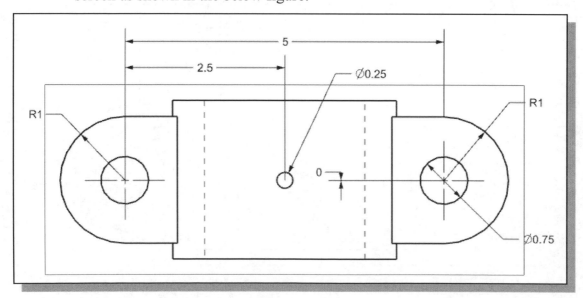

Delete Feature Dimensions

❖ The **Delete** command is a more intuitive approach to remove feature dimensions. The removed features are not stored in a temporary location, as the *Hide* command does. To redisplay any deleted feature dimensions, we will need to use the *Feature Parameters* command (p. 8-11).

1. Select a dimension and notice the **Delete** command is available.

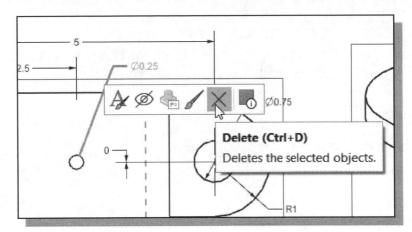

2. The *Delete* command can also be activated using the key combination, hit the [**Ctrl+D**] quick key once.

3. Select the **0.0** location dimension and the two center lines in the top view as shown.

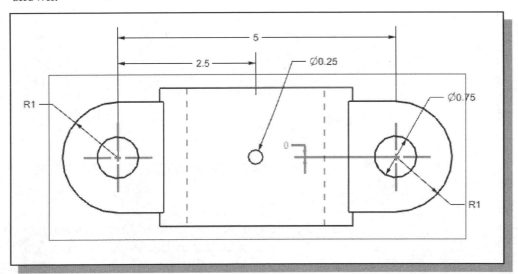

   4.   Click **OK** to accept the selections.

❖ Note another quick option to delete objects is to use the **Pre-Selection** approach; pre-select the objects first then hit the [**Delete**] key in the keypad.

Add Center Marks

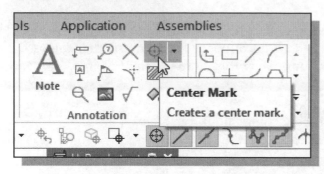

1. Select the **Center Mark** command through the *Dimension* toolbar.

2. Click once, with the left-mouse-button (MB1), on the two outer **arcs** and the small drill hole at the center in the top view as shown.

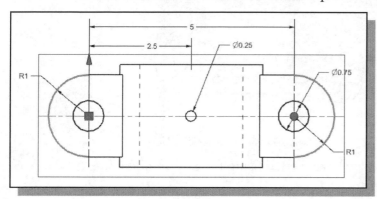

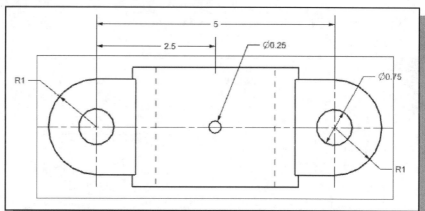

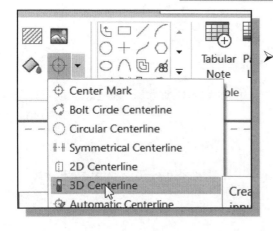

➢ Note that the **Center Mark** command is used to create center marks on circular views of cylindrical features. For edge views of cylindrical features, we can use the **2D Centerline** command. Note other centerline options are also available.

Add Additional Dimensions – Reference Dimensions

- Besides displaying the **feature dimensions**, dimensions used to create the features, we can also add additional **reference dimensions** in the drawing. *Feature dimensions* are used to control the geometry, whereas *reference dimensions* are controlled by the existing geometry. In the drawing layout, therefore, we can **add** or **delete** *reference dimensions*, but we can only hide the *feature dimensions*. One should try to use as many *feature dimensions* as possible and add *reference dimensions* only if necessary. It is also more effective to use *feature dimensions* in the drawing layout since they are created when the model was built. Note that additional *Drafting mode* entities, such as lines and arcs, can be added to drawing views. Before *Drafting mode* entities can be used in a reference dimension, they must be associated to a *drawing view*.

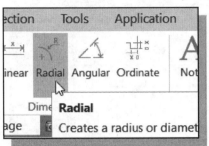

1. On your own, delete the diameter dimension in the front view.

2. Select the **Radial dimension** command through the *Ribbon Bar area* as shown.

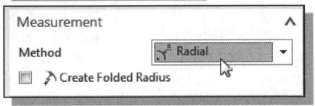

3. In the *radial Dimension*, set the *Measurement method* to **Radial** as shown.

4. Select the **bottom arc** in the front view.

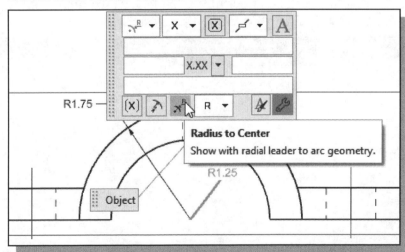

5. Place the dimension text in the front view as shown.

6. Click **Close** to accept the settings and exit the command.

Change the Dimension Appearance

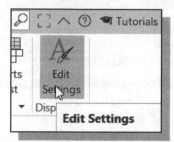

1. Click **Edit Settings** in the *Ribbon Bar area* as shown.

2. Select the two length dimensions *5* and *2.5*, by **clicking** on the dimensions with the left-mouse-button.

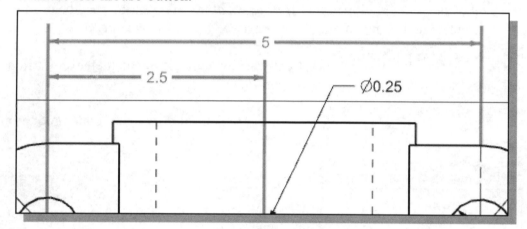

3. Click **OK** to accept the selections.

4. Change the display of the number of digits after the decimal point to **2** and turn on the **Show Trailing Zeros** option as shown.

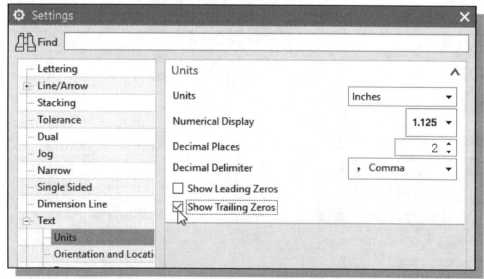

- On your own, examine the other options available in the *Settings dialog box*.

Associative Functionality – Modifying Feature Dimensions

- *NX's associative functionality* allows us to change the design at any level, and the system reflects the changes at all levels automatically.

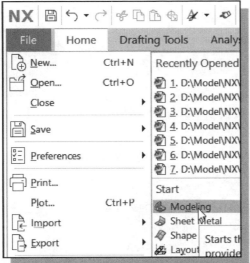

1. Select the **File** tab with a single click of the left-mouse-button (**MB1**) in the *Ribbon* bar.

2. Pick **Modeling** in the *Applications list* as shown in the figure to switch to the *Part Modeling Mode*.

3. Double-click, with the left-mouse-button (MB1), on the **Base Feature** to enter the *Edit Mode*.

4. Click **Sketch Section** in the *Extrude dialog box*.

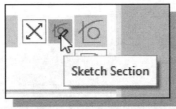

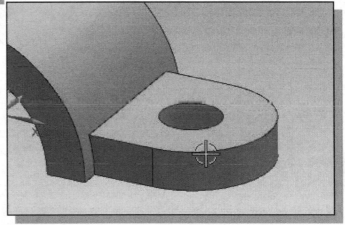

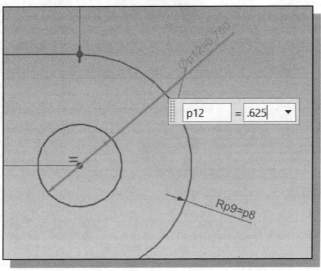

5. Double-click on the diameter **0.750** dimension of the base feature as shown in the figure.

6. In the *Edit Dimension* dialog box, enter **0.625** as the new diameter dimension.

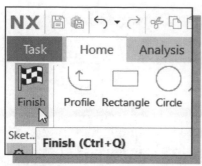

7. Click **Finish Sketch** to exit the NX sketcher mode and return to the *Feature option dialog window*.

8. Click **OK** to accept the changes.

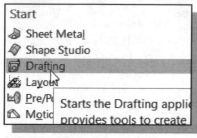

9. Select the **File** tab with a single click of the left-mouse-button (**MB1**) in the *Ribbon* bar.

10. Select **Drafting** from the *Applications list*.

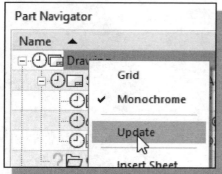

➢ In the *Part Navigator*, the little clock symbol indicates the drawing needs to be updated.

11. Right-mouse click on the **Drawing** item to bring up the option menu and choose **Update** as shown.

➢ Notice the diameter dimension in the top view is updated to 0.625. (On your own, adjust the display to displaying 3 digits if necessary.)

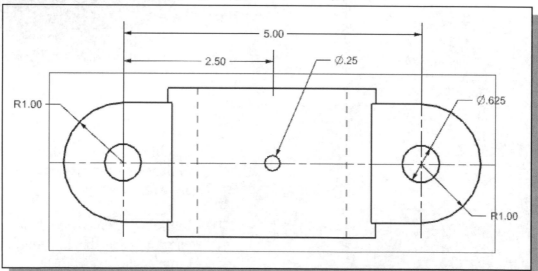

❖ One of the great advantages of *Parametric Modeling* is the ease in making modification of designs. The *associative functionality* is an impressive and important feature of *parametric modeling*. On your own, complete the U-Bracket drawing.

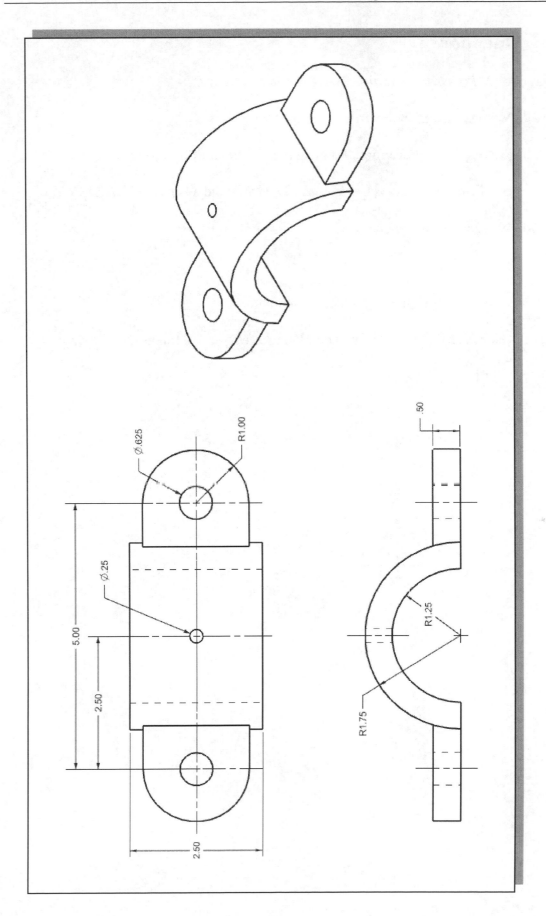

Review Questions:

1. What does *NX*'s *associative functionality* allow us to do?

2. How do we reposition a view in a drawing?

3. How do we display feature/model dimensions in the drafting mode?

4. What is the difference between a *feature dimension* and a *reference dimension*?

5. How do we reposition dimensions?

6. What is a *base view*?

7. Describe the general usage of the Hide command.

8. Create sketches showing the steps you plan to use to create the model shown on the next page:

Exercises: (Create the Solid models and the associated 2D drawings.)

1. **Slide Mount** (Dimensions are in inches.)

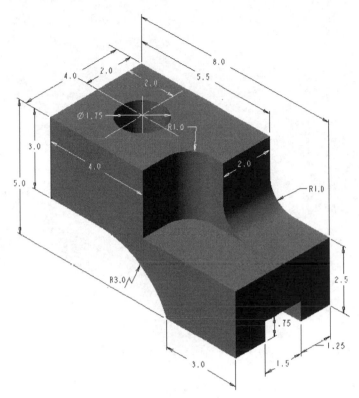

2. **Corner Stop** (Dimensions are in inches.)

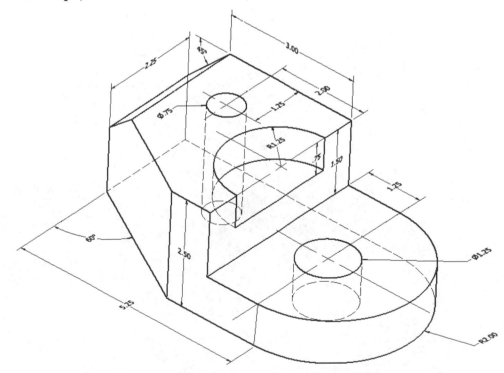

3. **Switch Base** (Dimensions are in inches.)

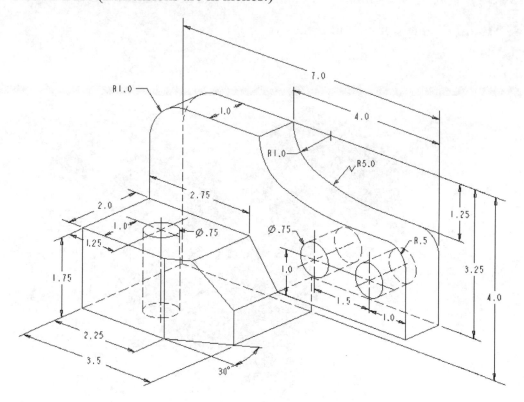

4. **Angle Support** (Dimensions are in inches.)

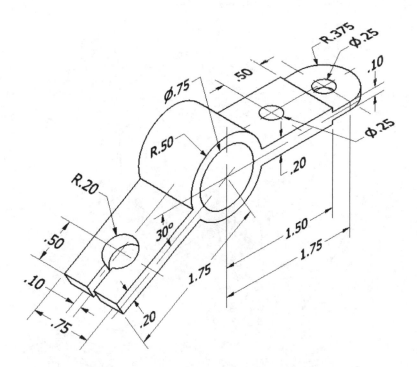

5.   **Block Base** (Dimensions are in inches. Plate Thickness: 0.25)

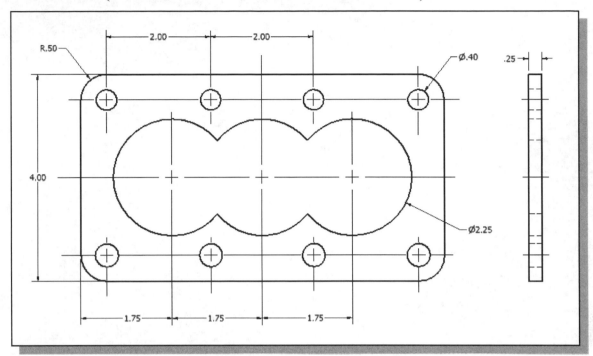

6.   **Shaft Guide** (Dimensions are in inches.)

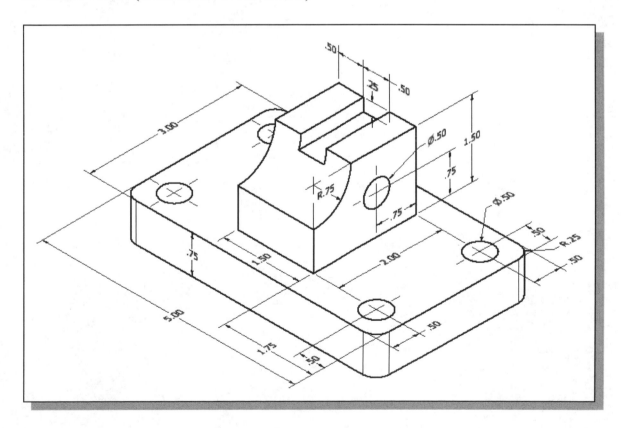

Notes:

Chapter 9
Datum Features and Auxiliary Views

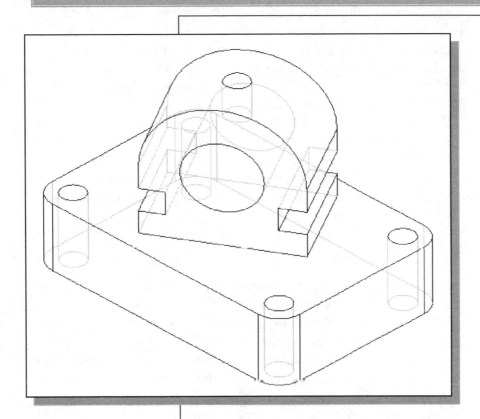

Learning Objectives

- ♦ **Understand the Concepts and the Use of Datum Features**
- ♦ **Using the Different Options to Create Datum Features**
- ♦ **Creating Auxiliary Views in 2D Drawing Mode**
- ♦ **Creating and Adjusting Centerlines**
- ♦ **Create A New Border and Title Block**
- ♦ **Create and use a template file in the 2D Drawing mode**

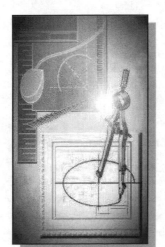

Datum Features

Feature-based parametric modeling is a cumulative process. The relationships that we define between features determine how a feature reacts when other features are changed. Because of this interaction, certain features must, by necessity, precede others. A new feature can use previously defined features to define information such as size, shape, location and orientation. *NX* provides several tools to automate this process. Datum Features can be thought of as user-definable datum, which are updated with the part geometry. We can create Datum Planes, axes, or points that do not already exist. Datum Features can also be used to align features or to orient parts in an assembly. In this chapter, the use of the **Offset** option and the **Angled** option to create new Datum Planes, surfaces that do not already exist, are illustrated. By creating parametric Datum Features, the established feature interactions in the CAD database assure the capturing of the design intent. The default Datum Features, which are aligned to the origin of the coordinate system, can be used to assist the construction of the more complex geometric features.

Auxiliary Views in 2D Drawings

An important rule concerning multiview drawings is to draw enough views to accurately describe the design. This usually requires two or three of the regular views, such as a front view, a top view and/or a side view. However, many designs have features located on inclined surfaces that are not parallel to the regular planes of projection. To truly describe the feature, the true shape of the feature must be shown using an **auxiliary view**. An *auxiliary view* has a line of sight that is perpendicular to the inclined surface, as viewed looking directly at the inclined surface. An *auxiliary view* is a supplementary view that can be constructed from any of the regular views. Using the solid model as the starting point for a design, auxiliary views can be easily created in 2D drawings. In this chapter, the general procedure of creating auxiliary views in 2D drawings from solid models is illustrated.

The Rod-Guide Design

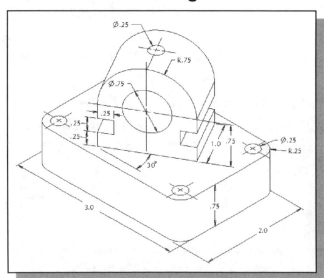

❖ Based on your knowledge of *NX* so far, how would you create this design? What are the more difficult features involved in the design? Take a few minutes to consider a modeling strategy and do preliminary planning by sketching on a piece of paper. You are also encouraged to create the design on your own prior to following through the tutorial.

Modeling Strategy

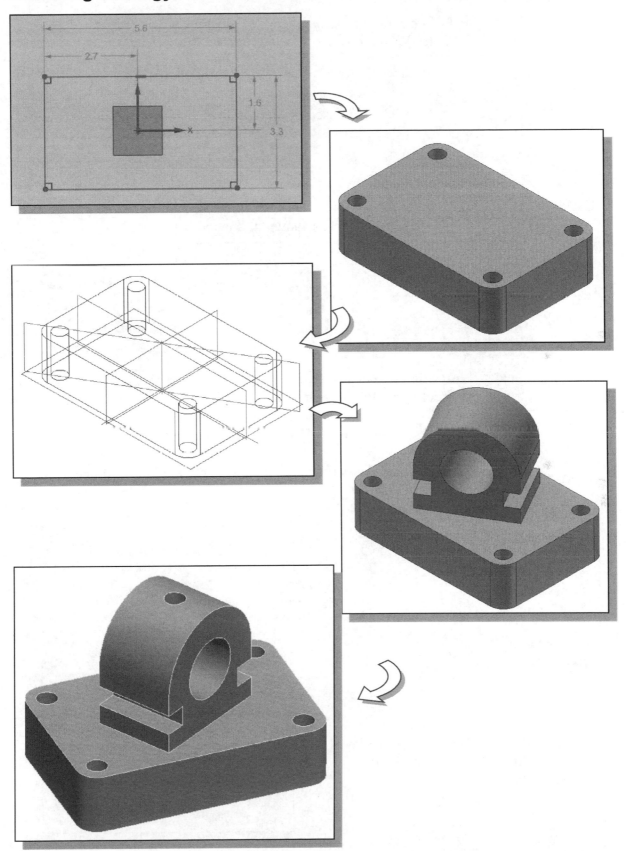

Start NX

1. Select the **NX** option on the *Start* menu or select the **NX** icon on the desktop to start *NX*. The *NX* main window will appear on the screen.

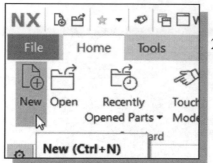

2. Select the **New** icon with a single click of the left-mouse-button (**MB1**) in the *Standard toolbar area*.

3. Select the **inches** units is set as shown in the below figure.

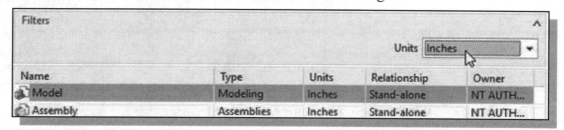

4. Select **Model** in the *Template list*. Note that the *Model template* will allow us to switch directly into the **Modeling task** as indicated in the *templates list*.

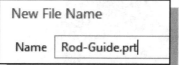

5. Enter **Rod-Guide** as the *New File Name*.

6. Click **OK** to proceed with the *New File* command.

Create the Base Feature

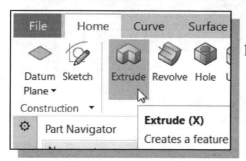

1. In the *Feature Toolbars* (toolbars aligned to the right edge of the main window), select the **Extrude** icon as shown.

2. Select the **XY plane** of the displayed Work Coordinate System to align the *Sketch Plane*.

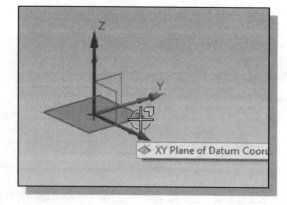

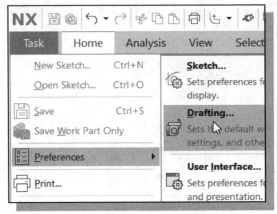

3. In the Menu pull down menu, choose **Preferences → Drafting** as shown.

4. Under *Dimension →Text*, set the *text orientation* to **Horizontal** test as shown.

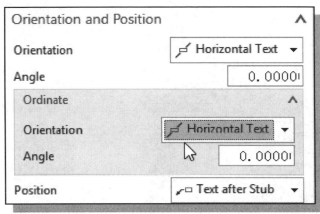

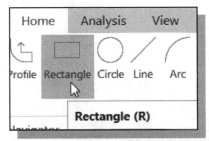

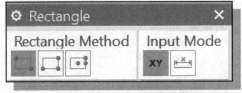

5. Select the **Rectangle** command by clicking once with the **left-mouse-button** on the icon in the *Sketch Curve* toolbar.

6. Notice the different options available to create rectangles; confirm the **By 2 Points** option is activated as shown.

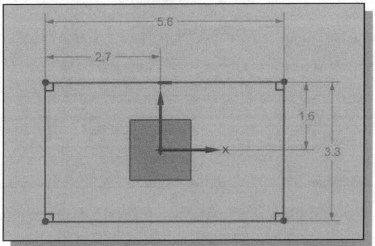

7. Create a **rectangle** of arbitrary size with the datum coordinate system near the center of the rectangle as shown. (Note the *Auto-Dimensioning* option will automatically maintain a fully defined sketch.)

8. On your own, use the **Rapid Dimension** command to create and adjust the 2D sketch as shown in the below figure.

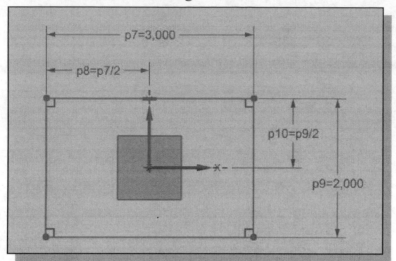

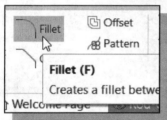

9. Select the **Fillet** command by clicking once with the **left-mouse-button (MB1)** on the icon in the *Sketch Curve* toolbar.

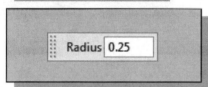

10. Enter **0.25** as the new radius of the fillet.

11. Create the **four rounded corners** and apply the proper constraints to make the sketch fully constrained, as shown in the figure below.

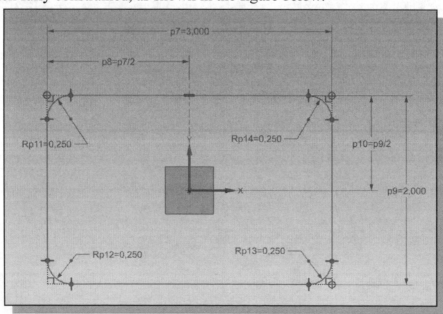

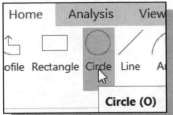

12. Select the **Circle** command by clicking once with the **left-mouse-button (MB1)** on the icon in the *Sketch Curve* toolbar.

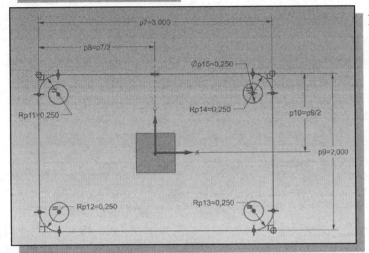

13. Create **four circles** (diameter **0.25**) aligned to the centers of the four arcs as shown. (Hint: Use the *Equal Radius* constraint on the circles.)

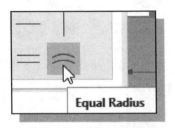

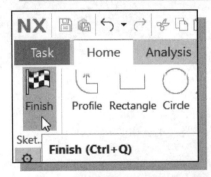

14. Click **Finish Sketch** to exit the *NX sketcher* mode and return to the *Feature option dialog window*.

15. In the *Extrude* pop-up window, enter **0.75** as the extrusion distance.

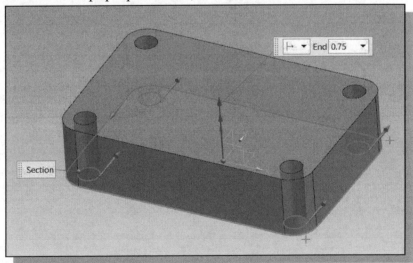

16. Click the **OK** button to proceed with creating the feature.

Create an Angled Datum Plane

1. In the *Part Features* toolbar, select the **Datum Plane** command by left-clicking the icon.

❖ Note that the template files we have used so far all contain a datum coordinate system that was created through the **Datum CSYS** command.

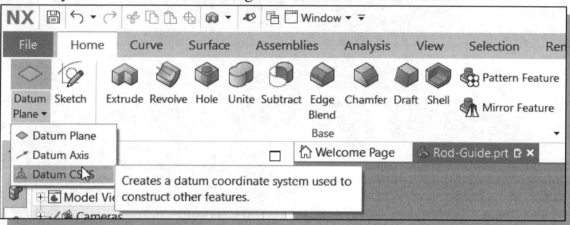

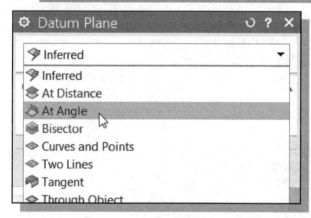

❖ In the *Datum Plane* dialog box, different options are available to create datum planes. Note that different components are needed for each option.

2. Select the **At Angle** option by clicking the corresponding icon as shown in the figure.

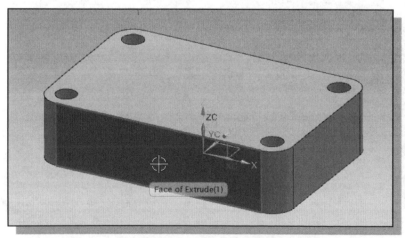

3. Inside the *Graphics* window, select the **Front Face** of the base feature as the first reference of the new *Datum Plane*.

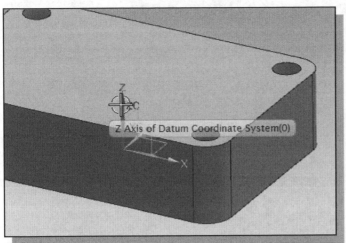

4. Inside the *Graphics* window, left-click once to select the **Z Axis** as the second reference of the new Datum Plane.

5. In the *Angle* pop-up window, enter **-30** degrees as the rotation angle for the new *Datum Plane* as shown in the below figure.

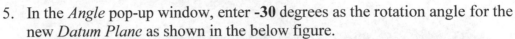

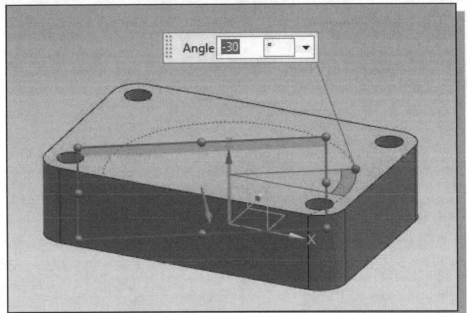

❖ Note that the *angle* is measured relative to the selected reference plane.

6. On your own, examine the other settings for the datum plane, such as the **Plane Orientation** option.

7. Click the **OK** button to accept the setting and create the angled datum plane.

Create an Extruded Feature Using the Datum Plane

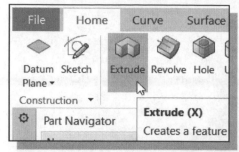

1. In the *Feature Toolbars* (toolbars aligned to the right edge of the main window), select the **Extrude** icon as shown.

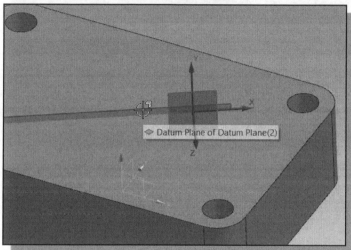

2. Select the newly created **datum plane** as shown.

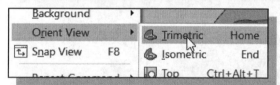

3. In the display toolbar, select **Trimetric** to reset the display angle of the model; then use the dynamic rotate option to orient the display as shown in the below figure.

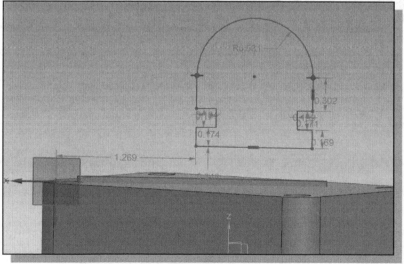

4. On your own, create a 2D rough sketch, above the base feature as shown in the figure, by using the profile command. Note the line segments are aligned to each other and the arc is tangent to the adjacent vertical edges.

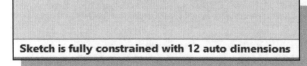

❖ Notice the message "*Sketch is fully constrained with 12 auto dimensions*" is shown.

Apply Proper Constraints

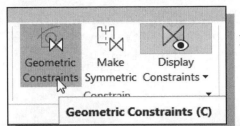

1. Select the **Constraints** command in the *Sketch Constraints* toolbar.

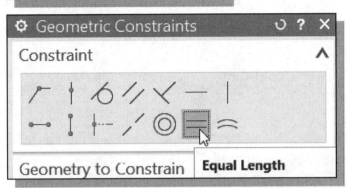

2. On your own, apply the **Equal Length** constraint to match the shorter line segments on both sides of the 2D sketch as shown.

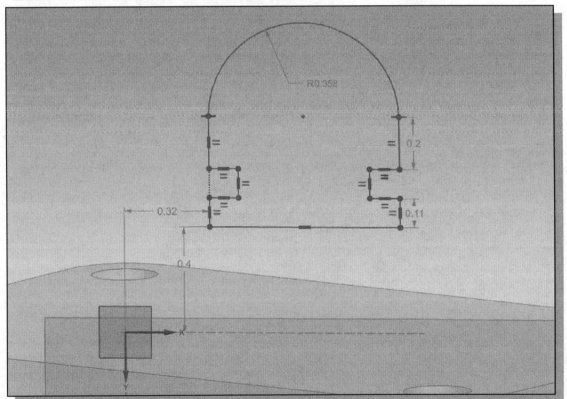

❖ Notice the message "*Sketch is fully constrained with 5 auto dimensions*" is displayed. The number of dimensions necessary to fully describe the sketch is reduced as constraints were applied.

> Sketch is fully constrained with 5 auto dimensions

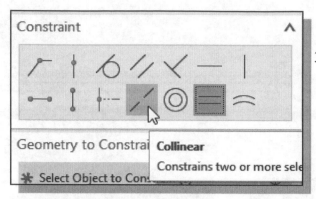

3. If necessary, add the **Collinear** constraint to align the two sets of outer vertical line segments.

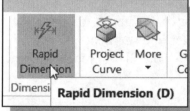

4. Select the **Rapid Dimension** command in the *Constraints* toolbar as shown.

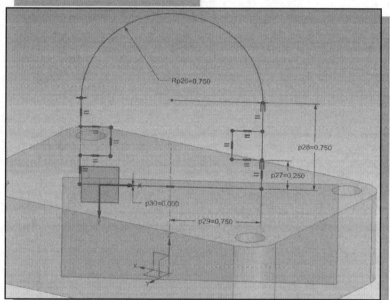

5. On your own, crcatc/adjust the five dimensions as shown in the figure. Note the center of the arc is aligned to the Z-Axis of the default coordinate system as shown.

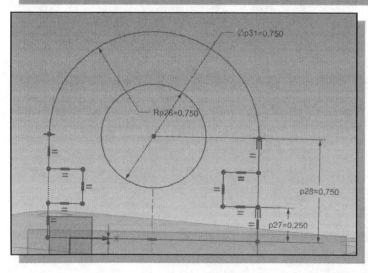

6. On your own, add a **0.75 circle** to complete the 2D sketch as shown in the figure.

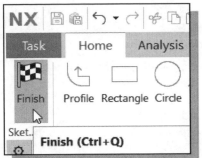

7. Click **Finish Sketch** to exit the NX sketcher mode and return to the *Feature option dialog window*.

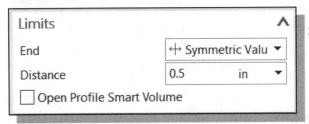

8. In the *Extrude* pop-up window, select Symmetric value and enter **0.5** as the extrusion distance.

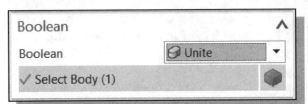

9. In the *Extrude* pop-up window, set the *Extrude* option to **Unite** as shown.

10. Click **OK** to create the extrusion feature.

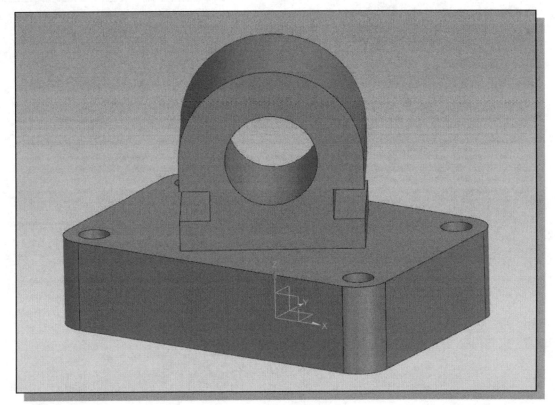

Create an Offset Datum Plane

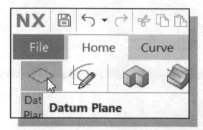

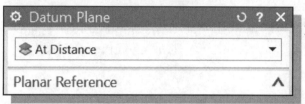

1. In the *Part Features* toolbar, select the **Datum Plane** command by left-clicking the icon.

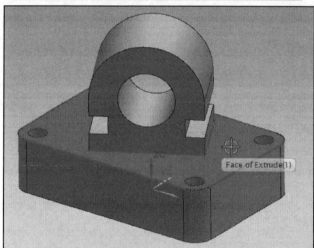

2. Select the **At Distance** option by selecting the corresponding type as shown in the figure.

3. Inside the *Graphics* window, select the ***Top Face*** of the base feature as the reference of the new *Datum Plane*.

4. Enter **0.75** as the offset distance and create only **one** datum plane as shown.

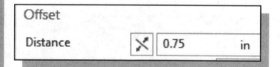

5. Click **OK** to create the offset datum plane.

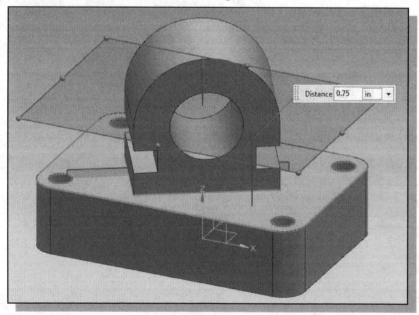

Creating a hole feature using the new Datum Plane

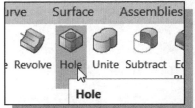

1. In the *Feature Operation* toolbar (the toolbar that is located to the right side of the *Form Feature Toolbar*), select the **Hole** command by releasing the left-mouse-button on the icon.

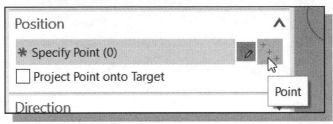

2. In the *Hole dialog box*, click **Sketch Section** to enter the sketch mode in defining the center point.

3. Rotate the display so that the edges of the new datum plane are roughly parallel to the edges of the screen.

4. Pick the **new datum plane** we just created as shown.

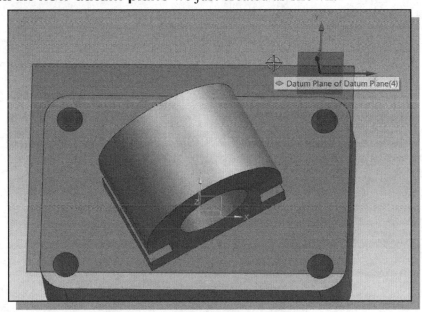

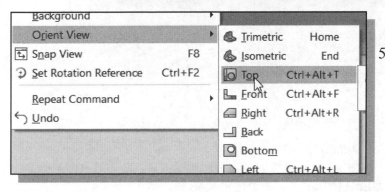

5. Right-mouse-click to bring up the option menu, set the display to the **top view** as shown.

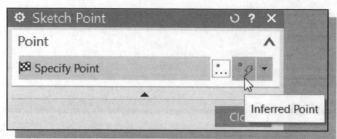

6. In the *Sketch CSYS dialog box*, set the plane method to **Inferred** as shown.

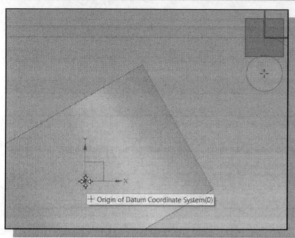

7. Select the **Sketch Origin** as the center point of the *hole feature*.

8. Click **Close** to accept the selection.

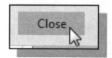

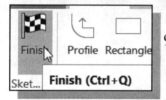

9. Click **Finish Sketch** to exit the NX sketcher mode and return to the *Hole Feature dialog box*.

10. Set the hole direction option to **Along Vector** and click Reverse direction to set direction to align the **Z** axis. Enter **0.25** as the diameter of the hole.

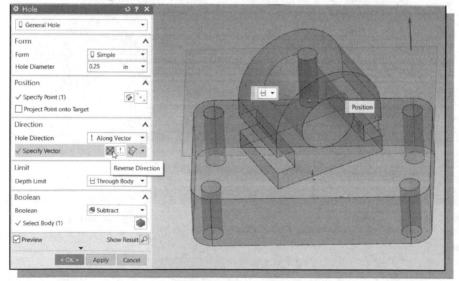

11. Click **OK** to create the *Hole* feature.

12. Click **Save** to save the design, **Rod-Guide.prt**

Create a Title Block Template

❖ Several options are available in NX to create drawing templates. In this section, we will examine the use of Export/Import commands to create drawing templates.

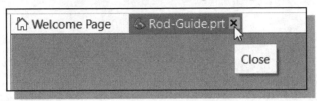

1. Close the *Rod-Guide* model by clicking on the lower [**X**] next to the NX part navigator window as shown.

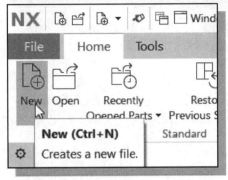

2. Start a new file by clicking the **New** icon with the left-mouse-button (**MB1**) in the *Standard toolbar area*.

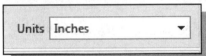

3. In the *New Part File* window, set the units to **Inches** as shown.

4. Select **Blank** in the *Template list*. Note that the *Blank template* contains no pre-defined settings.

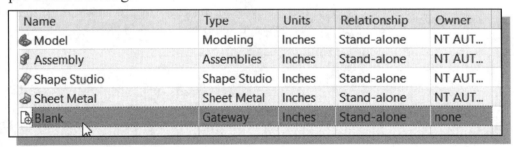

Name	Type	Units	Relationship	Owner
Model	Modeling	Inches	Stand-alone	NT AUT...
Assembly	Assemblies	Inches	Stand-alone	NT AUT...
Shape Studio	Shape Studio	Inches	Stand-alone	NT AUT...
Sheet Metal	Sheet Metal	Inches	Stand-alone	NT AUT...
Blank	Gateway	Inches	Stand-alone	none

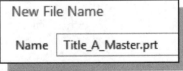

5. In the *New File Name* section, enter **Title_A_Master** as the File *Name*.

6. Click **OK** to proceed with the *New Part File* command.

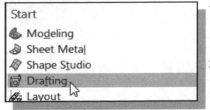

7. Select the **File** tab with a single click of the left-mouse-button (**MB1**) in the *Ribbon* bar area.

8. Pick **Drafting** in the pull-down list as shown in the figure.

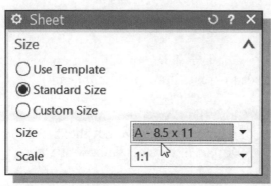

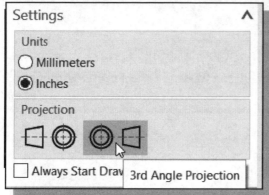

9. Select **A-8.5X11** from the *sheet size* list.

10. Confirm the **Scale** is set to **1:1**, which is *Full Scale*.

❖ Note that *Sheet1* is the default drawing sheet name that is displayed above the *sheet size* list.

11. In the settings area, confirm the *Projection* type is set to **Third Angle of projection** as shown.

12. **Uncheck** the *Automatically Start Base View Command* option as shown.

13. Click **OK** to accept the settings and proceed with the creation of the drawing sheet.

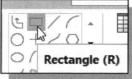

14. Create a rectangle and adjust the size of the rectangle to **10.25 X 7.75** as shown in the below figure. (Hint: Use the **Rapid Dimension** command to add the dimensions.)

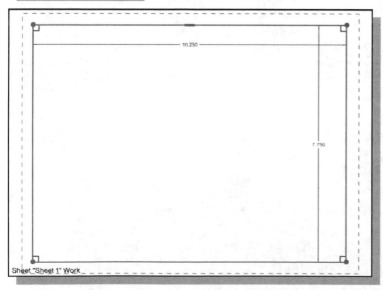

15. On your own, click and drag one of the edges of the rectangle to reposition the rectangle so that it is roughly at the center of the drawing sheet.

❖ Note that any sketches created in the drafting mode are still parametric sketches, which means we can apply different types of constraints to assure the proper geometric properties of the 2D sketches.

16. Switch **on** the **Continuous Auto Dimensioning** option in the *Sketch* toolbar.

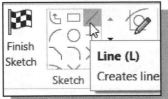

17. Select the **Line** command by clicking once with the left-mouse-button as shown.

18. On your own, create and adjust the horizontal and vertical lines with the dimensions as shown. (Hint: Apply proper **constraints** first.)

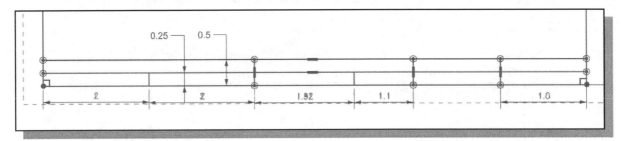

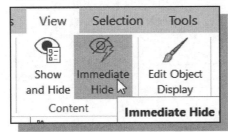

19. Select the **Immediate Hide** command in the *View* toolbar or use the quick key [**Ctrl+B**] to bring up the *Hide dialog box*.

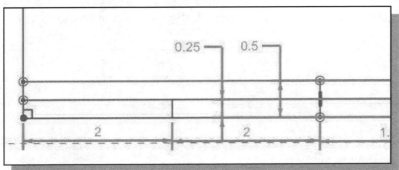

20. Hide **all of the dimensions** by clicking with the left-mouse-button on each of the dimension text.

21. Click **OK** to accept the selection and proceed to hide the selected items.

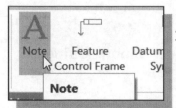

22. Click **Note** in the toolbar as shown.

23. Click on the **Arrow** icon to expand the list of options.

24. Pick the **Annotation Setting** option in the settings area as shown.

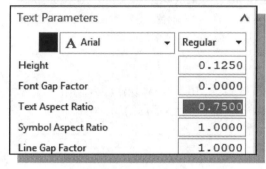

25. Enter **0.125** as the *Character size* and 0.75 as the new *Text Aspect Ratio*.

26. Set the color to black.

27. Click **Close** to accept the settings.

28. Enter your school name, or company name, in the ***annotation editor*** *box* and place the text in the title block as shown.

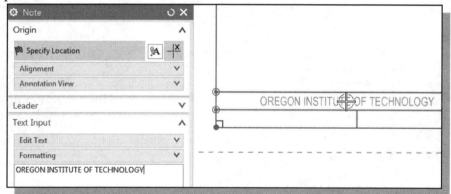

29. On your own, repeat the above steps and enter the text as shown in the figure below.

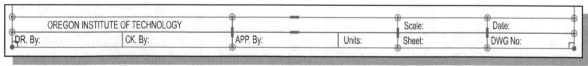

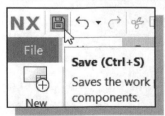

30. Save the title block design by clicking **Save** in the standard toolbar as shown.

❖ Note that this is the **master file**; additional modifications can be applied and a new template can be generated.

Use the Export File Command

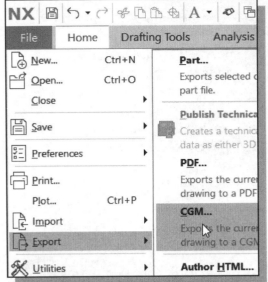

1. Select the **Export CGM** command through the *File* pull-down menu:
   [File] → [Export] → [CGM]

❖ Note the different options available, such as PNG, Jpeg, STL, IGES and DXF/DWG. You are encouraged to try using some of these formats.

2. Click on the **folder** icon and set the file name to **Title_A_Master_Sheet1**.

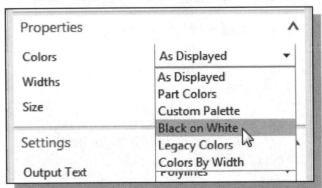

3. Click **OK** to accept the file name and location.

4. Click on the triangle icon and under the properties options, set the *Colors* option to **Black on White** as shown in the figure.

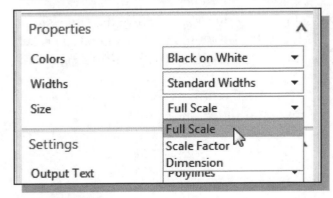

5. Confirm the *Scale factor* is set to **Full Scale**.

6. Click **OK** to accept the settings and save the CGM file.

Reopen the Rod Guide Design

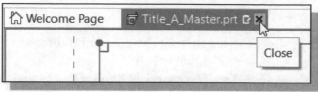

1. Close the *Title_A_Master* file by clicking on the lower [**X**] near the upper left corner of the NX main window as shown.

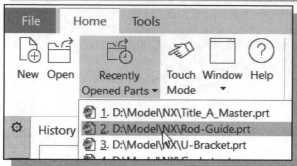

2. Reopen the **Rod Guide** design through the *Open a Recent Part* icon as shown.

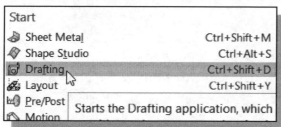

3. Click on the **File** icon in the *Ribbon* toolbar area to display the available options.

4. Select **Drafting** from the *Applications list*.

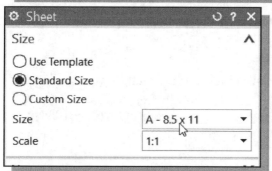

5. Select **A-8.5X11** from the *sheet size* list.

6. Confirm the **Scale** is set to **1:1**, which is ***Full Scale***.

➢ Note that ***Sheet 1*** is the default drawing sheet name that is displayed in the *Drawing Sheet Name* area.

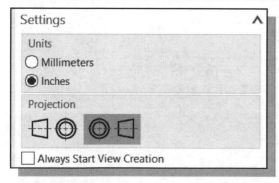

7. In the settings area, confirm the *Units* are set to **Inches** and the *Projection* type is set to **Third Angle of projection** as shown.

8. Confirm the *Always Start Drawing View* option is turned off as shown.

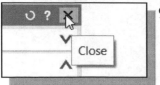

9. Click **OK** to accept the settings and proceed with the creation of the drawing sheet.

10. Click **Close** to exit the *base view wizard*.

Import the Title Block

1. Select the **Import CGM** command through the *File* pull-down menu:
   [File] → [Import] → [CGM]

2. Enter or select **Title_A_Master_sheet 1.cgm** in the *Import CGM* window as shown.

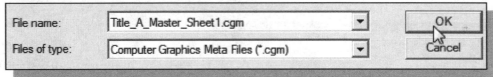

3. Click **OK** to proceed with importing the selected CGM file.

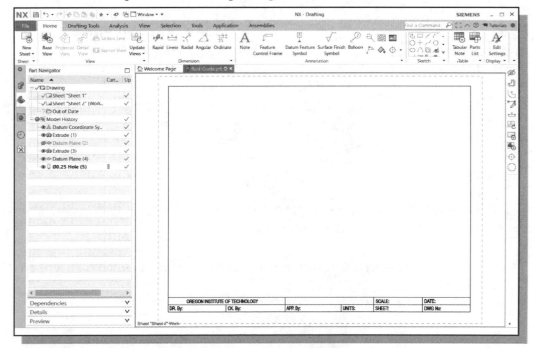

❖ The **Import** and **Export** commands provide a fairly flexible way for us to reuse title block and borders. You are encouraged to try the other file formats and achieve the same result as shown in the above figure.

Add a Base View

❖ In *NX Drawing Mode*, the first drawing view we create is called a **base view**. A *base view* is the primary view in a drawing; other views can be derived from this view. When creating a *base view*, NX allows us to specify one of the principal views to be shown. By default, *NX* will treat the *XZ plane* as the front view of the solid model. Note that there can be more than one *base view* in a drawing.

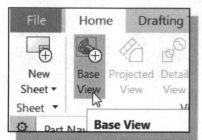

1. Click on the **Base View** icon in the *Drawing Layout toolbar* to create a new base view.

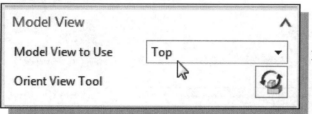

2. In the *Base View* option toolbar, set the view option to **Top** as shown.

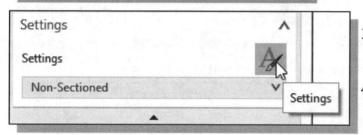

3. In the *Settings* option list, click the **Settings** icon as shown.

4. In the *settings* box, click the **Hidden Lines** tab as shown.

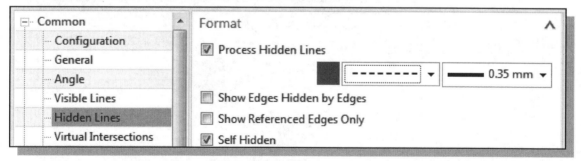

5. Choose **Dashed** as the line type to be used for *hidden lines* in the Base view.

6. Click on the **Visible Lines** tab and select **0.70mm** as the line type for the Base View.

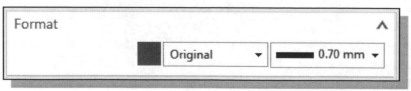

7. Click **OK** to accept the settings and exit the *View Style* command.

8. Place the **Base View**, the *Top view* of the *Rod-Guide* model, near the upper left corner of the drawing sheet. **Do not** exit the *Project View* option.

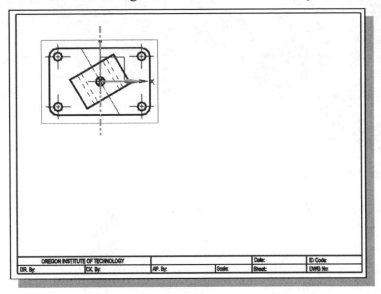

Create an Auxiliary View

❖ In *NX Drawing Mode*, the **Projected View** command is used to create standard views such as the *top* view and *front* view. For non-standard views, the **Auxiliary View** command is used. *Auxiliary* views are created using orthographic projections. Orthographic projections are aligned to the base view and inherit the base view's scale and display settings.

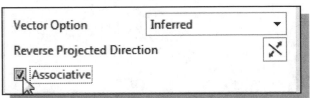

1. Switch on the **Associative** option as shown.

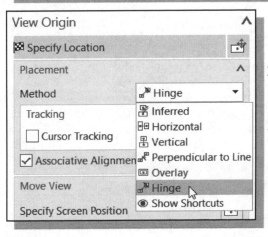

2. Confirm the **Associative Alignment** option is switched **on**.

3. Set the *Placement Method* to **Hinge** as shown.

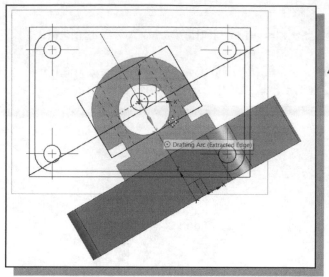

4. Select the **front edge** of the top portion of the solid model to set the hinge reference as shown.

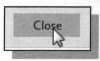

5. Click **Close** to end the *Projected View* command.

6. On your own, reposition the *Auxiliary* view by dragging and dropping with the **left-mouse-button** as shown in the below figures.

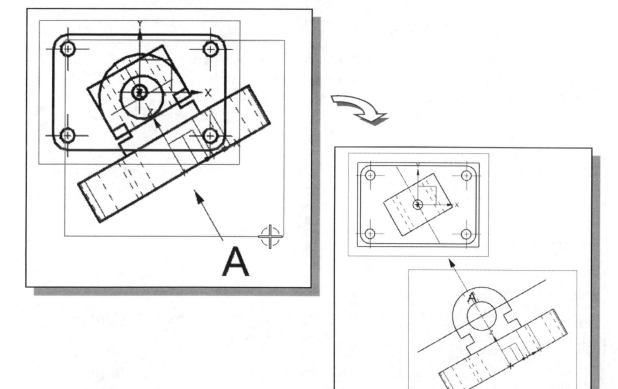

❖ In NX, creating an *Auxiliary* view is very similar to creating a standard projected view.

Turn Off the Datum Planes and Reference Labels

➤ The display of the datum planes, datum coordinate system and reference labels can be turned off in the drafting mode.

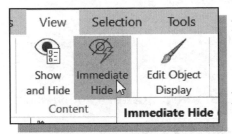

1. Select the **Immediate Hide** command in the *View* toolbar or use the quick key **[Ctrl+B]** to bring up the *Hide dialog box*.

➤ The *Hide* command can be used to temporarily turn off the display of objects.

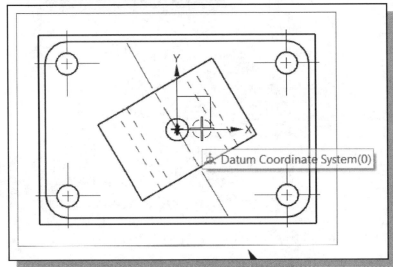

2. Select the **datum coordinate system** and the **Drafting point** near the center of the top view as shown.

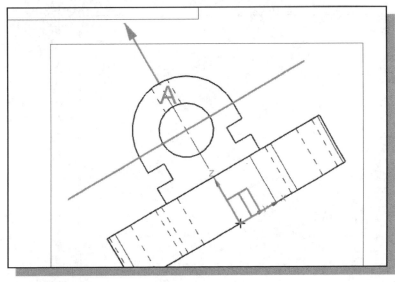

3. Select the **Reference Arrow/Plane, drafting point** and the **Reference label** and **View label** in the auxiliary view as shown.

4. Click **OK** to accept the selections and proceed to hide the selected objects.

Turn Off the Display Borders Option

❖ The display of the borders around the 2D views can be turned off by setting the drafting preferences. Note that some of the drafting commands can reset this setting.

1. Select **Drafting** in the *Preferences* menu list as shown.

2. Turn off the **Display Borders** option under *View Workflow* as shown.

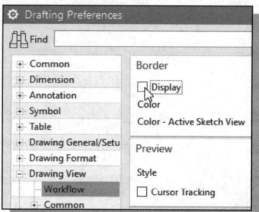

3. Click **OK** to accept the settings and exit the *Drafting Preferences* command.

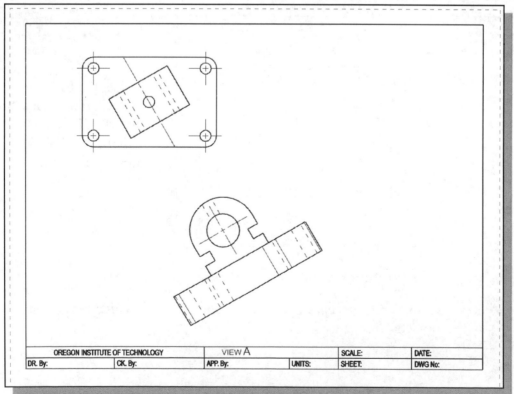

Add another Base View

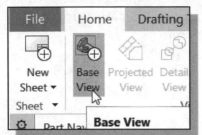

1. Click on the **Base View** in the *Drawing Layout toolbar* to create a new base view.

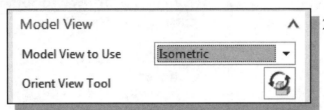

2. Select **Isometric** in the Model view list, and place an isometric view to the right side of the drawing sheet as shown.

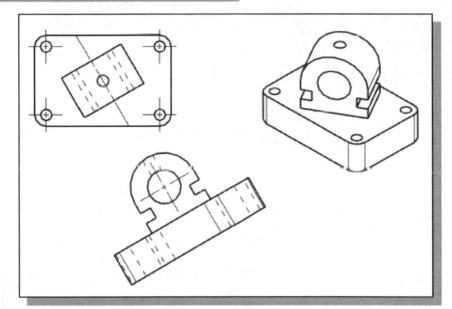

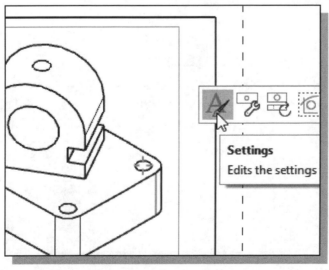

3. Move the cursor on the **Isometric view** until you see the border highlighted.

4. **Left-mouse-click** once on the *Isometric view* to bring up the option list.

5. Select the **Settings** option as shown.

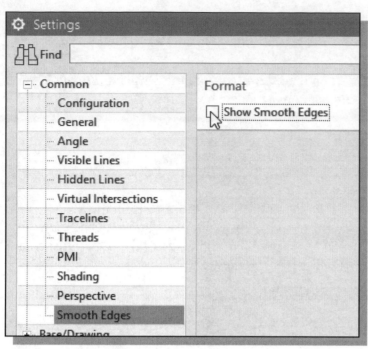

6. In the *Settings dialog box*, expand the **Common** list item.

7. Left-mouse-click (MB1) the **Smooth Edges** tab as shown.

8. Turn off the **Show Smooth Edges** option as shown.

9. Click **OK** to accept the settings.

10. On your own, repeat the above steps to turn off the edges in the *Aux. view*.

Drawing Display Option

1. Select **Visualization** in the *Preferences menu list* as shown.

2. Click on the **Color → Drawing Layouts** as shown.

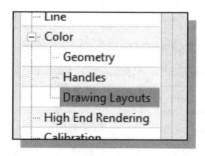

3. Activate the **Monochrome Display** and the **Show Widths** options as shown.

4. Confirm the **Background** color to **white**.

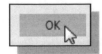

5. Click **OK** to accept the settings.

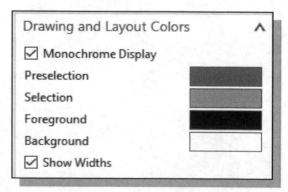

Display Feature Dimensions

By default, feature dimensions are not displayed in 2D views in *NX*.

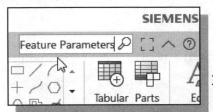

1. Enter **Feature Parameters** in the **Command Finder** box as shown.

2. Select **Feature Parameters** in the **Command Finder** dialog box.

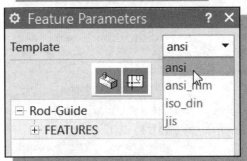

3. In the *Feature Parameters* window, notice the **ANSI** standard is used for the dimension settings.

❖ Two steps are required: (1) select the features and (2) select the view to display the dimensions associated with the feature.

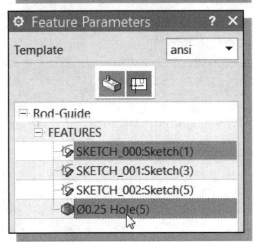

4. Expand the *FEATURES* list and choose the **Base** feature (Sketch1) and the Simple **Hole** feature (Hole 5) as shown.

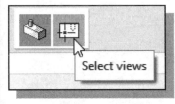

5. Click the **Select Views** icon to proceed to the next step.

6. Inside the graphics window, select the **Top View** by clicking once with the **left-mouse-button (MB1)** as shown.

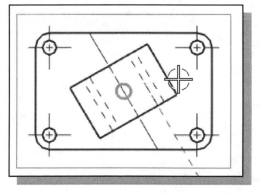

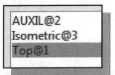

❖ Note that the corresponding view name is highlighted as soon as we click on a view in the graphics window.

7. Click **Apply** to accept the settings.

8. Left-mouse-click once the **Select features** button as shown.

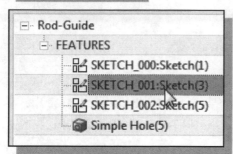

9. Select **second extrude feature** by left-clicking once in the features list.

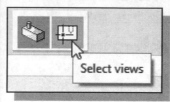

10. Click the **Select Views** icon to proceed to the next step.

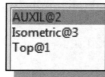
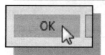

11. Inside the feature parameters window, select the **Aux View** by clicking once with the **left-mouse-button (MB1)** as shown.

12. Click **OK** to accept the settings and exit the *feature parameters* window. Click Close to exit *Command Finder*.

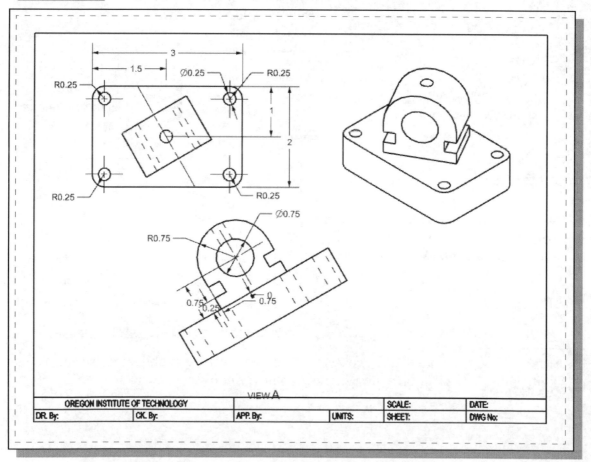

Delete and Add Dimensions

- We will use the **Delete** command to remove the unwanted feature dimensions.

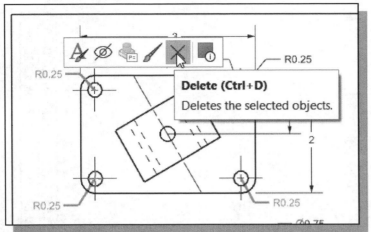

1. Select one of the radius dimensions of the top view and select **Delete** through the *icon* menu as shown.

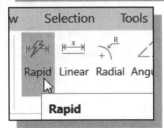

2. Select the **Rapid dimension** in the toolbar.

3. On your own, create the necessary dimensions in the *Aux. View* as shown in the below figure

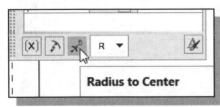

4. For the radius and diameter dimensions, toggle the **Radius to Center** option as shown.

5. Also set the *Dimension Side 1* to **Out** under the *Dimension Settings: Line/Arrow → Arrowhead.*

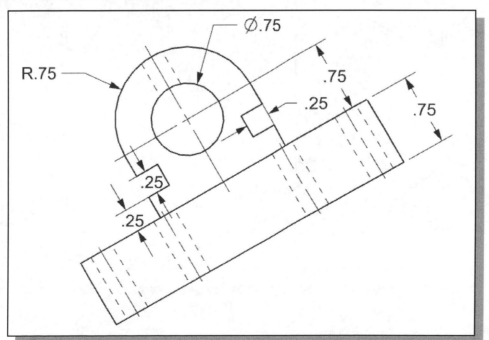

6. On your own, create the necessary dimensions and center line to complete the *rod guide drawing* as shown in the below figure.

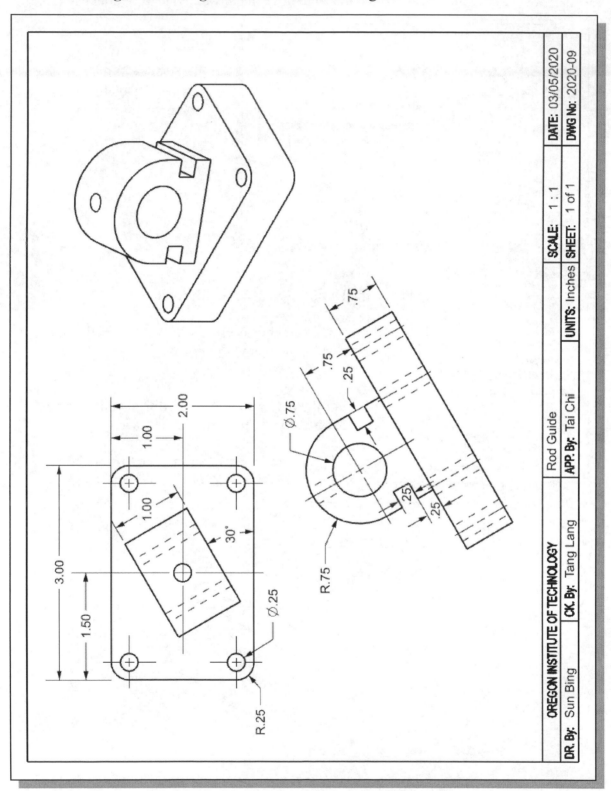

OREGON INSTITUTE OF TECHNOLOGY		Rod Guide	SCALE: 1 : 1	DATE: 03/05/2020
DR. By: Sun Bing	CK. By: Tang Lang	APP. By: Tai Chi	UNITS: Inches	SHEET: 1 of 1
				DWG No: 2020-09

Review Questions:

1. What are the different types of Datum Features available in *NX*?

2. Why are Datum Features important in parametric modeling?

3. Describe the purpose of auxiliary views in 2D drawings?

4. What are the required elements to create an auxiliary view?

5. What are the advantages of using a template file?

6. Describe the differences between the 1st angle projection and 3rd angle projection used in drafting standards.

7. Describe the general procedure to display feature dimensions in the drafting mode of NX.

Exercises: (Create the Solid Models and the associated 2D drawings.)

1. **Rod Slide** (Dimensions are in inches.)

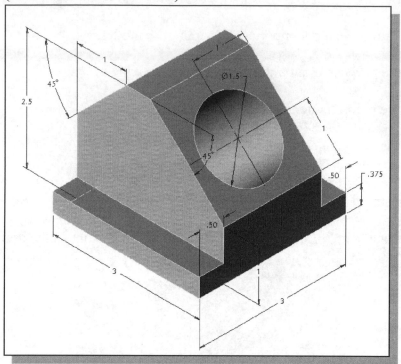

2. **Anchor Base** (Dimensions are in inches.)

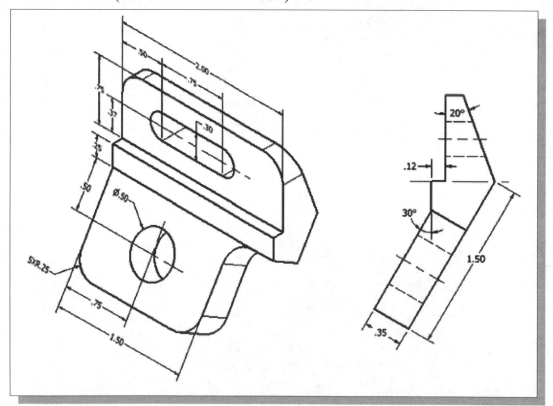

3. **Bevel Washer** (Dimensions are in inches.)

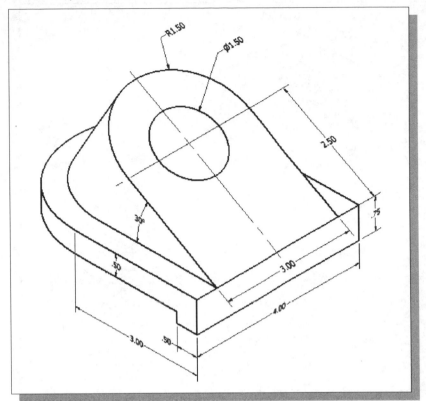

4. **Angle V-Block** (Dimensions are in inches.)

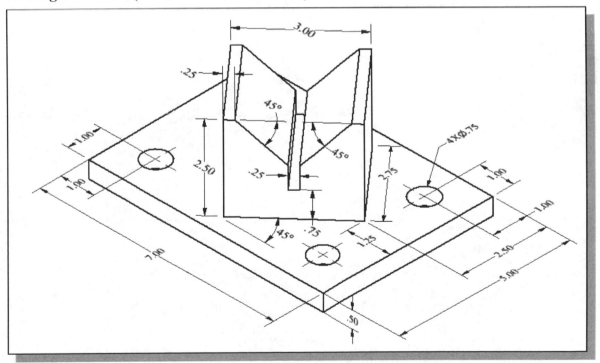

5. **Angle Support** (Dimensions are in millimeters.)

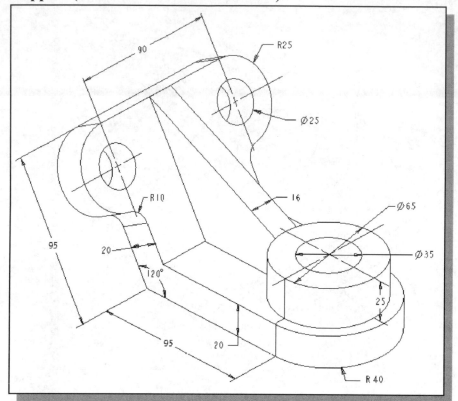

6. **Jig Base** (Dimensions are in millimeters.)

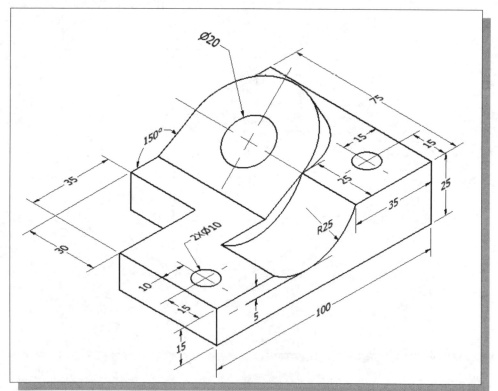

Chapter 10
Introduction to 3D Printing

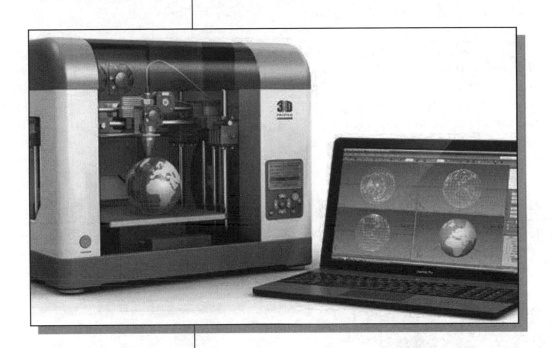

Learning Objectives

- ♦ **Understand the History and Development of 3D Printing**
- ♦ **Be aware of the Primary types of 3D Printing Technologies**
- ♦ **Be able to identify the commonly used Filament types for Fused Filament Fabrication**
- ♦ **Understand the general procedure for 3D Printing**

What is 3D Printing?

3D Printing is a type of *Rapid Prototyping* (RP) method. Rapid prototyping refers to the techniques used to quickly fabricate a design to confirm/validate/improve conceptual design ideas. 3D printing is also known as **"Additive Manufacturing"** and construction of parts or assemblies is usually done by addition of material in thin layers.

Prior to the 1980s, nearly all metalworking was produced by machining, fabrication, forming, and mold casting; the majority of these processes require the removal of material rather than adding it. In contrast to the *Additive Manufacturing* technology, the traditional manufacturing processes can be described as **Subtractive Manufacturing**. The term *Additive Manufacturing* gained wider acceptance in the 2010s. As the various additive processes continue to advance and become more mature, it is quite clear that material removal will no longer be the main manufacturing process in the very near future.

The basic principle behind *3D printing* is that it is an additive process. 3D printing is a radically different manufacturing method based on advanced technology that create parts directly, by adding material layer by layer at the sub millimeter scale. One way to think about 3D Printing is the additive process is really performing "**2D printing over and over again.**"

A number of limitations exist to the traditional manufacturing processes, which has been based on human labor and made by hand ideology, including the expensive tooling, designing of fixtures, and the assembly of parts. The *3D printing* technology provides a way to create parts with complex geometric shapes quite easily using thin layers. The traditional *subtractive manufacturing* can also be quite wasteful as excess materials are cut and removed from large stock blocks, while the *3D printing* process only uses the material needed for the parts. *3D printing* is an enabling technology that encourages and drives innovation with unprecedented design freedom while being a tooling-less process that reduces costs and lead times. The relatively fast turnaround time also makes *3D printing* ideal for prototyping. Components with intricate geometry and complex features can also be designed specifically for *3D printing* to avoid complicated assembly requirements. *3D printing* is also an energy efficient technology that can provide better environmental friendliness in terms of the manufacturing process itself and the type of materials used for the product. There are quite a few different techniques to 3D print an object. *3D Printing* brings together two fundamental innovations: the manipulation of objects in the digital format and the manufacturing of objects by addition of material in thin layers.

The term **3D-printing** originally refers only to the smaller 3D printers with moveable print heads similar to an inkjet printer. Today, the term **3D-printing** is used interchangeably with **Additive Manufacturing**, as both refer to the technology of creating parts through the process of adding/forming thin layers of materials.

Development of 3D Printing Technologies

The earliest 3D printing technology was first invented in the 1980s; at that time it was generally called **Rapid Prototyping** (**RP**) technology. This is because the process was originally conceived as a fast and time-effective method for creating prototypes for product development in industry. In 1981, Dr. Hideo Kodama of *Nagoya Municipal Industrial Research Institute* invented two methods of creating three-dimensional plastic models with photo-hardening polymer through the use of a UV Laser. In 1986, the first US patent for **stereolithography** apparatus (**SLA**) was issued to Charles Hull, who first invented his SLA machine in 1983. Chuck Hull went on to co-found *3D Systems Corporation*, which is one of the largest companies in the 3D printing sector today. Chuck Hull also designed the **STL** (**ST**ereo**L**ithography) file format, which is widely used by 3D printing software performing the digital slicing and infill strategies common to the additive manufacturing processes. The first available commercial RP system, the **SLA-1** by *3D Systems* (as shown in the figure below), was made available in 1987.

The 1980s also mark the birth of many RP technologies worldwide. In 1989, Carl Deckard of *University of Texas* developed the **Selective Laser Sintering (SLS)** process. In 1989, Scott Crump, one of the founders of *Stratasys Inc.*, also created the **Fused Deposition Modeling (FDM)**. In Europe, Hans Langer started *EOS GmbH* in Germany; the company focuses on the **Laser Sintering (LS)** process. The *EOS systems Corp.* also developed the **Direct Metal Laser Sintering (DMLS)** process. Today, *3D Systems*, *EOS* and *Stratasys* are still the main leaders in the *Additive Manufacturing* industry.

During the 1990s, the *3D printing* sector started to show signs of distinct diversification with two specific areas of emphasis which are much more clearly defined today. First, there was the high end of 3D printing, still very expensive systems, which were geared towards part production for relatively complex designs. For example, in 1995, *Sciaky Inc.* developed an additive welding process based on its proprietary **Electron Beam Additive Manufacturing (EBAM)** technology. Many *RP* system companies, such as *Solidscape*, *ZCorporation*, *Arcam* and *Objet Geometries* were all launched in the 1990s. At the other end of the spectrum, some of the 3D printing system manufacturers started to develop smaller desktop systems in the 1990s.

The idea of creating low-cost desktop 3D printers also intrigued many technology professionals and hobby enthusiasts during the late 1990s. In 2004, a retired professor, Dr Adrian Bowyer (person on the left in the below photo), started the **RepRap** (*Replication Rapid-Prototyper*) project of an open source, self-replicating 3D printer (**RepRap 1.0 - Darwin**). This sets the stage for what was to come in the following years. It was around 2007 that the open source *3D printing* movement started gaining visibility and momentum. In January of 2009, the first commercially available open source 3D printer, the **BFB RapMan** 3D printer, became available. The *Makerbot Industries* also came out with their **Makerbot** 3D printer in April of 2009. Since then, a host of low-cost desktop 3D printers have emerged each year.

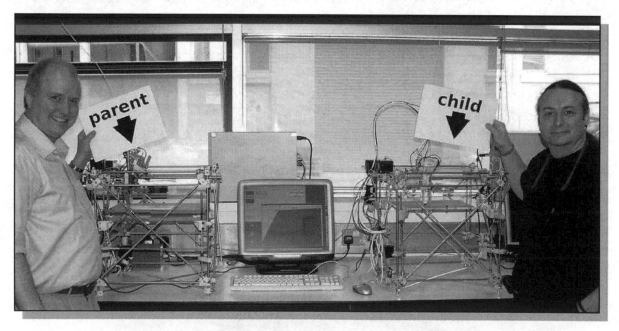

In the beginning of the 2010s, alternative 3D printing processes, such as using the **Polymer Resins** material, became available at the desktop level of the market. The **B9 Creator** by *B9Creations*, using **Digital Light Processing (DLP)** technology, came first in June of 2012, followed by the **Form 1** desktop printer by *Formlabs Inc.* Both 3D printers were launched via KickStarter's crowd-funding website and both enjoyed huge success. 2012 was also the year that many different mainstream media noticed the exciting 3D printing technology, which dramatically increased awareness and uptake to the general public. 2013 was also a year of significant growth and consolidation. One of the most notable moves was the acquisition of Makerbot by Stratasys. In 2016, the new developments in 3D printing concentrate more on multi-color, multi-material using single or multiple extruders and new technologies to shorten the 3D printing time.

As a result of the market divergence, the price of desktop 3D printers continues to go down each year. Today, very capable fully assembled desktop 3D printers, such as *Robo3D R1+, Prusa I3 MK2*, can be acquired for under $1000. Fully assembled smaller desktop 3D printers, such as Xyz-printing's *DA Vinci mini 3D printer* and M3D's *Micro 3D* can be acquired for less than $350. Unassembled desktop 3D printer kits can even be acquired for under $200.

Another trend that happened in the 2010s is the availability of **3D printing Services**. 3D printing services are growing quite rapidly in the US. For example, many public libraries, especially in California, are now providing 3D printing services to the general public and **UPS** started its worldwide 3D printing services in May of 2016. This trend is spreading throughout the US, with many more companies planning to provide 3D printing services in the very near future. It is now quite feasible, and perhaps more economical, to 3D print designs without owning or ever touching a 3D printer, but understanding of the technology is still needed to increase productivity.

As the exponential adoption rate continues on all fronts, more and more technologies, materials, applications, and online services will continue to emerge. It is predicted that the development of 3D printing will continue in the years to come and 3D printing will eventually become the mainstream manufacturing method in industries and in homes.

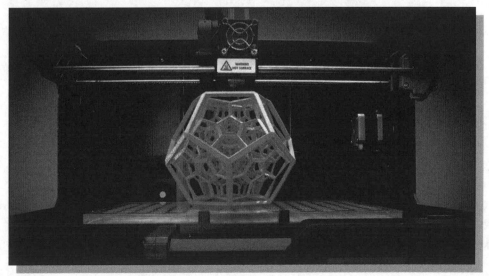

Primary types of 3D Printing processes

There are quite a few different techniques to 3D print an object. The different types of 3D printers each employ a different technology that processes different materials in different ways. For example, some 3D printers process powdered materials (nylon, plastic, ceramic and metal), which utilize a light/heat source to sinter/melt/fuse layers of the powder together in the defined shape. Others process polymer resin materials and again utilize a light/laser to solidify the resin in thin layers. **Stereolithography (SLA or SL), Fused Deposition Modeling (FDM or FFF)** and **Laser Sintering (LS or SLS)** represent the three primary types of 3D printing processes; the majority of the other 3D printing technologies are variations of the three main types.

Stereolithography

Stereolithography (**SLA** or **SL**) is widely recognized as the first 3D printing process; it was certainly the first to be commercialized. *SLA* is a laser-based process that works with photopolymer resins. The photopolymer resins react with the laser and cure to form a solid in a very precise way to produce very accurate parts. It is a complex process, but simply put, the photopolymer resin is held in a container with a movable platform inside. A laser beam is directed in the X-Y axes across the surface of the resin according to the 3D data supplied to the machine. The resin hardens precisely as the laser hits the designated area. Once the current layer is completed, the platform within the container drops down by a fraction (in the Z axis) and the subsequent layer is traced out by the laser. This 2D layer tracing continues until the entire object is completed.

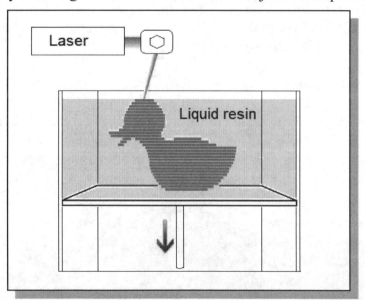

Because of the nature of the SLA process, support structures are needed for some parts, specifically those with overhangs or undercuts. These support structures need to be removed once the part is created. Many 3D printed objects using SLA need to be further cleaned and/or cured. Curing involves subjecting the part to intense light in an oven-like machine to fully harden the resin. SLA is generally accepted as being one of the most accurate 3D printing processes with excellent surface finish.

Fused Deposition Modeling (FDM) / Fused Filament Fabrication (FFF)

3D printing utilizing the extrusion of thermoplastic material is probably the most popular 3D printing process. The original name for the process is **Fused Deposition Modeling (FDM)**, which was developed in the early 1990s and is a trade name registered by *Stratasys*. However, a similar process, **Fused Filament Fabrication (FFF)**, has emerged since 2009. A majority of desktop 3D printers, both open source and proprietary, utilize the FFF process that is a more basic extrusion form of FDM.

The FDM and FFF processes work by melting plastic filament that is deposited, via a heated extruder, one layer at a time, onto a build platform according to the 3D data supplied to the 3D printer. Each layer hardens as it cools down and bonds to the previous layer.

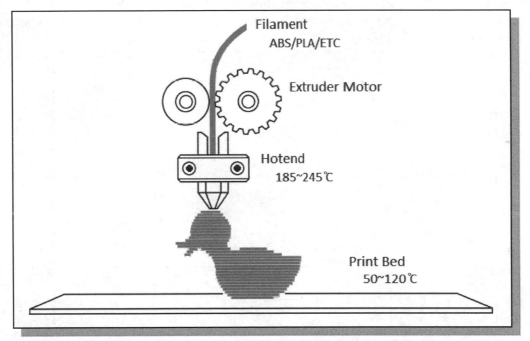

Stratasys has developed a range of proprietary industrial grade materials for its FDM process that are suitable for production applications. However, the most common materials for both FDM and FFF 3D printers are **ABS (Acrylonitrile Butadiene Styrene)** and **PLA (Polylactic Acid)**. The FDM and FFF processes require support structures for any applications with overhanging geometries. This generally entails a second, typically water-soluble or breakaway material, which allows support structures to be easily removed once the print is complete.

The FDM and FFF printing processes can be slow for large parts or parts with complex geometries. The layer to layer adhesion can also be a problem, resulting in parts that warp or separate easily. The surface finish of FDM and FFF printed parts might appear a bit rough as the thin layers are generally visible. To improve the appearance, several options are feasible, such as using Acetone, Sanding and/or Spray paint.

Laser Sintering / Laser Melting

Laser Sintering (LS) or **Selective Laser Sintering (SLS)** creates tough and geometrically intricate parts using a high-powered CO_2 laser to fuse/sinter/melt powdered thermoplastics. The main advantage of SLS *3D printing* is that as a part is made, it remains encased in powder; this eliminates the need for support structures and allows for very complex 3D geometries to be 3D printed. SLS can be used to produce very strong parts as exceptional materials such as Nylon and metal powders are commonly used.

Laser sintering refers to a laser-based 3D printing process that works with powdered materials. The laser is traced across a powder bed of tightly compacted powdered material, according to the 3D data provided to the machine, in the X-Y axes. As the laser interacts with the powdered material it sinters and fuses the particles to each other forming a solid. As each layer is completed the powder bed drops incrementally and a roller is used to compact the powder over the top surface of the bed prior to the next pass of the laser for the subsequent layer.

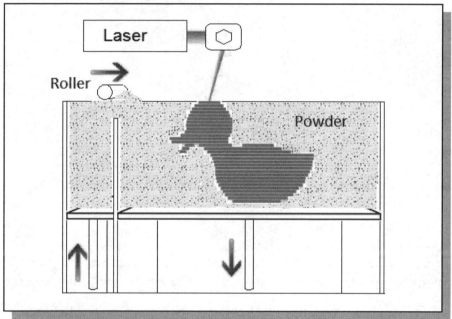

The build chamber is completely sealed as it is necessary to maintain a precise temperature during the process specific to the melting point of the powdered material of choice. One of the key advantages of this process is that the powder bed serves as an in-process support structure for overhangs and undercuts, and therefore complex shapes that could not be manufactured in any other way become possible with this process. Because of the high temperatures required for laser sintering, cooling can take a long time. Porosity is also a common issue with this process; an additional metal infiltration process may be required to improve mechanical characteristics.

Laser sintering can process plastic and metal materials, although metal sintering does require a much higher-powered laser and higher in-process temperatures. Parts produced with this process are much stronger than parts made with SLA or FDM, although generally the surface finish and accuracy is not as good.

Primary 3D Printing Materials for FDM and FFF

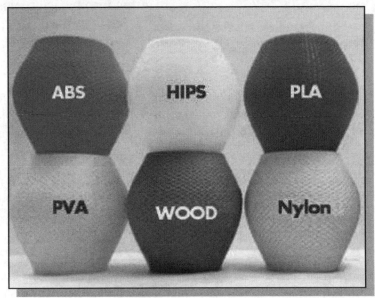

ABS (Acrylonitrile Butadiene Styrene)

ABS is a popular choice for 3D printing. It is a strong thermoplastic that is among one of the most widely used plastics. It is tough with mild flexibility, making it more durable to stress and has a higher heat resistance of up to 200 degrees Fahrenheit. However, this material has a tendency to shrink, which can affect the accuracy of designs. ABS has a pretty high melting point, and can experience warping if cooled while printing. Because of this, ABS objects are printed typically on a heated surface. ABS also requires ventilation when in use, as the fumes can be unpleasant. The aforementioned factors make ABS printing difficult for hobbyist printers, though it's the preferred material for professional applications.

PLA (Polylactic Acid)

PLA is a staple and it is becoming one of the most popular choices for 3D printing with good reason. Aside from the fact that is a biodegradable thermoplastic derived from renewable resources such as corn starch, tapioca roots, chips or starch, or sugarcane, *PLA* is a very rigid material that is easy to use for 3D printing, and it is able to withstand a good amount of impact and weight. It also has a glossier finish than ABS and in most scenarios PLA is the preferred material for 3D printing large objects. The main disadvantage of PLA is it's not as heat resistant as ABS, so it should not be placed in environments that exceed 140 degrees Fahrenheit.

Flexible (Thermoplastic Elastomer)

Flexible material is for applications that require incredible rubbery flex in their applications. Flexible filament goes beyond bending; it is more like rubber. When it comes to Flexible filament, it's all about finding a balance between flexibility (softness) and printability. This softness is sometimes indicated with a *Shore* value (like 85A or 60D). Higher Shore value means less flexibility. Harder filaments (less flexible) are easier to 3D print with compared to softer, more flexible filaments.

PETG (Polyethylene Terephthalate)

PETG is a material that is similar to *PLA*, with more attractive characteristics such as being generally a tougher and denser material, and good heat resistance of up to 190 degrees Fahrenheit. It claims to have the strength of *ABS*, while printing as easily as *PLA*.

HIPS (High Impact Polystyrene) and PVA (Polyvinyl Alcohol)

HIPS and *PVA* are relatively new materials that are growing in popularity for their dissolvable properties. They are used for creating support material. Their ability to dissolve under certain liquids means that they can be easily removed. These materials can be hard to print with because they don't stick well to the build plates. Be sure to not print *PVA* too hot either, as it can turn into tar and jam the extruder.

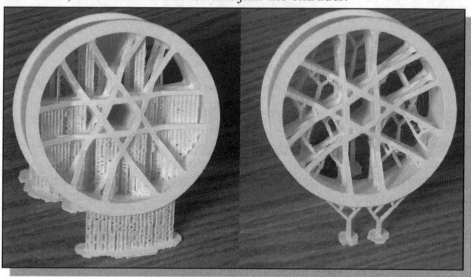

Wood Fiber

Wood Fiber filament contains a mixture of recycled wood and binding polymer. Thus, a 3D printed object can look and smell like real wood. Due to its wooden nature, it's difficult to tell that the object is 3D printed. Using Wood filament is similar to using a thermoplastic filament like ABS or PLA. However, a 3D object having a wooden-like appearance can be created with this material.

From 3D model to 3D printed Part

To create a 3D printed part, it all starts with making a virtual design of the object. This virtual design may be created with a computer-aided design (CAD) package, via a 3D scanner, or by a digital camera and photogrammetry software. 3D scanning and photogrammetry software can process the collected digital data on the shape and appearance of a real object and create a digital 3D model. The 3D virtual design can generally be modified with 3D CAD packages, allowing verification of the virtual design before it is 3D printed.

Once the virtual design is verified, the 3D data will then be transferred to the 3D printing software. There is a multitude of file formats that 3D printing software supports. However, the most popular are the STL file format and the OBJ file format. The STL file format is the most commonly used file format for 3D printing. Most CAD software has the capability of exporting models in the STL format. The STL file contains only the surface geometry of the modeled object. The OBJ file format is considered to be more complex than the STL file format as it is capable of displaying texture, color and other attributes of the three-dimensional object. However, the STL file format holds the top spot for 3D printing, as this file format is simpler to use, and most CAD packages work better with STL files than OBJ files.

Once the 3D data of the virtual design is transferred into the 3D printing software, further examination and/or repair can be performed if necessary. The 3D printing software will also process the imported 3D data by the special software known as a **Slicer**, which converts the model into a series of thin layers and produces a G-code file containing instructions tailored to a specific type of 3D printer. G-code is the common name for the most widely used numerical control (NC) programming language. It is used mainly in computer-aided manufacturing to control automated machine tools. The generated G-code file can be sent to the 3D printer and create the 3D printed part.

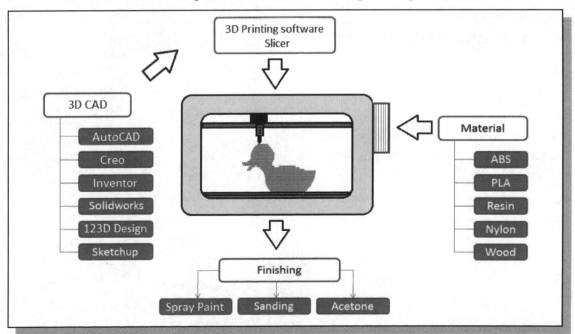

Start NX

1. Select the **NX** option on the *Start* menu or select the **NX** icon on the desktop to start *NX*. The *NX* main window will appear on the screen.

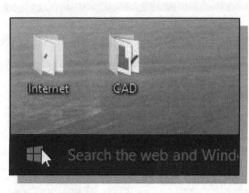

2. Select the **Rod-Guide.prt** file in the displayed *History palette* with a single click of the left-mouse-button.

➢ The ***History Palette*** of the *resource bar* provides a quick way to retrieve recently worked on parts.

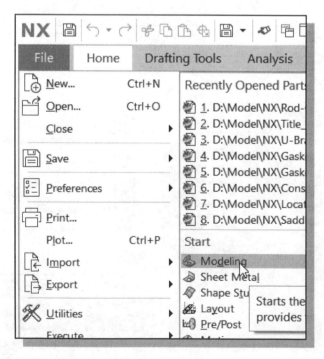

3. On your own, switch to the **Modeling** application through the *File menu* as shown.

Export the Design as an STL file

3D printers' slicer program generally can accept a 3D model with the STL or OBJ file formats. NX allows us to save the 3D model in STL format through the **Export** menu.

1. In the *File Menu,* select the **[Export]** → **[STL]** command as shown.

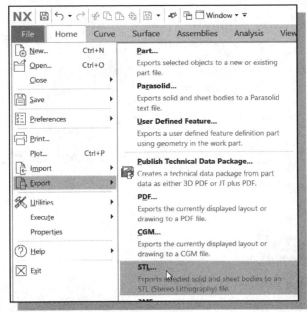

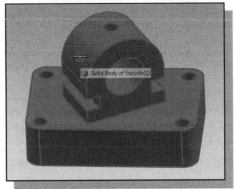

2. Select the Rod-Guide design with the left-mouse-button as shown.

3. Choose a folder to export the STL file.

- Note that the STL file can be in machine-code (Binary) or regular Text format; slicer software will accept both formats.

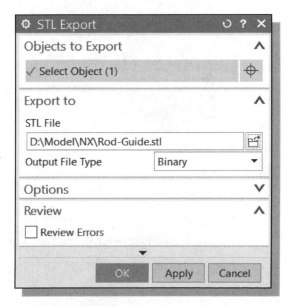

4. Click **Apply** to proceed with saving the file.

5. Click **Cancel** to exit the Export STL command.

Using the 3D Printing software to create the 3D Print

To 3D print the model, we will open the STL file in the 3D printing software. We will use **Matter Control** to demonstrate the procedure. Note that *Matter Control* (Freeware) supports quite a few desktop 3D printers. The procedures illustrated here are also applicable to other similar software.

1. Start the **Matter Control** software.

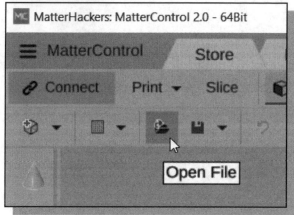

2. In the *Quick access menu*, select **Open File** as shown.

3. On your own, switch to the saved STL file folder and select the Rod-Guide.STL file as shown.

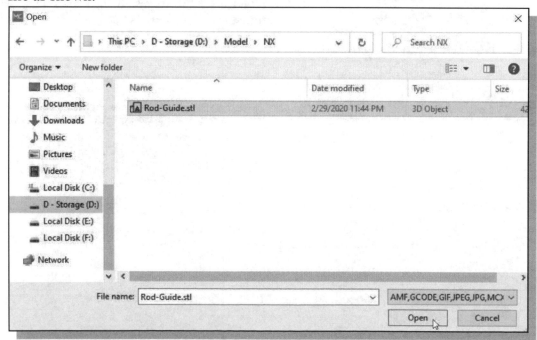

4. Click **Open** to import the STL file into *Matter Control*.

5. Once the STL file is imported into the program, the STL model is displayed in the *View* window. Note that the model is imported with the incorrect size; it is very small as shown in the below figure. Most 3D Printers expect the model to be in millimeters—the Rod-Guide design was created in inches.

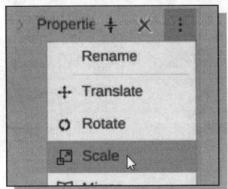

6. To the right-side of the main window, click on **Scale** under the properties option menu as shown.

7. Set the *Scale ratio* to **inches to mm** as shown. Note that 1inch equals 25.4 mm.

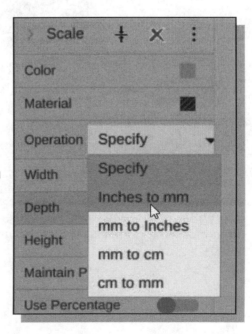

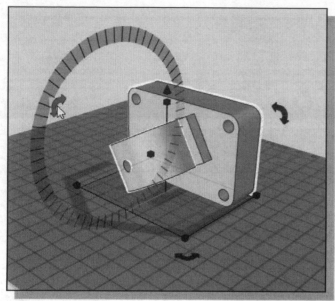

8. To **Rotate** the design, move the cursor on the curve-arrow icon and drag with the left mouse button to perform the rotate operation.

9. On your own, experiment with rotating the design. Set the orientation as shown.

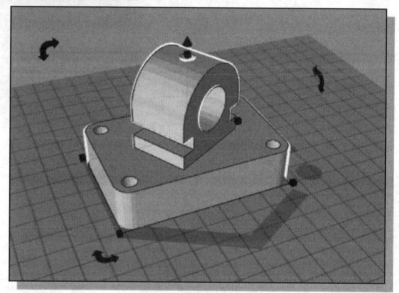

• Note the straight-arrows icons can be used to move/translate the design in X/Y/Z directions.

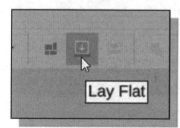

10. Click on the Lay Flat command to align the bottom face of the design to the print-bed.

11. On your own, examine the different display option, such as **Home** and **View Cube**, near the upper right corner of the main window.

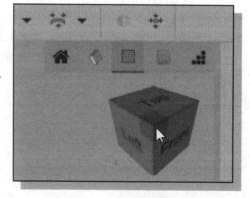

12. Use the right-mouse-button on the design to display the option list; note the different options available to modify the design.

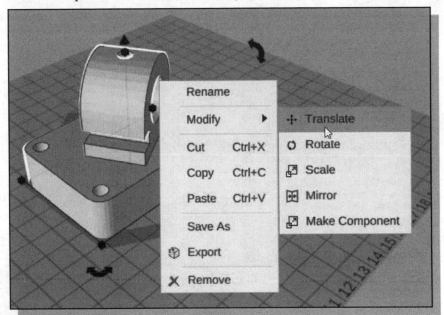

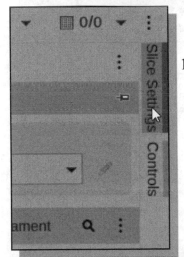

13. Near the upper right corner of the main window, click **Slice Settings** to review the Slicer program settings.

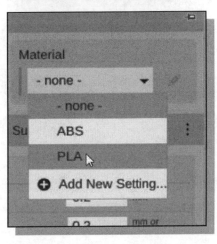

14. In the **Material list**, choose **PLA** as the type of filament properties. Note that the related properties, such as the diameter, extruder temperature, and bed temperature can be adjusted to match the actual filament being used.

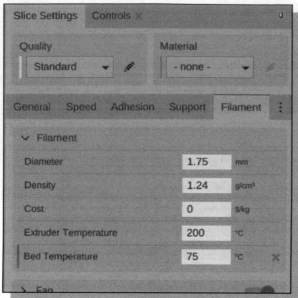

15. Note the different settings are organized as groups, which can be used to adjust the 3D printer's settings and the commands to directly control the movements of the extruder of the printer.

• Note that the temperatures are different based on the types of filaments, as well as the different brands and the type of printer bed. It may be necessary to do some testing and/or experimenting when a new roll of filament is used.

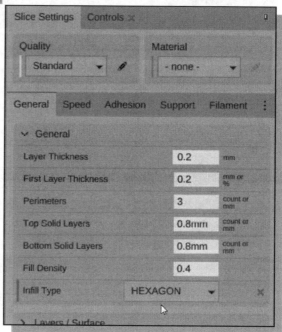

16. Switch to the **General** Tab, and examine the Layer control settings, such as the layer thickness, top layer and bottom layer thickness.

17. Also note the settings for **Infill Type**, the **Fill Density**.

18. Click on the **Speed** tab and examine the related settings to control the 3D printing speed.

19. Click Slice to process the 3D model, which includes slicing and generating the G-code for the specific 3D printer.

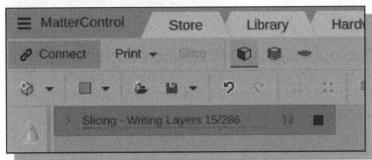

20. Matter Control will display a status bar showing the slicing progress.

- Note that Matter Control allows the use of other Slicers, such as Cura or Slic3r, which are the two extremely popular open source slicer programs.

21. On your own, drag the vertical slider to review the thin layers generated by the slicer.

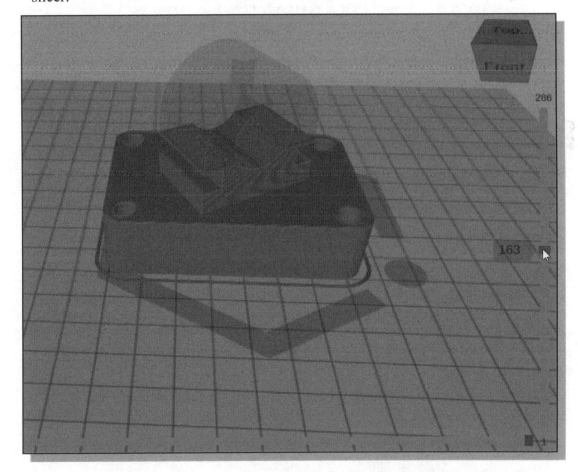

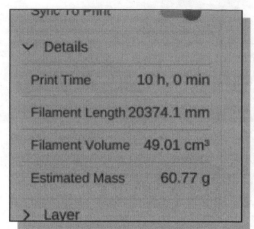

- Note that with the current settings, it will take about 10 hours to complete the print using 20374.1 mm of filament.

22. Click **Connect -→ Print** to start the printing of the 3D model.

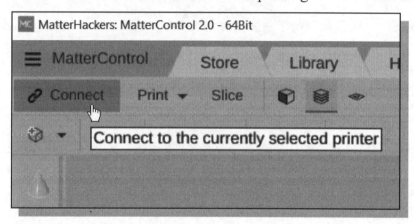

23. *Matter Control* will now begin printing; note the Print time for the model is displayed in the status window.

Questions:

1. What is the main difference between the **Additive Manufacturing** and the traditional **Subtractive Manufacturing** technologies?

2. Which 3D printing process is recognized as the first 3D printing process?

3. Describe the general procedure to create a 3D printed part.

4. What are the three primary types of 3D printing processes?

5. Which 3D printing process is the most popular 3D printing process?

6. What is the main advantage of using PLA over ABS for FFF process?

7. Which are the most popular file formats used for 3D printing?

8. What is the main function of a **Slicer** program?

Notes:

Chapter 11
Symmetrical Features in Designs

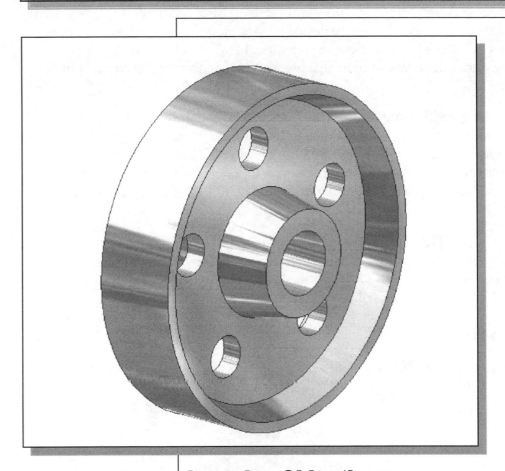

Learning Objectives

♦ **Create Revolved Features**
♦ **Use the Mirror Feature Command**
♦ **Import a pre-made Border and Title Block**
♦ **Create Circular Patterns**
♦ **Use NX's Associative Functionality**
♦ **Using the Basic Tools to create Symmetrical Features**

Introduction

In parametric modeling, it is important to identify and determine the features that exist in the design. *Feature-based parametric modeling* enables us to build complex designs by working on smaller and simpler units. This approach simplifies the modeling process and allows us to concentrate on the characteristics of the design. Symmetry is an important characteristic that is often seen in designs. Symmetrical features can be easily accomplished by the assortments of tools that are available in feature-based modeling systems, such as *NX*.

The modeling technique of extruding two-dimensional sketches along a straight line to form three-dimensional features, as illustrated in the previous chapters, is an effective way to construct solid models. For designs that involve cylindrical shapes, shapes that are symmetrical about an axis, revolving two-dimensional sketches about an axis can form the needed three-dimensional features. In solid modeling, this type of feature is called a ***revolved feature***.

In *NX*, besides using the **Revolve** command to create revolved features, several options are also available to handle symmetrical features. For example, we can create multiple identical copies of symmetrical features with the **Feature Pattern** command or create mirror images of models using the **Mirror Feature** command. We can also use *construction geometry* to assist the construction of more complex features. In this lesson, the construction and modeling techniques of these more advanced options are illustrated.

A Revolved Design: *PULLEY*

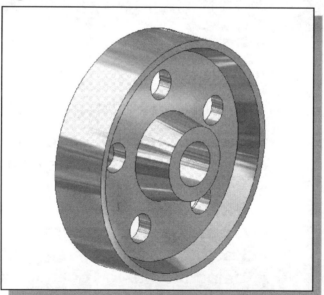

❖ Based on your knowledge of *NX*, how many features would you use to create the design? Which feature would you choose as the **base feature** of the model? Identify the symmetrical features in the design and consider other possibilities in creating the design. You are encouraged to create the model on your own prior to following through the tutorial.

Modeling Strategy – A Revolved Design

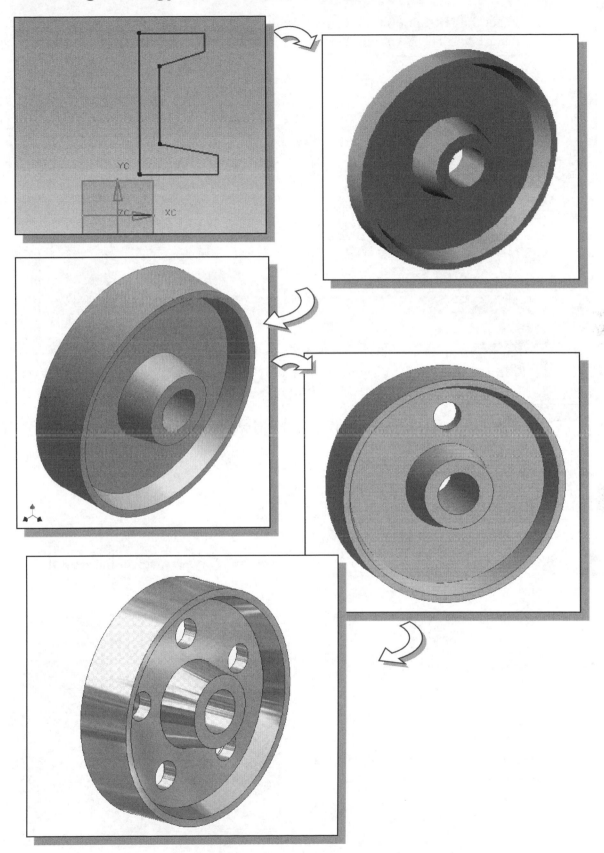

Start NX

1. Select the **NX** option on the *Start* menu or select the **NX** icon on the desktop to start *NX*. The *NX* main window will appear on the screen.

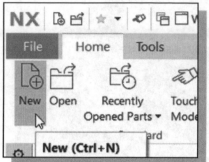

2. Select the **New** icon with a single click of the left-mouse-button (**MB1**) in the *Standard toolbar area*.

3. Select the **inches** units as shown in the below figure.

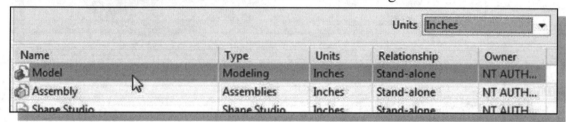

	Units	Inches			▼
Name	Type	Units	Relationship	Owner	
Model	Modeling	Inches	Stand-alone	NT AUTH...	
Assembly	Assemblies	Inches	Stand-alone	NT AUTH...	
Shape Studio	Shape Studio	Inches	Stand-alone	NT AUTH...	

4. Select **Model** in the *Template list*. Note that the *Model template* will allow us to switch directly into the **Modeling task** as indicated in the *templates list*.

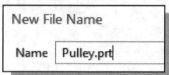

5. Enter **Pulley** as the *New File Name*.

6. Click **OK** to proceed with the New File command.

Create the Base Feature

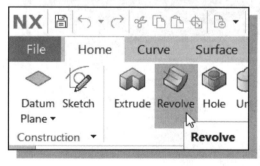

1. In the *Feature Toolbars,* select the **Revolve** icon as shown.

❖ The revolved feature requires a 2D sketch and an axis of rotation.

2. Select the **XY** plane of the displayed Datum Coordinate System.

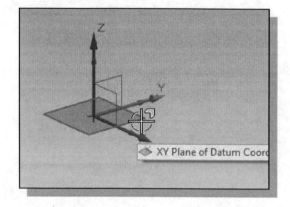

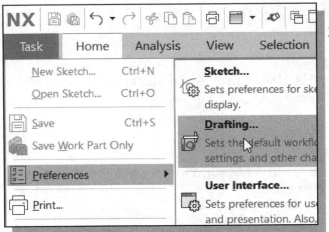

3. In the *Task tab*, select
    Preferences → Drafting as
    shown.

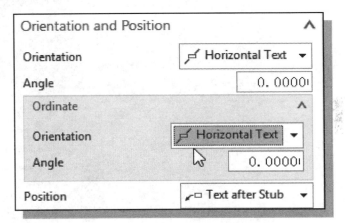

4. Under **Dimension →Text** list,
    set the *Text orientations* to
    Horizontal as shown.

5. Click **OK** to accept the
    settings.

6. Create a closed-region rough sketch as shown below. (Note that the *Pulley* design
    is symmetrical about a horizontal axis as well as a vertical axis; this property
    allows us to simplify the 2D sketch as shown below.)

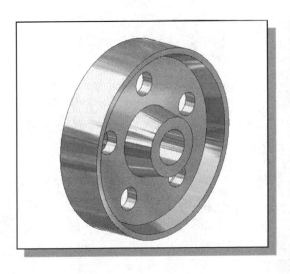

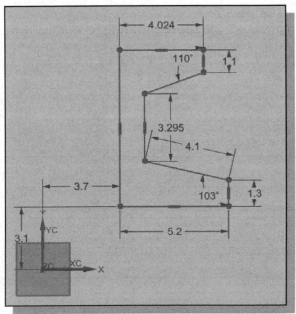

Completing the Sketch and Creating the Feature

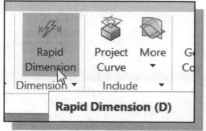

1. Select the **Rapid Dimension** command in the *Sketch* toolbar as shown.

2. On your own, create and adjust the **dimensions** as shown below. (Hint: Modify the smaller numbers first.)

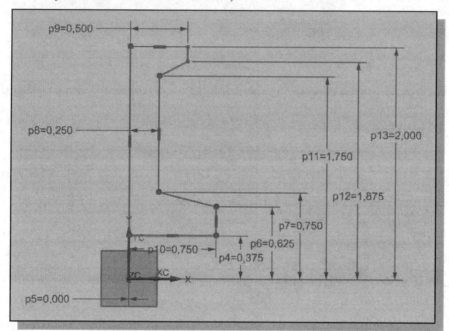

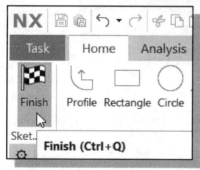

3. Click **Finish** to exit the NX sketcher mode and return to the *Feature option dialog window*.

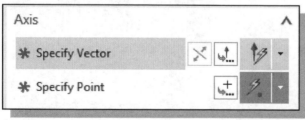

4. In the *Revolve* feature dialog box, the **Specify Vector** option is activated. NX expects the definition of a rotational axis.

❖ For a revolved feature, there are two required elements: a 2D sketch and an axis of rotation.

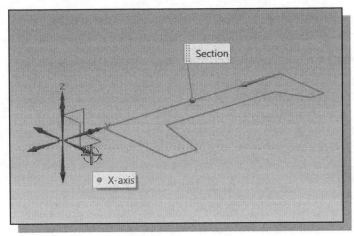

5. Select the **X-axis** of the DCS/WCS to set the rotation vector.

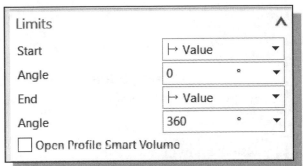

6. Select the **origin** of the DCS/WCS to set the rotation axis point.

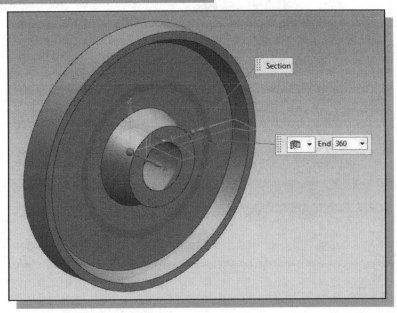

7. Confirm the start angle is set at **0** and the end angle at **360**, in the *Angular Limits area* as shown.

8. Click **OK** to create the revolved feature.

❖ Note that although it is possible to create the above revolved feature using the extrude command, it is more difficult. It is very important for a designer to recognize the symmetrical features in the design and use the most effective options to create them.

Mirroring Features

- In *NX*, features can be mirrored to create and maintain complex symmetrical features. We can mirror a feature about a work plane or a specified surface. We can create a mirrored feature while maintaining the original parametric definitions, which can be quite useful in creating symmetrical features. For example, we can create one quadrant of a feature then mirror it twice to create a solid with four identical quadrants.

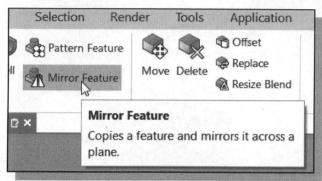

1. Select the **Mirror Feature** command in the *Ribbon Toolbar*.

➤ The **Mirror Feature** command can be used to create copies of features that are parametrically linked to the original copy.

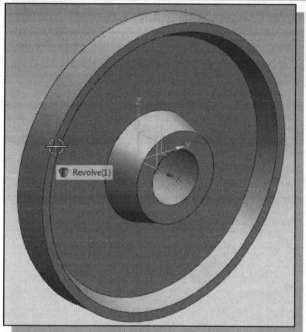

2. Select the **Revolved feature** in the graphics window as shown.

3. Use the middle-mouse-button (**MB2**), or click **OK** to proceed with the Mirror Feature command.

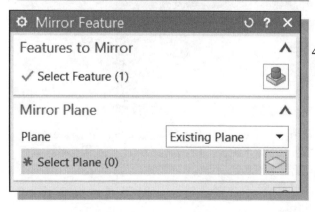

4. Confirm the **Existing Plane** option is set and the Select Plane option is activated as shown. In the message area, the message "*Select mirror plane*" is displayed.

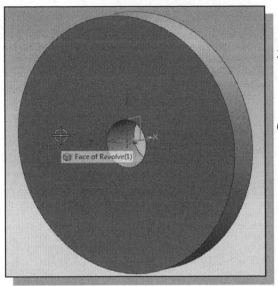

5. On your own, dynamically rotate the solid model so that we are viewing the *back surface* as shown.

6. Select the **back surface** as the planar surface about which to mirror.

7. Click on the **OK** button to accept the settings and create a mirrored feature.

8. Select **Wireframe with Dim Edges** in the *View Toolbar* in the *Ribbon bar* area, or press-down on MB2 and select the icon, to adjust the display of the model on the screen.

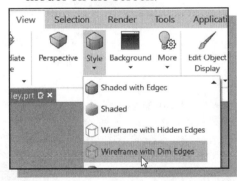

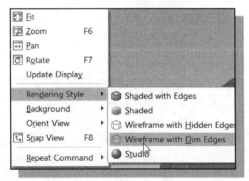

❖ Note the *Mirror Feature* command has created a separate feature.

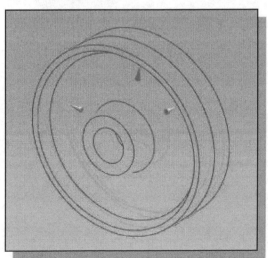

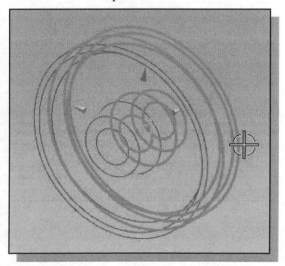

Join the Two Solid Features

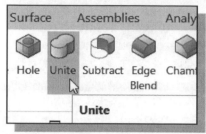

1. Activate the **Unite** command by clicking the left-mouse-button on the icon as shown.

❖ The *Unite* command can be used to join two separate bodies into one object.

❖ Note the first selection option, *Target Body*, is activated as shown.

2. On your own, select the **original revolved feature** as the *Target Body*.

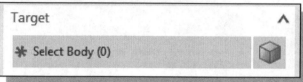

❖ Note the second selection option, *Tool Body*, is activated as shown.

3. On your own, select the **mirrored feature** as the *Tool Body*.

4. Do not activate any of the options listed under Settings, as we want to create one solid model.

5. Click on the **OK** button to accept the settings and create a mirrored feature.

6. On your own, use the 3D-Rotate command to dynamically rotate the solid model and view the resulting solid.

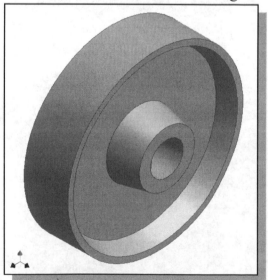
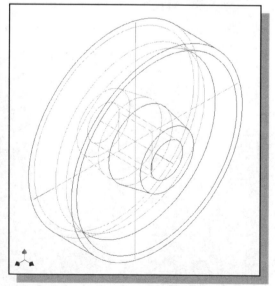

Create a Pattern Leader

- The *Pulley* design requires the placement of five identical holes on the base solid. Instead of creating the five holes one at a time, we can simplify the creation of these holes by using the **Circular Array** command, which allows us to create duplicate features. Prior to using the Circular Array command, we will first create a *pattern leader*, which is a regular extruded feature.

1. In the *Form Feature Toolbars* (toolbars aligned to the right edge of the main window), select the **Extrude** icon as shown.

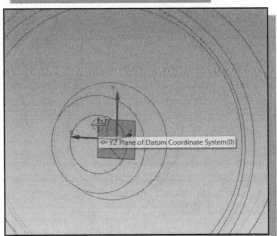

2. Select the vertical plane, **YZ plane** of the datum coordinate system, that is through the center of the model as the sketch plane as shown.

3. Click **OK** to accept the selection of the *Sketch Plane*.

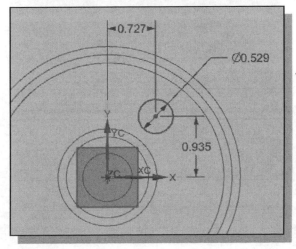

4. Create a **circle** near the upper right corner as shown in the figure.

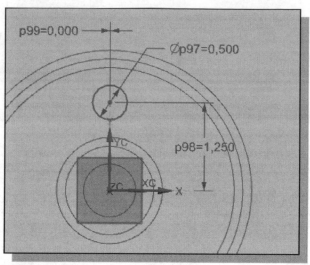

5. On your own, modify the **three dimensions** as shown.

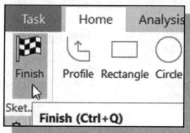

6. Click **Finish Sketch** to exit the *NX sketcher mode* and return to the *Feature option dialog window*.

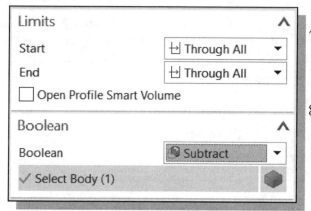

7. Inside the *Extrude* dialog box, select the **Subtract** operation, and set both *Limits* to **Through All** as shown.

8. Click on the **OK** button to accept the settings and create the cut feature.

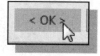

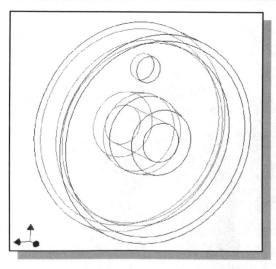

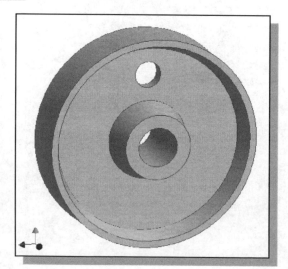

Circular Array

In *NX*, existing features can be easily duplicated with the two array commands, **rectangular array** and **circular array**. The arrayed features can be parametrically linked to the original feature, which means any modifications to the original feature are also reflected in the arrayed features. Three elements are needed for creating a circular array: (1) The rotation axis about which the instances are generated, (2) The total number of instances in the array (including the original feature) and (3) The angle between the instances.

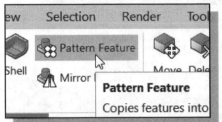

1. Select the **Pattern Feature** command through the feature toolbar.

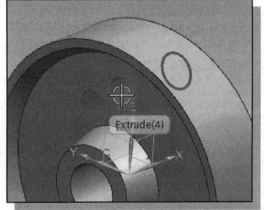

2. In the **Pattern Feature** dialog box, select the **Circular** command by clicking the left-mouse-button on the icon as shown.

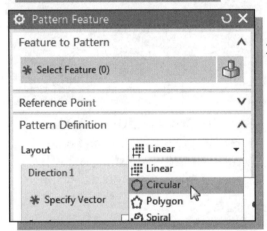

3. Choose the **last feature**, the *Extruded* feature, in the *Instance* window as shown.

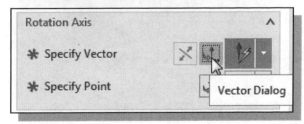

4. Select **Vector Dialog** in the option list to proceed with the Circular Array command.

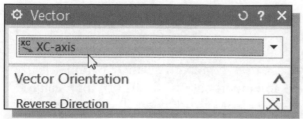

5. Select the *XC-Axis* as the datum axis, the axis of rotation, as shown in the figure.

6. Click **OK** to proceed to the next step of the circular array command.

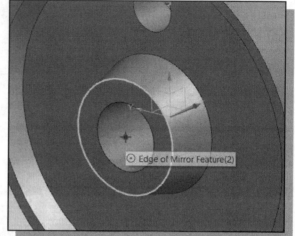

7. Select the *Center Point* of the front circle as the reference point as shown in the figure.

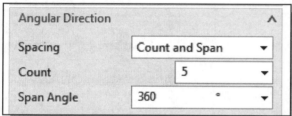

8. Select the **Count and Span** *Spacing method* in the dialog box and enter **5** as the total number of instances and **360** degrees as the span angle as shown.

9. Click **OK** to proceed to the next step of the circular array command.

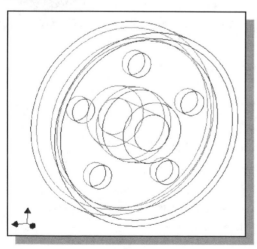

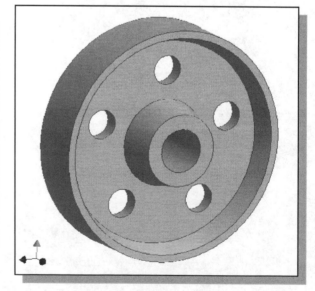

Create a New Drawing in the Drafting Mode

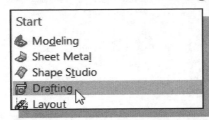

1. Click on the **File** icon in the *Ribbon* toolbar area to display the available options.

2. Select **Drafting** from the *Applications list*.

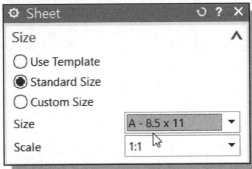

3. Select **A-8.5X11** from the *sheet size* list.

4. Confirm the **Scale** is set to **1:1**, which is *Full Scale*.

➢ Note that *Sheet 1* is the default drawing sheet name that is displayed in the *Drawing Sheet Name* area.

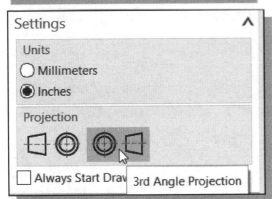

5. In the settings area, confirm the *Units* are set to **Inches** and the *Projection* type is set to **Third Angle of projection** as shown.

6. **Uncheck** the *Always Start View Creation Command* option.

7. Click **OK** to accept the settings and proceed with the creation of the drawing sheet.

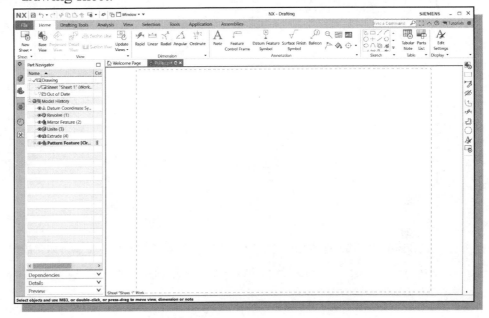

Import the Predefined Title Block

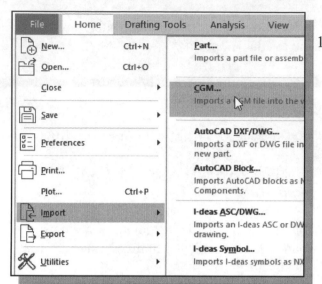

1. Select the **Import CGM** command through the *File* pull-down menu: **[File] → [Import] → [CGM]**

2. Enter or select **Title_A_Master_sheet 1.cgm** in the *Import CGM* window as shown.

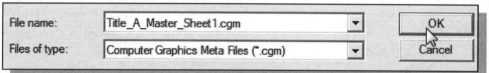

3. Click **OK** to proceed with importing the selected CGM file.

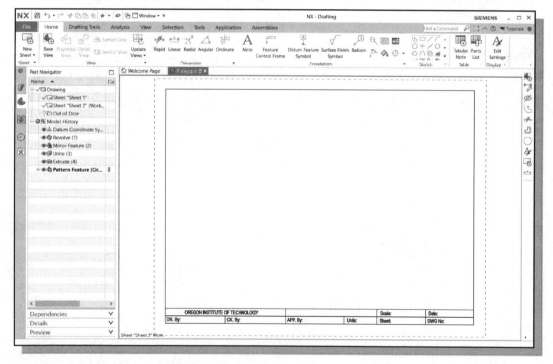

❖ Note that the **Import** and **Export** commands provide a fairly flexible way for us to reuse *title block* and borders.

Create 2D Views

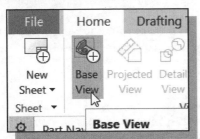

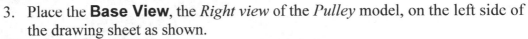

1. Click on the **Base View** icon in the *Drawing Layout toolbar* to create a new base view.

2. In the *Base View* option toolbar, set the view option to **Right** as shown.

3. Place the **Base View**, the *Right view* of the *Pulley* model, on the left side of the drawing sheet as shown.

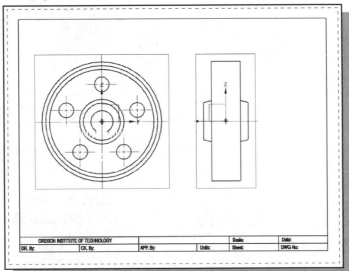

4. On your own, create a **side view** to the right side as shown in the above figure; click once with the middle-mouse-button to end the create view wizard.

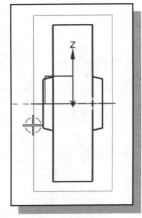

❖ Note the default display of views does not show any hidden lines. And as we realize a section view is more suitable for our design, we will delete this projected view and create a *section view*.

5. Select the **Projected View** by clicking with the left-mouse-button (MB1) on the view border as shown.

6. Click the [**Delete**] icon or hit the [**Delete**] key, in the keypad, once to erase the view completely.

Add a Section View

❖ Two elements are needed when creating a section view: (1) The parent view, where the projection is derived from, and (2) a cutting plane line identify where the virtual cut is to occur.

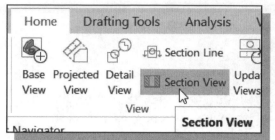

1. Select the **Section View** command through the *Drawing layout toolbar* as shown.

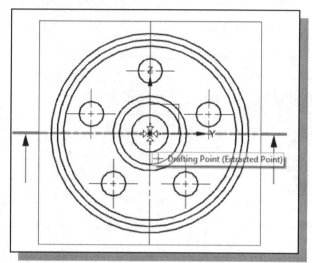

2. Inside the graphics window, select the **center** of the base view to align the center of the cutting plane line as shown.

3. Select the **Settings** option in the Settings list as shown.

4. Set the *Section Line → Type* option to use the **Arrows away from Line** for the cutting-plane line.

5. Set the *Line-type* to **Phantom** and *Line-width* to **0.70mm** as shown.

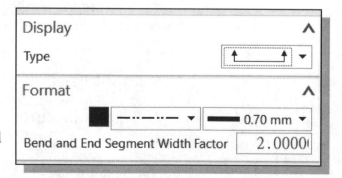

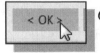

6. Click **OK** to accept the settings and proceed with the creation of the cutting-plane line.

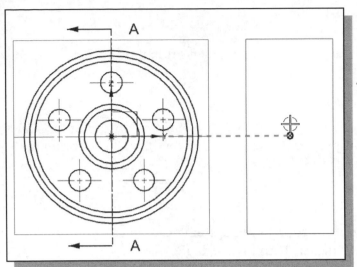

7. **Rotate** the *cutting-plane line* until it is vertical, by moving the cursor to the right of the base view.

8. Position the section view to the right side of the base view as shown.

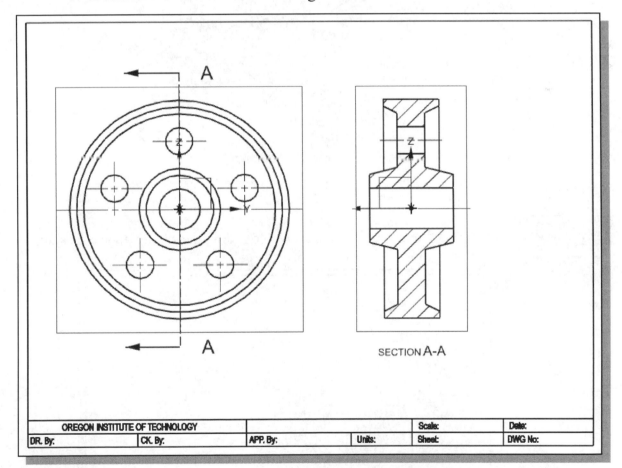

SECTION A-A

OREGON INSTITUTE OF TECHNOLOGY				Scale:	Date:
DR. By:	CK. By:	APP. By:	Units:	Sheet:	DWG No:

9. Click **Close** to end the section view option menu.

Turn Off the Datum Features

➢ The display of the datum coordinate system can be turned off in the *Drafting* mode or in the *Modeling* mode.

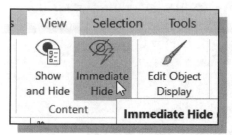

1. Select the **Immediate Hide** command in the *View* toolbar or use the quick key [**Ctrl+B**] to bring up the *Hide dialog box*.

2. On your own, select the **datum coordinate system** near the center of the base view as shown. Also select the **drafting points** in the views.

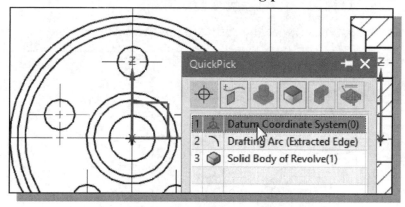

❖ Hint: Pick *Datum Coordinate System* in the *Quick Pick* window if necessary.

3. Click **OK** to accept the selection. Your drawing should appear as shown in the figure.

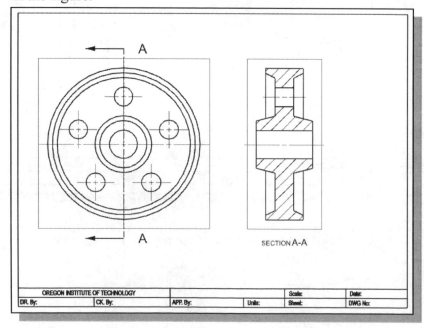

Add Dimensions

❖ One main advantage of *Parametric Modeling* over the traditional CAD system is the flexibility and multiple options available for different situations. For example, for the pulley design, instead of using many of the feature dimensions, we will use reference dimensions in the 2D drawing. All of the 2D drawing views are still associated to the 3D model and any changes to the solid model will be reflected in the 2D drawing instantly.

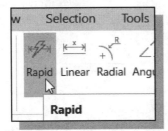

1. Select the **Rapid Dimension** command in the *Dimension toolbar* as shown.

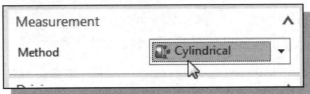

2. Select the **Cylindrical Dimension** command by choosing the corresponding icon in the *Measurement Method* list.

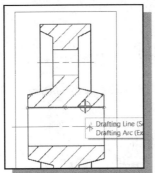

3. Select the **horizontal line** that is **above** the centerline as shown.

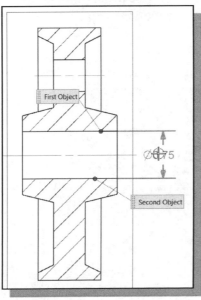

4. Select the **horizontal line** that is **below** the centerline as shown.

5. Place the dimension to the **right side** of the section view.

❖ The **Cylindrical dimension** command automatically adds the diameter symbol in front of the dimension value.

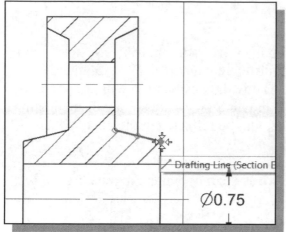

6. Select the **endpoint** of the top right corner on the hub as shown.

7. Select the corresponding second point on the hub and place the dimension to the right side as shown.

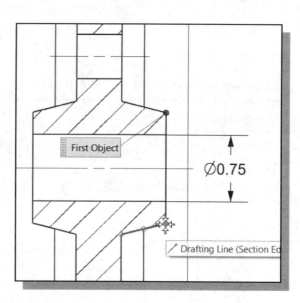

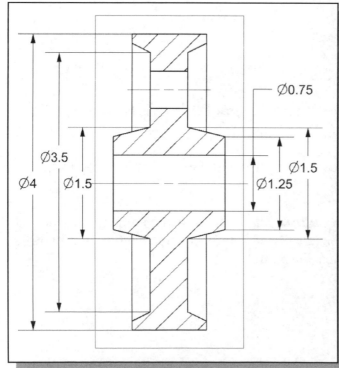

8. On your own, repeat the above steps and create the other **cylindrical dimensions** in the section view as shown.

❖ Hint: Use the drag and drop option to reposition the dimensions.

Adjust the Display of the Views

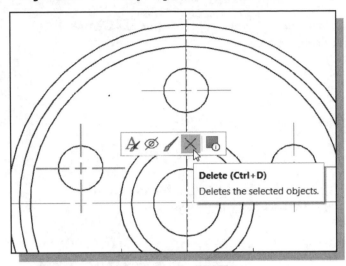

1. Select the **centerline** of any of the hole patterns by clicking once with the left-mouse-button.

2. Click the [**Delete**] icon to remove the center mark.

3. Repeat the above steps and remove the center marks on the five holes.

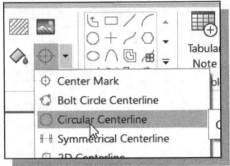

4. In the *Annotation toolbar*, select the **Circular Centerline** command as shown.

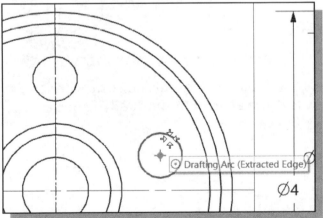

5. Select one of the five holes by clicking once with the **left-mouse-button** on the circle.

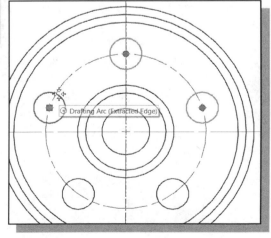

6. Going counterclockwise, select the next two holes to create a circular center line that passes through the associated center point as shown.

7. Click **OK** to accept the selections and create the centerline as shown.

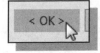

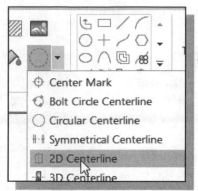

8. In the *Annotation toolbar*, select the **2D Centerline** command as shown.

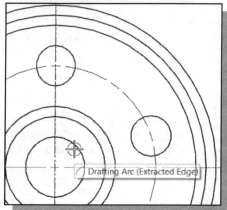

9. Select the **hole** aligned to the origin of the coordinate system as shown.

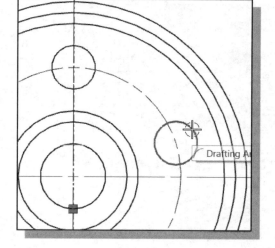

10. Select one of the five holes by clicking once with the **left-mouse-button** on the circle.

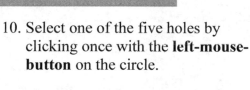

11. In the 2D centerline dialog box, switch on the **Set Extension Individually** option as shown.

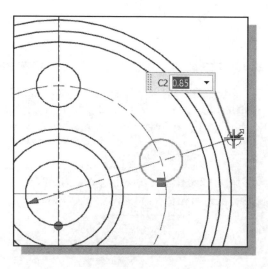

12. On your own, adjust the length of the centerline by dragging the two associated endpoints. (Set one end to 0.85 and the other end 0.0)

13. Click **Apply** to accept the settings and create the centerline.

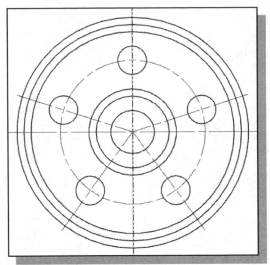

14. On your own, repeat the above steps and create the additional centerlines in the primary view as shown.

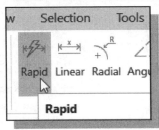

15. Select the **Rapid dimension** command as shown.

16. Set the *Measurement Method* to **Inferred** as shown.

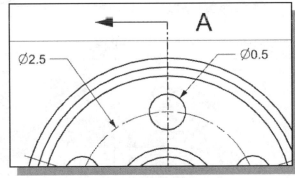

17. Create the two **diameter dimensions** in the primary view as shown.

18. Set the *Measurement Method* to **Angular** as shown.

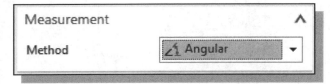

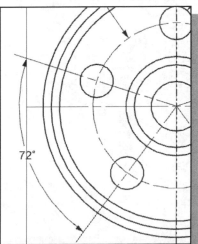

19. On your own, create the **angular dimension** as shown.

20. Click **Close** to end the *Rapid dimension* command.

Turn Off the Display Borders Option

❖ The display of the borders around the 2D views can be turned off by setting the drafting preferences.

1. Select **Drafting** in the *Preferences* menu as shown.

2. Turn off the **Display Borders** option under the *Drawing View list* as shown.

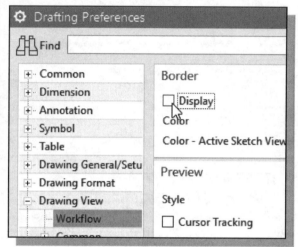

3. Click **OK** to accept the settings and exit the *View Style* command.

4. On your own, complete the dimensions/views as shown in the figure.

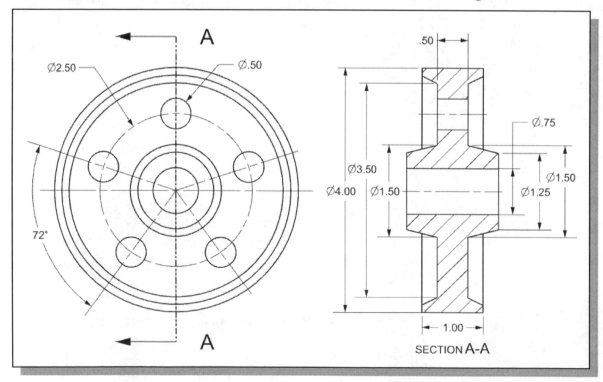

SECTION A-A

Associative Functionality – A Design Change

- *NX's associative functionality* allows us to change the design quickly, and the system reflects the changes at all levels automatically. We will illustrate the associative functionality by changing the circular pattern from five holes to six holes.

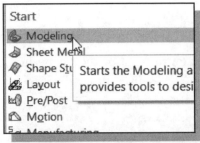

1. Switch back to the *Modeling module* by selecting **Modeling** in the *Start List* as shown.

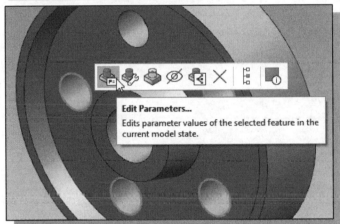

2. Left-click on one of the **arrayed features** to bring up the option menu.

3. Select **Edit Parameters** in the pop-up menu to switch to the edit mode.

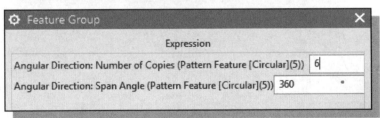

4. Adjust the *Count option* to **6** as shown in the figure.

5. Click **OK** to accept the settings and proceed with the updating of the solid model.

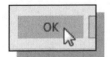

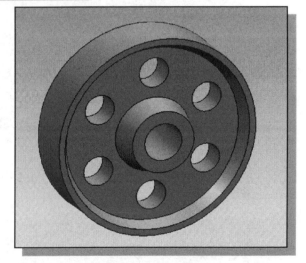

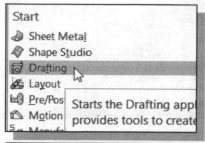

6. On your own, switch back to the **drafting mode**.

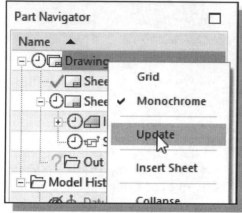

- Notice the **clock** icon in front of the *Drawing* item. By default, the drawing will not be updated automatically; the clock icon simply means an update is needed.

7. Right-mouse-click on **Drawing** and select **Update** in the pop-up menu as shown.

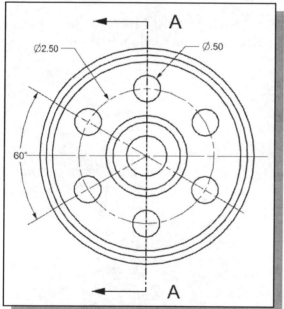

❖ Note that even though the views are updated correctly, we are still missing a centerline.

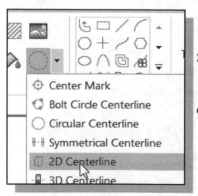

8. Select the **2D Centerline** command as shown and create the necessary center line.

9. On your own, complete the drawing as shown on the next page.

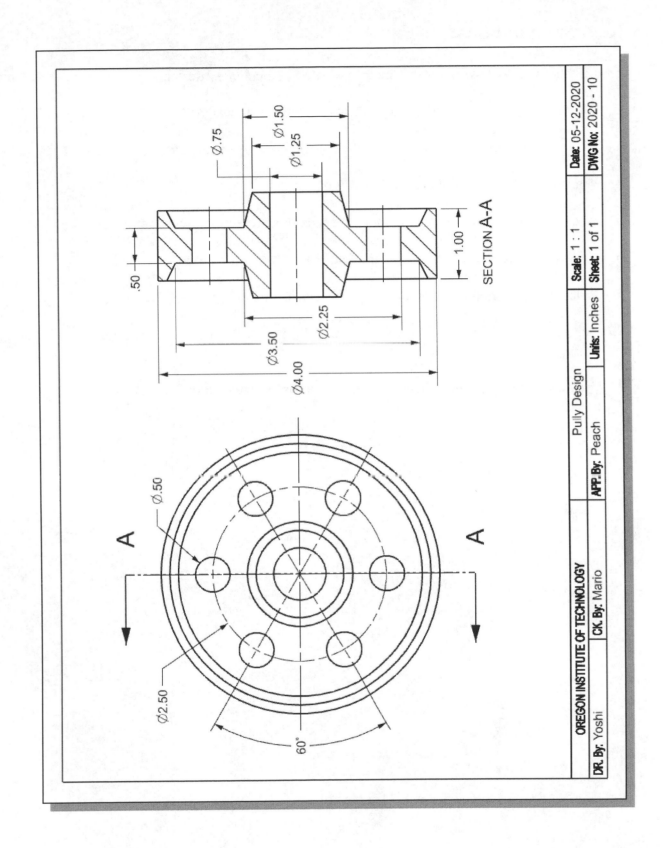

Ø.75

Ø1.50

Ø1.25

.50

Ø2.25

Ø3.50

Ø4.00

1.00

SECTION A-A

A

A

Ø.50

Ø2.50

60°

OREGON INSTITUTE OF TECHNOLOGY

Pully Design

Scale: 1 : 1

Date: 05-12-2020

DWG No: 2020 - 10

APP. By: Peach

Units: Inches

Sheet: 1 of 1

DR. By: Yoshi

CK. By: Mario

Review Questions:

1. List the different symmetrical features created in the *Pulley* design.

2. Why is it important to identify symmetrical features in designs?

3. Describe the steps required in using the **Mirror Feature** command.

4. When and why should we use the **Array** option?

5. What are the required elements in order to generate a sectional view?

6. How do you re-use a pre-defined title block in the drafting mode?

7. What are the general steps required to create a section view in NX?

8. Describe the different options available in NX to create centerlines in the drafting mode.

9. Create sketches showing the steps you plan to use to create the model shown on the next page:

Exercises: (All dimensions are in inches.)

1. **Shaft Support** (Dimensions are in inches.)

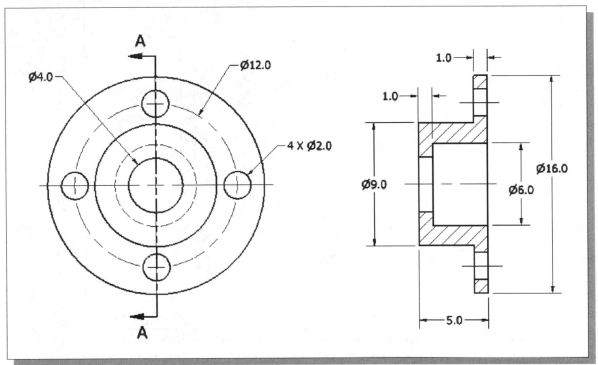

2. Plate thickness: 0.125 inch

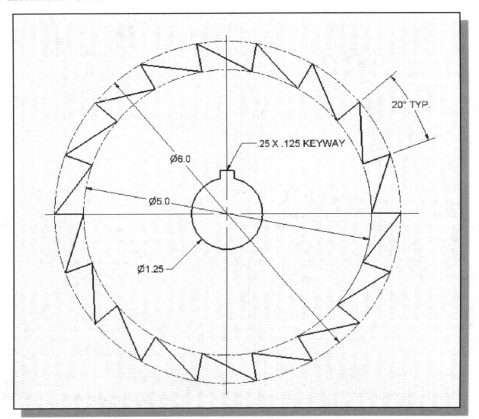

3. **Geneva Wheel** (Dimensions are in inches.)

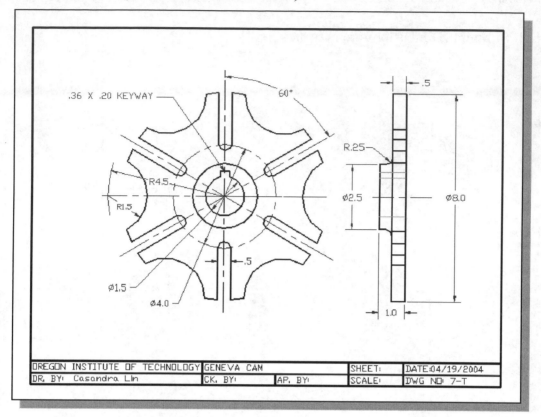

4. **Support Mount** (Dimensions are in inches.)

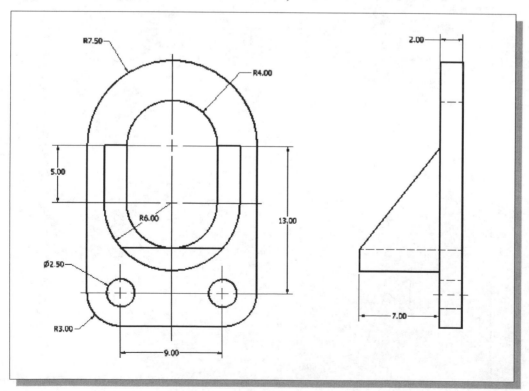

5. **Hub** (Dimensions are in inches.)

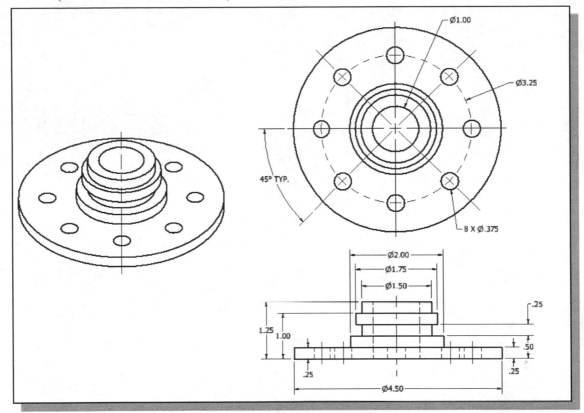

6. **Switch Base** (Dimensions are in inches.)

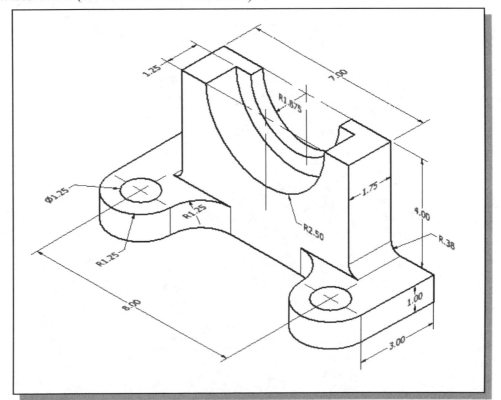

Notes:

Chapter 12
Advanced 3D Construction Tools

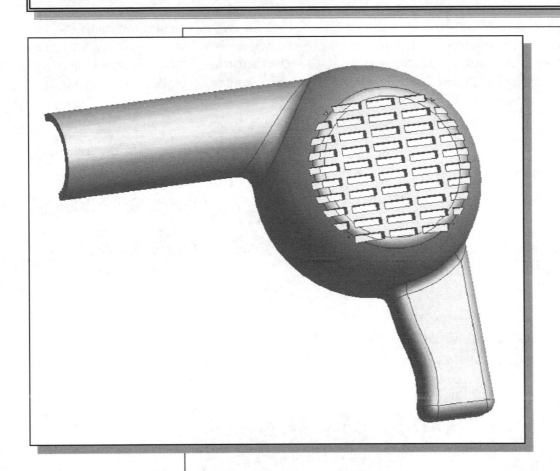

Learning Objectives

♦ **Understand the uses of the available 3D Construction Tools**

♦ **Set up Multiple Datum Planes**

♦ **Create Swept Features**

♦ **Use the Rectangular Array command**

♦ **Use the Shell Command**

♦ **Create 3D Rounds & Fillets**

Introduction

NX provides an assortment of three-dimensional construction tools to make the creation of solid models easier and more efficient. As demonstrated in the previous lessons, creating **extruded** features and **revolved** features are the two most common methods used to create 3D models. In this next example, we will examine the procedures for using the **Sweep** command, the **Shell** command, the **Rectangular Array** command, and also for creating **3D rounds** and **fillets** along the edges of a solid model. These types of features are common characteristics of molded parts.

The **Sweep** option is defined as moving a cross-section through a path in space to form a three-dimensional object. To define a sweep in *NX*, we define two sections: the trajectory and the cross-section.

The **Shell** option is defined as hollowing out the inside of a solid, leaving a shell of specified wall thickness.

The **Edge Blend** command allows us to create 3D rounds and fillets along the edges of a solid model. The *Edge Blend* command is one of the placed feature commands available in *NX*.

A Thin-Walled Design: Dryer Housing

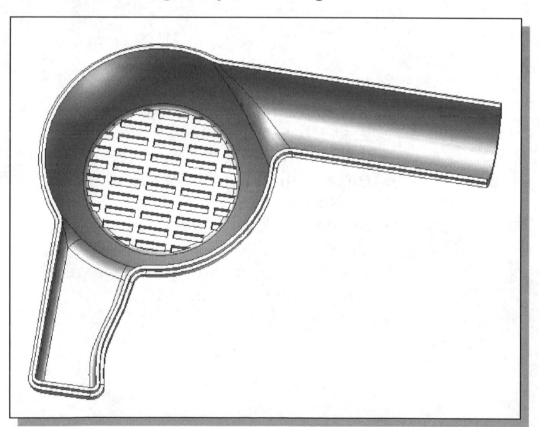

Modeling Strategy

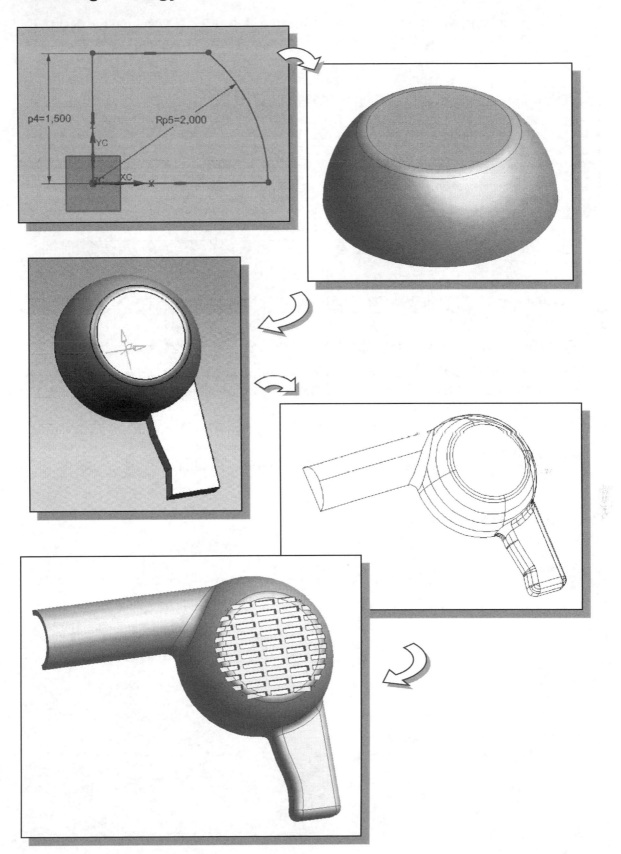

Start NX

1. Select the **NX** option on the *Start* menu or select the **NX** icon on the desktop to start *NX*. The *NX* main window will appear on the screen.

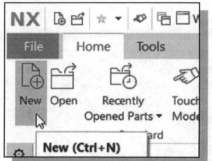

2. Select the **New** icon with a single click of the left-mouse-button (**MB1**) in the *Ribbon toolbar area.*

3. Select the **inches** units as shown in the below figure.

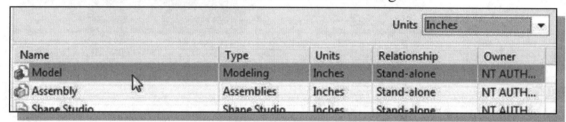

				Units	Inches ▼
Name		**Type**	**Units**	**Relationship**	**Owner**
Model		Modeling	Inches	Stand-alone	NT AUTH...
Assembly		Assemblies	Inches	Stand-alone	NT AUTH...
Shape Studio		Shape Studio	Inches	Stand-alone	NT AUTH

4. Select **Model** in the *Template list*. Note that the *Model template* will allow us to switch directly into the **Modeling task** as indicated in the *templates list*.

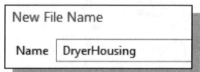

5. Enter **DryerHousing** as the *New File Name*.

6. Click **OK** to proceed with the New File command.

Create the Base Feature

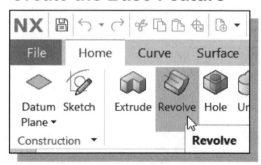

1. In the *Feature Toolbars,* select the **Revolve** command as shown.

2. Set the sketch plane to the **YZ** plane of the displayed Datum Coordinate System.

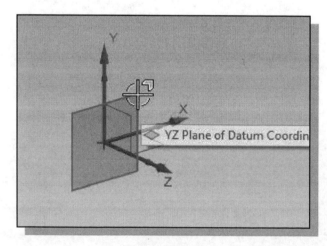

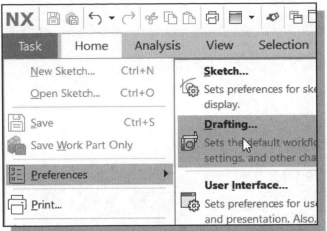

3. In the *Task tab*, select
    Preferences → Drafting as
    shown.

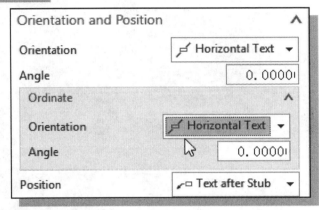

4. Under Dimension →Text list, set the
    Text orientations to **Horizontal** as
    shown.

5. Click **OK** to accept the
    settings.

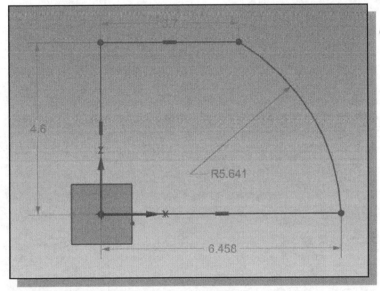

6. On your own, using the line
    and arc commands, start at
    the origin of the datum
    coordinate system and
    create the 2D sketch as
    shown. Note that all line
    segments are either
    horizontal or vertical. Do
    not be overly concerned
    with the displayed
    dimension values.

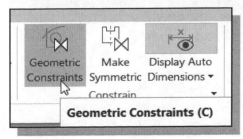

7. Select the **Geometric Constraints** command
    in the *Sketch Constraints* toolbar.

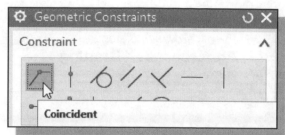

8. On your own, apply the **Coincident** constraint to align the center point of the arc to the origin of the datum coordinate system as shown.

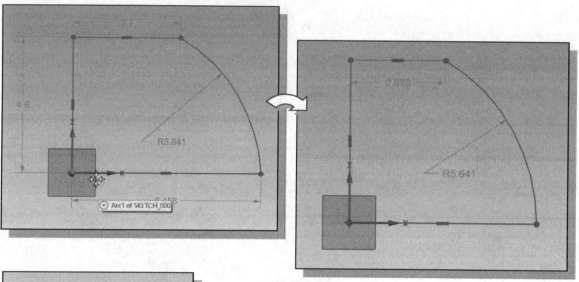

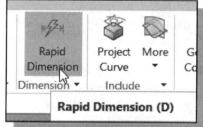

9. Select the **Rapid Dimension** command in the *Sketch Constraints* toolbar as shown.

10. On your own, create and modify the **two dimensions** as shown.

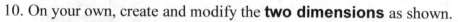

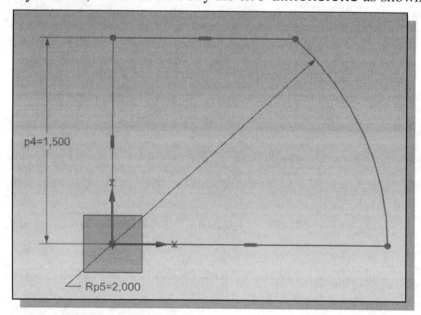

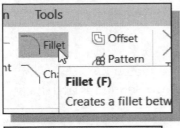

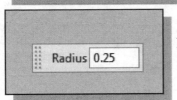

11. Click on the **Fillet** icon in the *2D Sketch Curve toolbar*.

12. In the *Radius* box, set the *radius* to **0.25**.

13. Select the top horizontal line and the arc to create a **rounded corner** as shown.

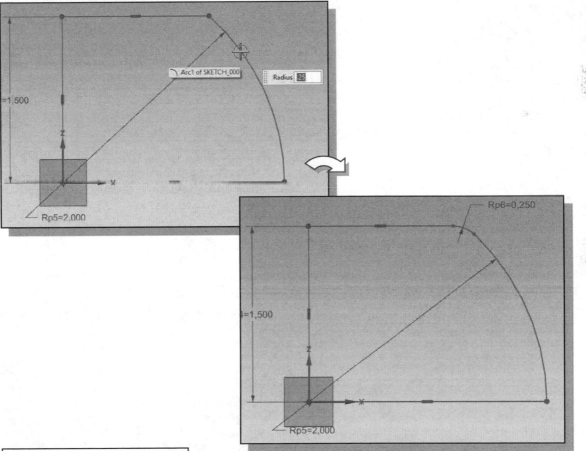

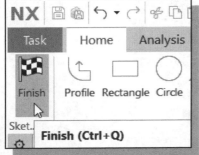

14. Click **Finish Sketch** to exit the NX sketcher mode and return to the *Feature option dialog window*.

Create a Revolved Feature

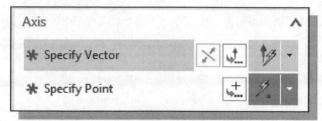

1. In the *Revolve* dialog box, the **Specify vector** option is activated as shown.

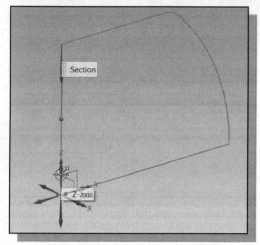

2. Select the **Vertical Edge** or the **Z Datum Axis** as the axis of rotation as shown.

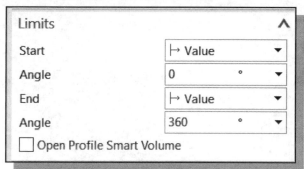

3. Confirm the start angle is set at **0** and the end angle at **360**, in the *Angular Limits area* as shown.

4. Click on the **OK** button to accept the settings and create the revolved feature.

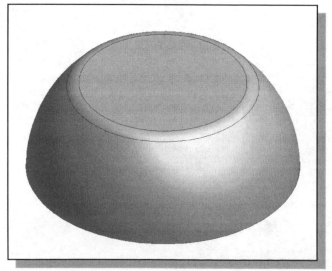

Create the Dryer Handle

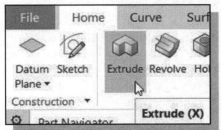

1. In the *Form Feature Toolbars,* select the **Extrude** command as shown.

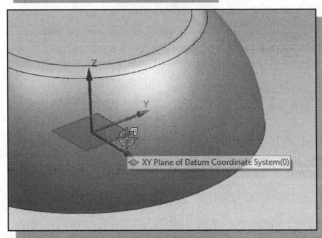

2. Select the **XY plane** as the *Sketch Plane.*

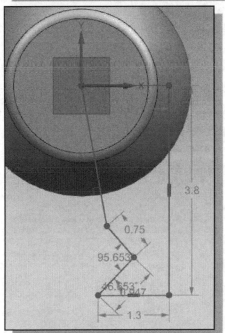

3. Using the **Profile** command, create the 2D sketch as shown. Note there are three inclined line segments on the left side of the sketch.

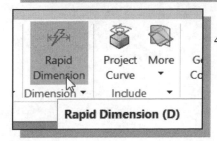

4. Select the **Rapid Dimension** command in the *Constraints* toolbar as shown.

5. On your own, create the **seven dimensions** as shown in the figure.

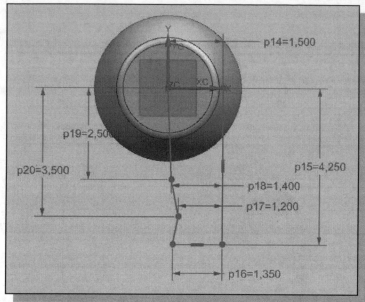

6. Click **Finish Sketch** to exit the NX sketcher mode and return to the *Feature option dialog window*.

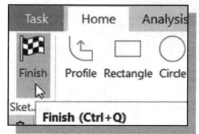

7. In the *Extrude* dialog box, set the extrude option to **Unite** and the *End limit* to **0.5** as shown.

8. Click **OK** to create the extrude feature.

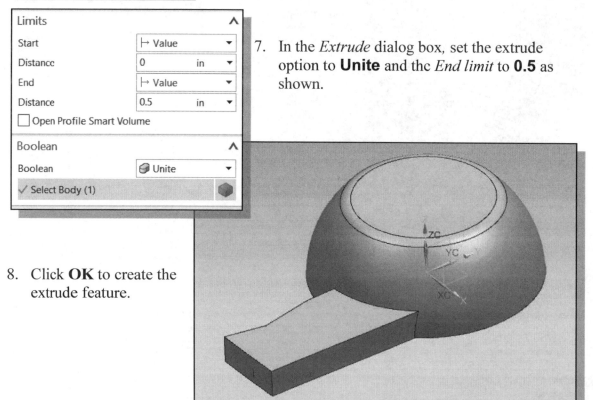

Create another Extruded Feature

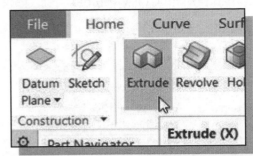

1. In the *Form Feature Toolbars Standard* toolbar, select the **Sketch** command by left-clicking once on the icon.

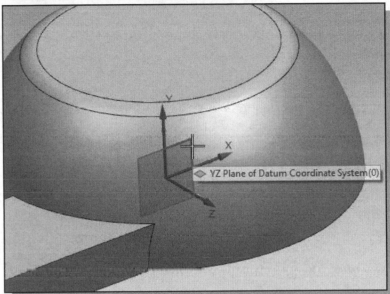

2. Select the **YZ** Plane to align the *Sketch Plane* as shown.

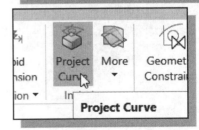

3. Select the **Project Curve** command by left-clicking once on the icon as shown.

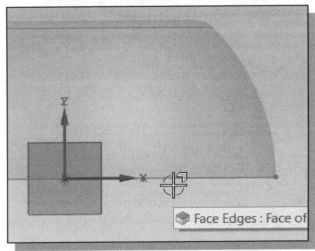

4. Select the *bottom edge* of the revolved feature as shown.

5. Click **OK** to accept the selection and proceed to create projected geometry.

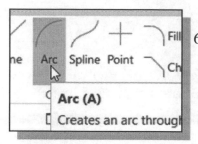

6. Activate the **Arc** command in the Curve toolbar.

7. Choose the **Arc by Center and Endpoints** option as shown.

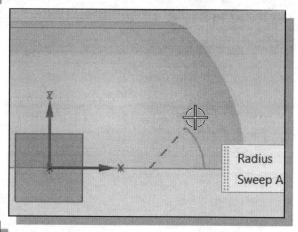

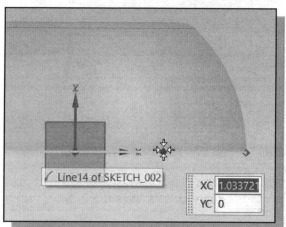

8. Align the **arc center** to the right of the origin of the coordinate system and on the bottom edge of the solid model.

9. Move the cursor toward the **right side** and click once with the left-mouse-button to align the end point to the bottom edge.

10. Move the cursor toward the **left side** and notice the **Radius** and **Sweep Angle** options. Click once with the left-mouse-button when the angle shows 180 degrees to align the end point to create an arc as shown.

11. Click once with the **middle-mouse-button** to end the **Arc** command.

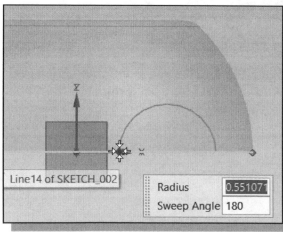

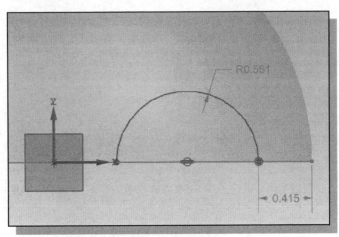

12. On your own, create a **line segment** connecting the two endpoints of the arc we just created.

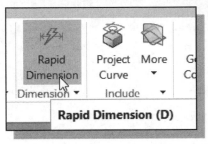

13. Select the **Rapid Dimension** command in the constraints toolbar as shown.

14. Create and modify the dimensions to fully constrain the 2D sketch as shown in the below figure.

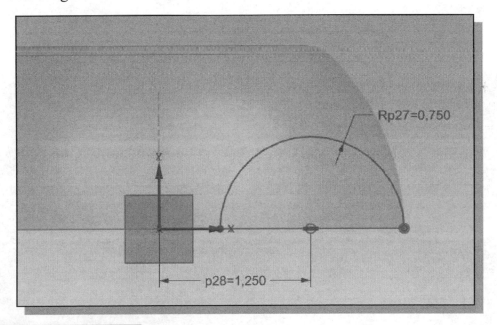

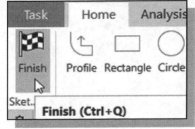

15. Click **Finish Sketch** to exit the NX sketcher mode and return to the feature dialog box.

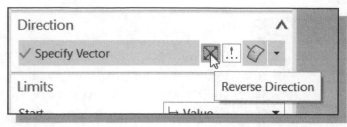

16. Click **Reverse Direction** to flip the extrusion direction as shown.

17. Enter **5.5** as the *End extrusion distance* as shown.

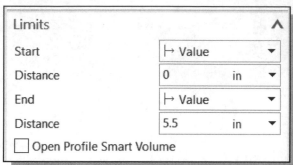

18. Set the *Boolean option* to **Unite** as shown.

19. Click **OK** to create the feature.

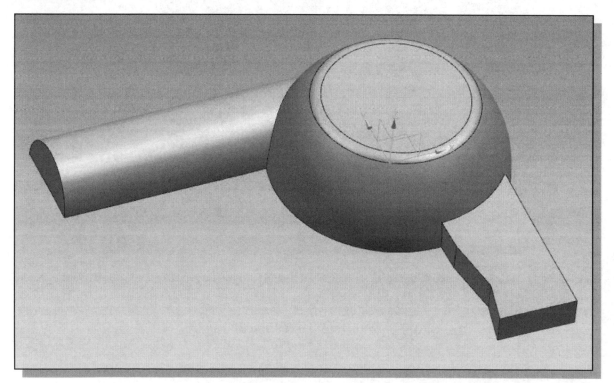

Create 3D Rounds and Fillets

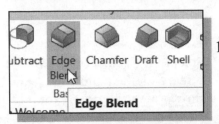

1. In the *Feature Operations* toolbar, select the **Edge Blend** command by left-clicking once on the icon.

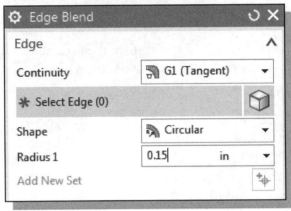

2. Note the Select Edge option is activated, also the default shape option is set to **Circular**.

3. Set the *Radius 1* option to **0.15** as shown in the above figure.

4. Select the **seven edges** as shown in the figure; note that six straight edges are of the handle section.

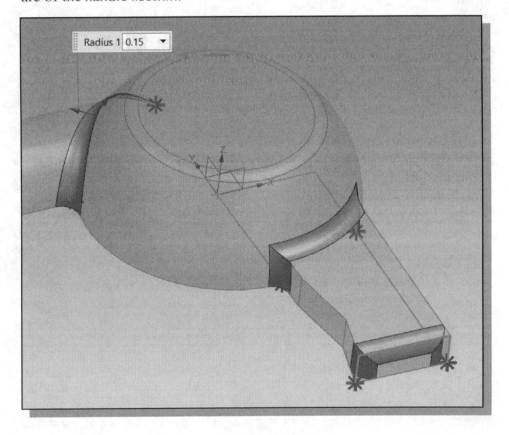

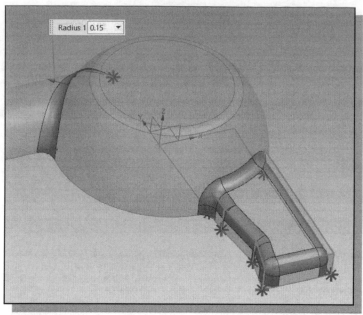

5. Select the adjacent **five edges** on the handle as shown in the figure. A total of thirteen edges are selected.

• Note the displayed rounded edges all have the same radius of **0.15**.

6. Click **OK** to accept the selections and proceed to create the second set of 3D rounds and fillets for our model.

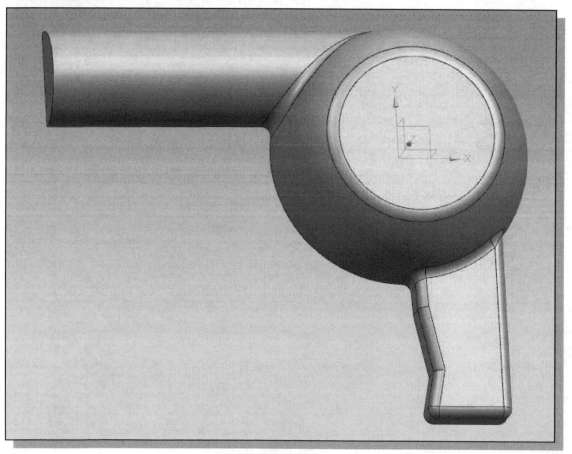

Create a Shell Feature

- The **Shell** command can be used to hollow out the inside of a solid, leaving a shell of specified wall thickness.

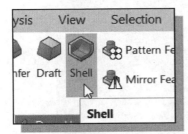

1. In the *Feature Operations* toolbar, select the **Shell** command by left-clicking once on the icon.

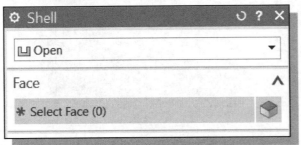

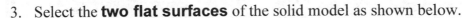

➢ Note that the **Open** option is activated by default.

2. On your own, use the **3D Rotation** option to display the back faces of the model as shown below.

3. Select the **two flat surfaces** of the solid model as shown below.

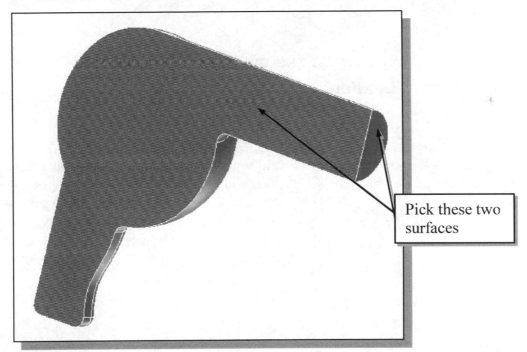

Pick these two surfaces

4. Set the *Thickness* option to a value of **0.125** as shown.

5. In the *Shell* dialog box, click on the **OK** button to accept the settings and create the shell feature.

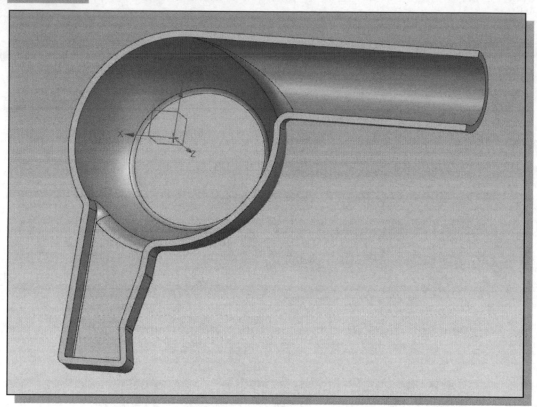

Create a Pattern Leader

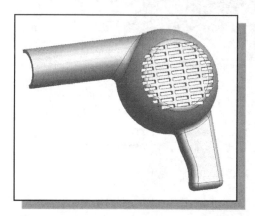

*   The *Dryer Housing* design requires the placement of identical holes on the top face of the solid. Instead of creating the holes one at a time, we can simplify the creation of these holes by using the **Rectangular Array** command to create duplicate features. Prior to using the command, we will first create a *pattern leader*, which is a regular extrude feature.

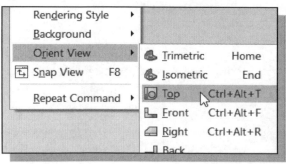

1. Select **Top View** in the *View toolbar* to adjust the display of the model on the screen.

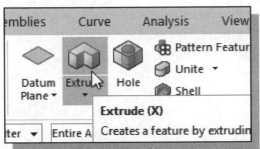

2. In the *Form Feature Toolbars,* select the **Extrude** icon as shown.

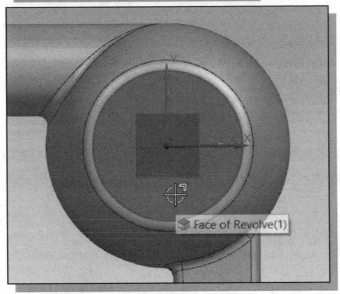

3. Pick the top face of the base feature, or use the select option and choose **Face of Revolve** in the *Quick Pick* box.

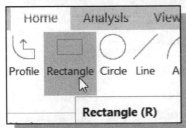

4. Activate the **Rectangle** command by clicking the corresponding icon as shown.

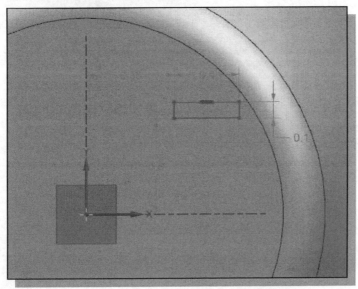

5. Create a **rectangle** of arbitrary size that is toward the right side of the coordinate system as shown.

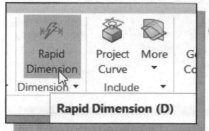

6. Select the **Rapid Dimension** command in the *Constraints* toolbar as shown.

7. On your own, create and modify the **dimensions** as shown in the below figure.

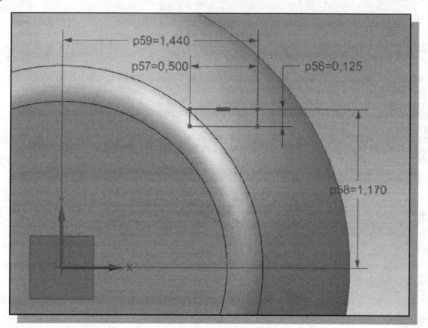

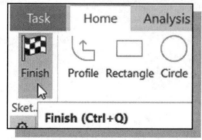

8. Click **Finish Sketch** to exit the NX sketcher mode and return to the *Feature option dialog window*.

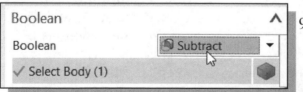

9. In the *Extrude* dialog box, set the *Boolean* option to **Subtract** as shown.

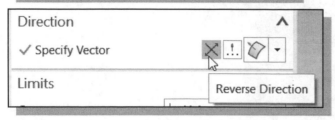

10. Click the **Reverse Direction** icon to flip the cut direction into the solid model.

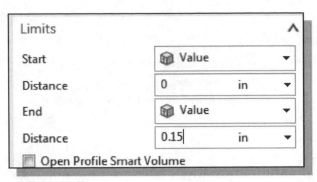

11. Set the *End limit Distance* to **0.15** as shown.

12. In the *Extrude* dialog box, click **OK** to proceed with creating the cut feature.

♦ Note that the pattern leader created is a fairly small cut on the solid model.

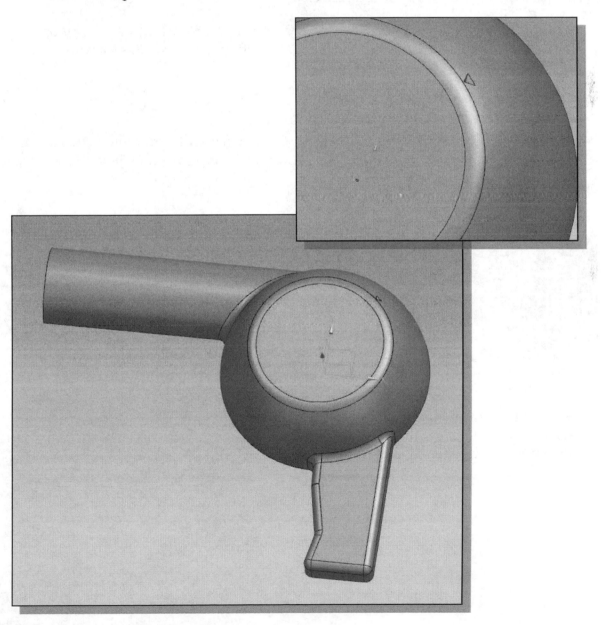

Creating a Rectangular Array

In *NX*, existing features can be easily duplicated. The **Array** commands allow us to create both rectangular and circular arrays of features. The arrayed features are parametrically linked to the original feature; any modifications to the original feature are also reflected on the arrayed features.

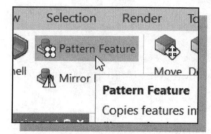

1. Select the **Pattern Feature** command in the Feature toolbar as shown.

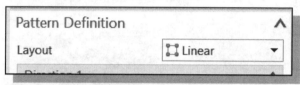

2. Select the **Rectangular Array** option as the pattern layout as shown.

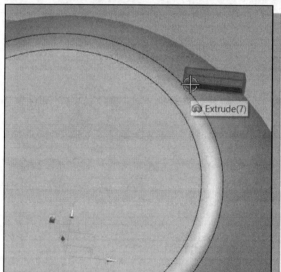

3. Choose the **last feature**, the *Extruded* cut feature, in the *graphics* window as shown.

4. Click once with the **Middle-mouse-button** to proceed to the next step.

5. Select the **X-axis** of the sketch coordinate system to set the first direction of the pattern copies.

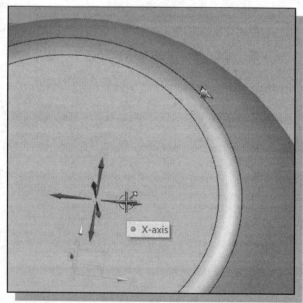

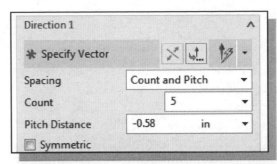

6. In the *Direction 1 option list*, enter **5** in the *Count* and **-0.58** in the *Pitch Distance* as shown.

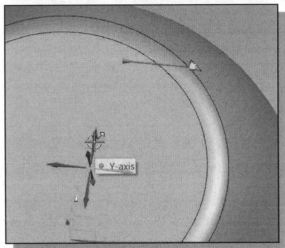

7. In the *Direction 2 option list*, switch on the **Use Direction 2** option.

8. Select the **Y-axis** of the sketch coordinate system to set the second direction of the pattern copies.

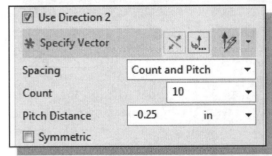

9. Enter **10** in the *Count* and **-0.25** in the *Pitch Distance* as shown.

10. Click **OK** in the option to proceed with the Rectangular Array command.

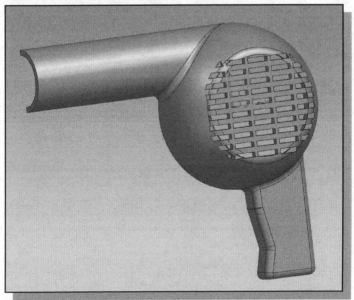

Create a Swept Cut Feature

❖ The **Sweep** operation is defined as moving a planar section through a planar (2D) or 3D path in space to form a three-dimensional solid object. The path can be an open curve or a closed loop but must be on an intersecting plane with the section. The **Extrude** operation, which we have used in the previous lessons, is a specific type of sweep. The **Extrude** operation is also known as a *linear sweep* operation, in which the sweep control path is always a line perpendicular to the two-dimensional section. Linear sweeps of unchanging shape result in what are generally called *prismatic solids* which means solids with a constant cross-section from end to end. In *NX*, we create a *swept feature* by defining a path and then a 2D sketch of a cross section. The sketched profile is then swept along the planar path. The Sweep Along guide operation is used for objects that have uniform shapes along a trajectory.

➢ **Define the Sweep Section**

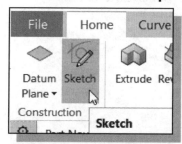

1. In the *Form Feature Toolbars Standard* toolbar, select the **Sketch** command by left-clicking once on the icon.

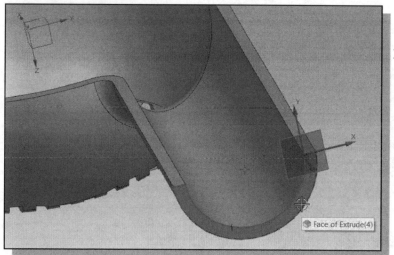

2. Select the plane on the **Circular End** of the model to align the *Sketch Plane* as shown.

3. Click **OK** to accept the selection of the *Sketch Plane*.

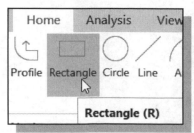

4. Activate the **Rectangle** command and choose the **By 2 points** option.

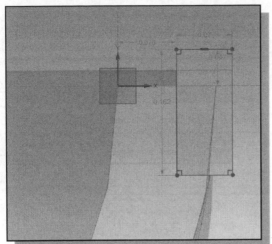

5. On your own, create a **rectangle** that is above the top right corner of the model as shown.

- Note that the rectangle is created a bit larger to assure proper operation in creating the feature.

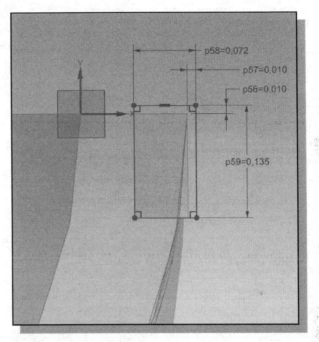

6. On your own, create and modify the **dimensions** associated to the sketch as shown in the figure.

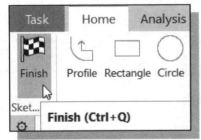

7. Click **Finish Sketch** to exit the NX sketcher mode and end the *Sketch option*.

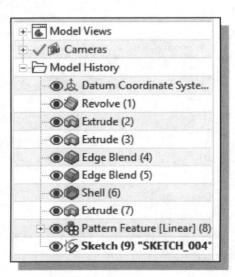

- Note the created sketch appears as an item in the Model History tree. The **Sweep** operation is defined as moving a planar section through a planar (2D) or 3D path in space to form a three-dimensional solid object. So far, we have created the 2D section, but still need to set up the sweep path.

➢ Create the Swept feature

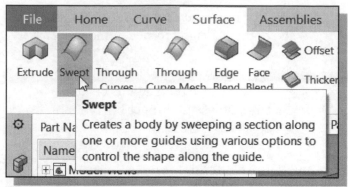

1. Select the **Swept** command in the *Surface tab* as shown.

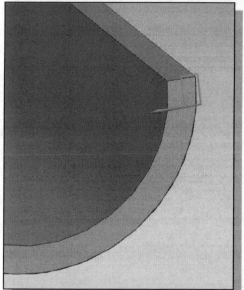

2. Select the **rectangle** as the *section string* as shown. (Hint: Use the dynamic *Zoom* function to assist the selection.)

3. Click on the **Guide** icon to activate the guide selection option.

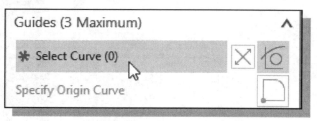

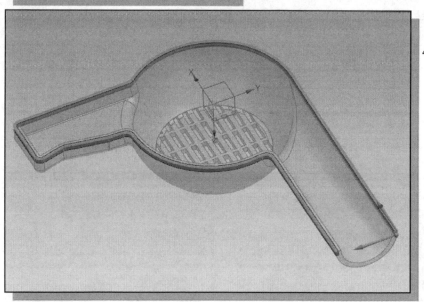

4. Select all the adjacent **line-Arc segments** connected to the top corner of the rectangle as shown. (Hint: Use the dynamic *Zoom* function to assist the selection.)

5. Click **OK** to accept the selections and create the swept feature.

➤ Note the swept feature is a separate solid from the dryer housing model; a *Boolean Subtract* is needed to combine the two solids.

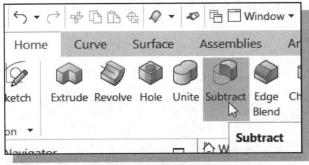

6. Click the **Subtract** icon to perform the cut operation of the swept feature from the existing model.

7. Select the **main body** of the dryer housing model as the *Target body*.

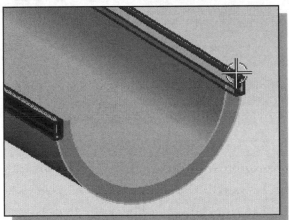

8. Select the **swept feature** as the *Tool body*.

9. In the *Subtract* dialog box, click **OK** to proceed with the operation and complete the model.

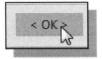

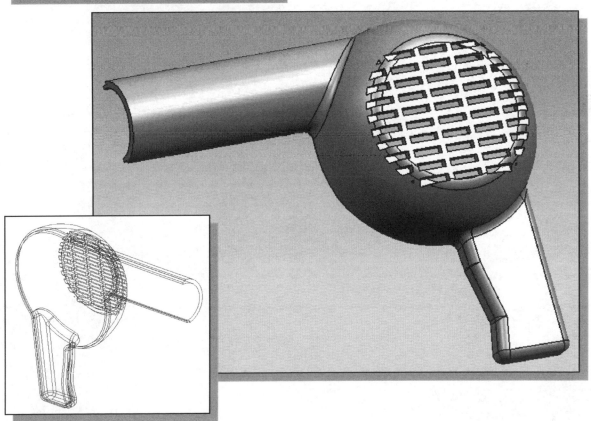

Review Questions:

1. Keeping the *History Tree* in mind, what is the difference between *cut with a pattern* and *cut each one individually*?

2. What is the difference between **Sweep** and **Extrude**?

3. What are the advantages and disadvantages of creating fillets using the **3D Fillets** command and creating fillets in the 2D profiles?

4. Describe the steps used to create the *Shell* feature in the lesson.

5. How do we modify the *Array* parameters after the model is built?

6. Describe the elements required in creating a *Swept* feature.

7. Create sketches showing the steps you plan to use to create the model shown on the next page:

Exercises:

1. **Motor Housing** (Dimensions are in inches.)

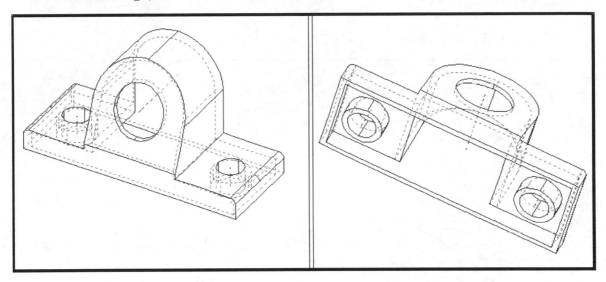

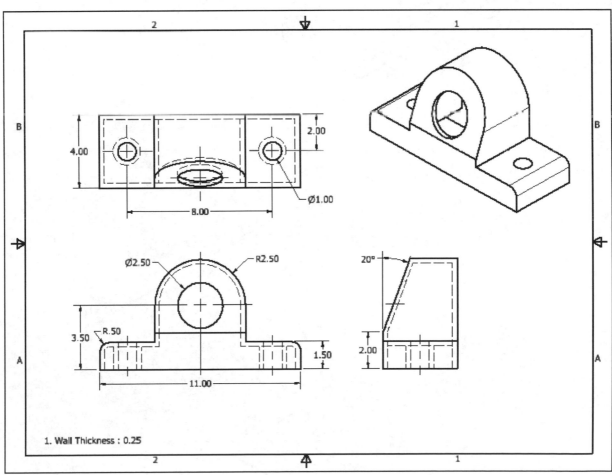

1. Wall Thickness : 0.25

2. Using the same dimensions given in the tutorial, construct the other half of the dryer housing. Plan ahead, and consider how you would create the matching half of the design. (Save both parts so that you can create an assembly model once you have completed chapter 14.)

3. **Index Key** (Dimensions are in inches.)

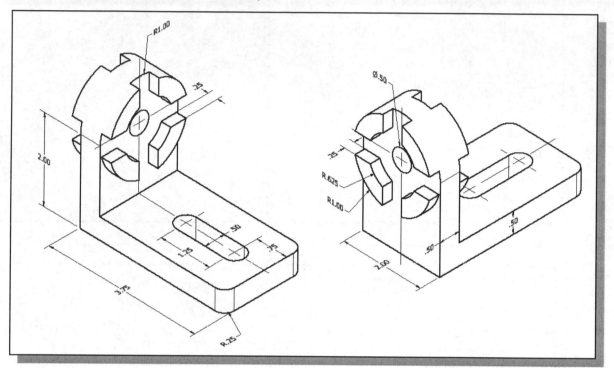

4. **Anchor Base** (Dimensions are in millimeters.)

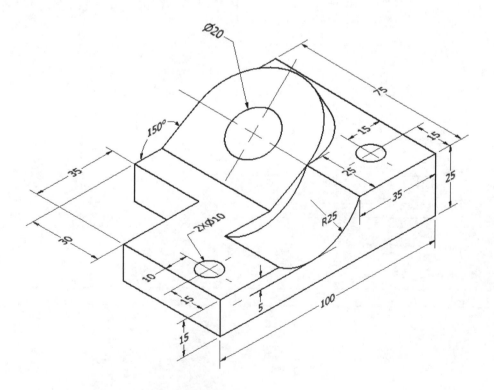

5. **Pivot Latch** (Dimensions are in inches.)

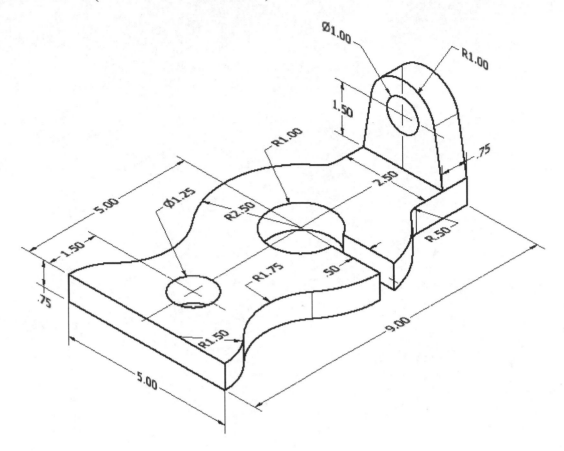

6. **Piston Cap** (Dimensions are in inches.)

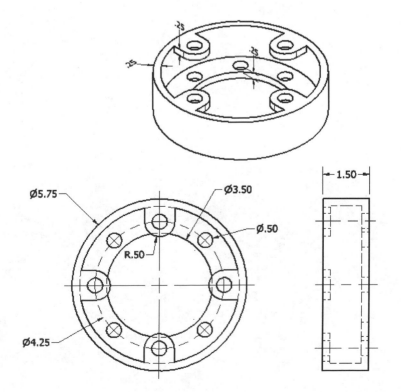

Notes:

Chapter 13
Basic Sheet Metal Designs

Learning Objectives

- ♦ **Understand the Sheet Metal Manufacturing Processes**
- ♦ **Understand the NX Sheet Metal Modeling Methodology**
- ♦ **Create Parts in the Sheet Metal Modeler Mode**
- ♦ **Utilize the NX Sheet Metal Tools to Create Bends and Flanges**
- ♦ **Create Flat Pattern Layouts**

Sheet Metal Processes

Sheet metal is one of the most commonly used materials in our everyday life. Sheet metal is simply a thin and flat piece of metal, which can be cut and bent into a variety of different shapes. The thicknesses of sheet metal can vary significantly, but the thickness is generally between 0.006" and 0.250".

Sheet metal is generally produced by reducing the thickness of a work piece by compressive forces applied through a set of rolls. This process is known as rolling and has been around since 1500 AD. Sheet metal is identified by the thickness, or gauge, of the metal and is generally available as flat pieces or in coils. The typical gauge of sheet metal ranges from 30 gauge to about 6 gauge; refer to Appendix A for more details. The higher the gauge number, the thinner the metal is. Aluminum, brass, copper, cold rolled steel, tin, nickel and titanium are some of the more commonly available sheet metal materials. Typical sheet metal applications are seen in cars, boats, airplanes, casing for electronic devices and many other things.

The main feature of sheet metal is its ability to be formed and shaped by a variety of processes, such as **bending** and **cutting**. Different processes can be used to achieve the desired shape and form. Some of the more commonly used sheet metal processes include:

Drawing
Drawing forms sheet metal into parts by using a punch, where the punch presses a sheet metal blank into a die cavity. This process is generally used to create shallow or deep parts with relatively simple shapes. **Soft punches** can also be utilized to create more arbitrary shapes. **Deep drawing** is generally done by making multiple steps; this process is known as *draw reductions*.

Stretch forming
Stretch forming is a process where the sheet metal is clamped around its edges and stretched over a die. This process is mainly used for the manufacturing of large parts with shallow contours, such as aircraft wings, or automotive door and window panels.

Spinning
Spinning is the process used to make axis-symmetric parts by applying a work piece to a rotating mandrel with the help of rollers. *Spinning* is commonly used to make cylindrical shapes, such as missile nose cones and satellite dishes.

Stamping
Stamping is the general term used to describe a variety of operations, such as bending, flanging, punching, embossing, and coining. The main advantage of stamping is its speed; designs containing simple or complex shapes can be formed at relatively high production rates.

Flanging
Flanging is a process used to strengthen different sections of a sheet metal part and also to form various shapes. This process is commonly used for a variety of parts, for example, aluminum cans for soft drinks.

Bending

Bending is a process by which sheet metal can be deformed by plastically deforming the material and changing its shape. The material is stressed beyond the yield strength but below the ultimate tensile strength. With this process, the surface area of the material does not change much. *Bending* usually refers to deformation about one axis.

Bending is a flexible process by which many different shapes can be produced. Standard die sets are used to produce a wide variety of shapes. The material is placed on the die and positioned in place with *stops* and *gauges*. The material is held in place with *hold-downs*. The upper part of the press, the ram, with the appropriately shaped punch descends and forms the v-shaped bend.

Bending is usually done using ***press brakes***. The lower die of the press contains a V-shaped groove. The upper part of the press contains a punch that will press the sheet metal down into the v-shaped die, causing it to bend.

The most commonly used modern *Bending* method is the ***air bending*** method, where a sharper die angle is used; for example, an 85 degree angle is used for a 90 degree bend. *Air Bending* is done with the punch touching the work piece, and the work piece not bottoming in the lower die. By controlling the push stroke of the upper punch, the metal is pushed down to the required bend angle. The groove width of the lower die is typically 8 to 10 times the thickness of the metal to be bent. The *press brake* can also be computer controlled to allow the making of a series of bends to assure a high degree of accuracy in manufactured parts.

Cutting

Cutting sheet metal can be done in various ways, from using a variety of hand tools to very large powered shears. Today, computer-controlled cutting is also available for very precise cutting. Most modern computer-controlled sheet metal cutting operations are ***CNC laser cutting*** and ***CNC punch press***.

CNC laser cutting is done by moving the laser beam over the surface of the sheet metal. The sheet metal is heated and then burnt by the laser beam. The quality of the edge can be extremely smooth. *CNC punching* is performed by moving the sheet metal between the computer-controlled punch. The top punch mates with the bottom die, cutting a simple shape, such as a square, circle, or hexagon from the sheet. An area can be cut out by making several hundred small square cuts around the perimeter. A *CNC punch* is less flexible than a laser for cutting compound shapes, but it is faster for repetitive shapes. A typical *CNC punch* has a choice of up to 60 tools in a ***turret***. A modern *CNC punch* can run as fast as 600 blows per minute. A *CNC punch* or a *CNC laser* machine can typically cut a blank sheet into the desired shapes in less than 15 seconds, with very high precision.

Sheet Metal Modeling

In reality, a sheet metal part is made from a piece of flat metal sheet of uniform thickness by cutting out a flat pattern and then folding it into the desired shape. To construct a computer sheet metal part, we can (1) simulate the actual production methods and start with a flat pattern layout to make the model; (2) use the building block approach which concentrates on the different sections of the formed 3D design; or (3) construct a solid model first, then convert it into a sheet metal model. All three methods are applicable in modern parametric modeling software such as *NX*.

Since the actual sheet metal manufacturing process requires a flat pattern layout, the accurate generation of the flat pattern layout in the computer modeling software is critical. The conversion between the 3D formed designs and 2D flat pattern layouts requires the using of the correct **K-Factor**, which can be used to determine the required **Bend Allowance**.

Bend allowance is the term used to describe how much material is needed between two panels to accommodate a given bend. Determining bend allowance is commonly referred to as **Bend Development**.

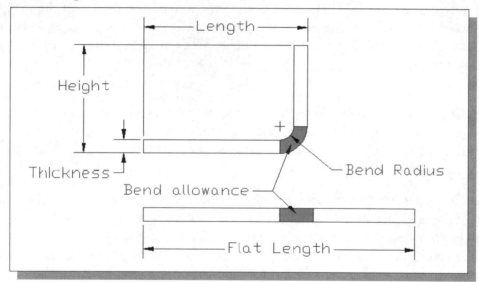

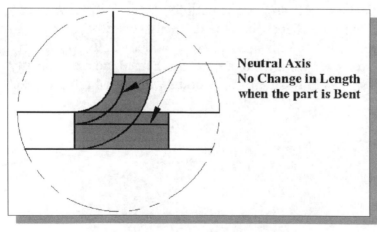

In sheet metal, the **Neutral Axis** is defined as the location where there is no change in length when the part is bent.

On the inside of the bend, above the neutral axis in the figure, the material is in compression, where the area below the neutral axis is in tension.

K-Factor (Neutral Factor)

The location of the neutral axis in a bend is called the ***k-factor***. Since the amount of inside compression is always less than the outside tension, the k-factor can never exceed **0.50** in practical use. To the other extreme, a reasonable assumption is that the k-factor cannot be less than **0.25**.

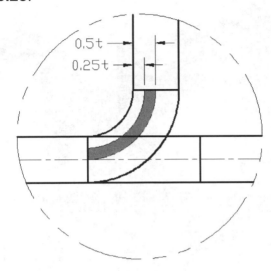

Several factors can change the k-factor, such as the type of bending (free vs. constrained), tool geometry, rate of bend, material (Mild Steel, Cold Rolled Steel, Aluminum, etc.), and even grain direction. With some grades of aluminum, the age of the material can also be a factor.

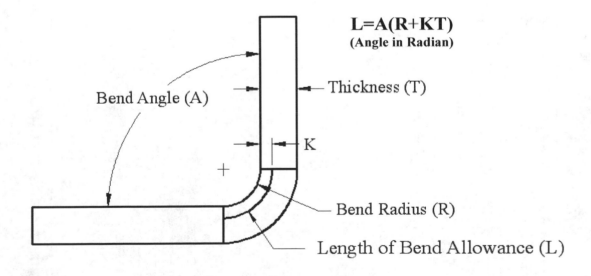

Sheet metal fabricators will typically have developed a k-factor table (usually through trial and error) to use. *NX* is set up to allow the k-factor to be added to create material specific profiles. By using the correct data, *NX* can be used to create fairly accurate and reliable flat patterns.

The Actuator Bracket Design

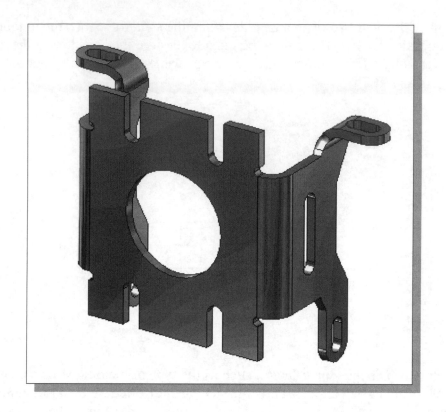

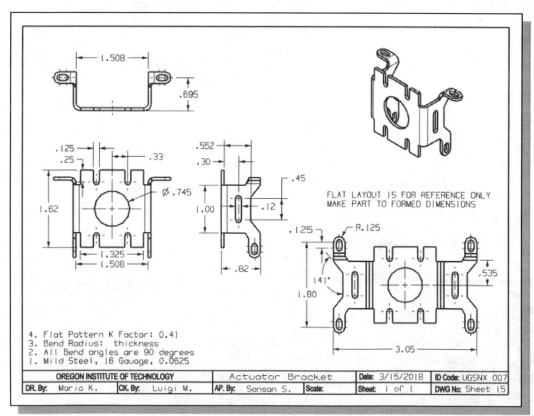

FLAT LAYOUT IS FOR REFERENCE ONLY
MAKE PART TO FORMED DIMENSIONS

4. Flat Pattern K Factor: 0.41
3. Bend Radius: thickness
2. All Bend angles are 90 degrees
1. Mild Steel, 16 Gauage, 0.0625

OREGON INSTITUTE OF TECHNOLOGY		Actuator Bracket		Date: 3/15/2018	ID Code: UGSNX 007
DR. By: Mario K.	CK. By: Luigi M.	AP. By: Sansan S.	Scale:	Sheet: 1 of 1	DWG No: Sheet 15

Start NX

1. Select the **NX** option on the *Start* menu or select the **NX** icon on the desktop to start *NX*. The *NX* main window will appear on the screen.

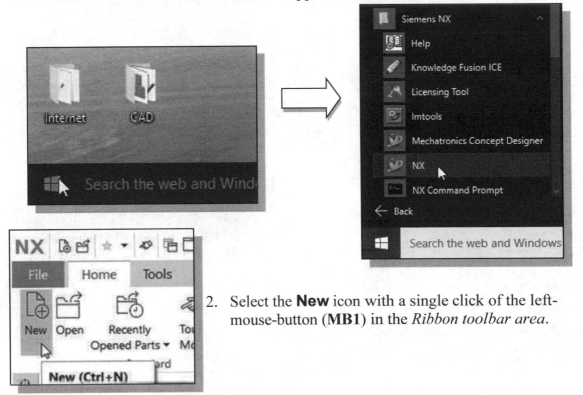

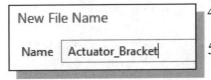

2. Select the **New** icon with a single click of the left-mouse-button (**MB1**) in the *Ribbon toolbar area*.

3. Select the **inches** units and select **Sheet Metal** in the *Template list*. Note that the *Sheet Metal template* will allow us to switch directly into the **Sheet Metal task**.

	Type	Units		
Model	Modeling	Inches	Stand-alone	NT AUT...
Assembly	Assemblies	Inches	Stand-alone	NT AUT...
Shape Studio	Shape Studio	Inches	Stand-alone	NT AUT...
Sheet Metal	Sheet Metal	Inches	Stand-alone	NT AUT...
Blank	Gateway	Inches	Stand-alone	none

New File Name

Name   Actuator_Bracket

4. Enter *Actuator_Bracket* as the *New File Name*.

5. Click **OK** to proceed with the New File command.

- Note the ribbon toolbar different commands available for sheet model designs.

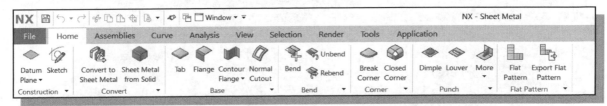

Sheet Metal Preferences

> ➢ The **Sheet Metal Preferences** command allows us to specify many of the sheet metal parameters, such as material thickness, Bend Radius and Cut Relief for the active sheet metal part.

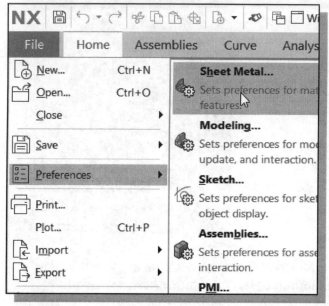

1. In the *Menu list*, select **Preferences → Sheet Metal** as shown.

❖ Note the *Default* sheet metal part properties setting is set to use **Value Entry**, which means the related sheet metal parameters are entered. Parameters can also be set by using the other two options: *Material selection* and *Tool selection*.

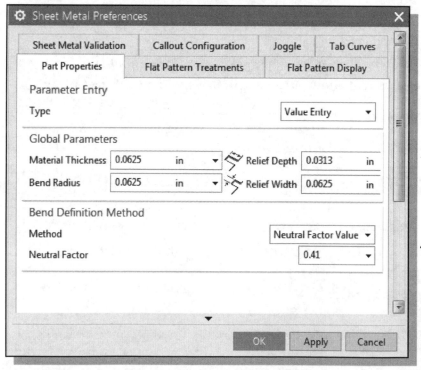

2. Enter **0.0625** as the *Material Thickness* and as the *Bend Radius* as shown.

3. Enter **0.0313** as the *Relief Depth* and **0.0625** as the *Relief Width* as shown.

4. Enter **0.41** as the *Neutral Factor (KFactor) Value* as shown.

5. Click **OK** to accept the settings and end the **Sheet Metal Preferences** command.

Create the Base Feature of the Design

❖ The main section of a sheet metal design is generally treated as the stationary portion of the design, to which all the other sections are added to form the final design. The main section is also typically the starting point of sheet metal modeling, and thus the base feature in parametric modeling.

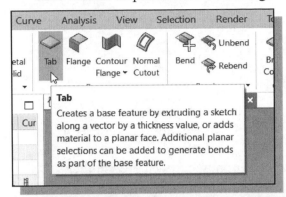

1. In the *Basic* toolbar select the **Tab** command by left-clicking once on the icon.

• The *Tab command* allows us to create the base feature by creating a flat 2D sketch.

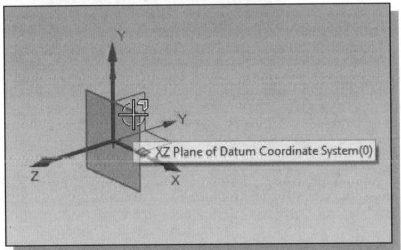

2. Select the **XZ Plane**, in the *graphics* area, to align the sketch plane.

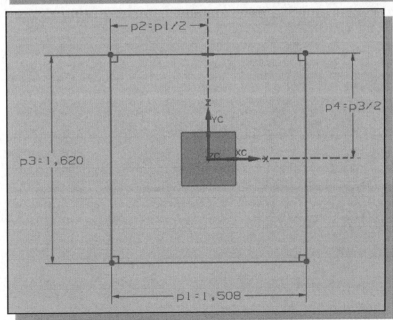

3. On your own, construct a **1.62 X 1.508** rectangle and apply the proper dimensions as shown. (Hint: Use the equations shown to align the rectangle to be centered at the origin.)

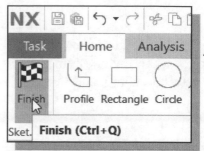

4. Select **Finish**, exiting the 2D Sketch mode and returning to the *Sheet Metal Tab* panel.

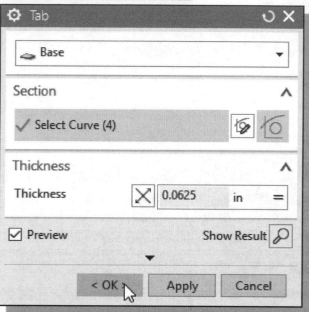

❖ In the *Tab* panel, the default thickness value, which was set through the sheet metal properties, is automatically entered in the *Thickness box*.

5. Click **OK** to accept the settings and proceed to creating the base feature of the design.

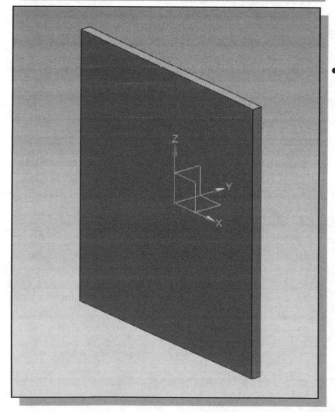

• The base feature of the design is constructed using the Tab command and the thickness was set through the *Sheet Metal Preferences* settings.

Create a Cut Feature with the Extrude Command

- In NX, the **Normal Cutout** command creates cut features perpendicular to the sheet metal surfaces.

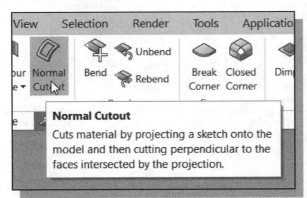

1. In the *Feature* toolbar, select the **Extrude** command in the *More* pull-down list as shown.

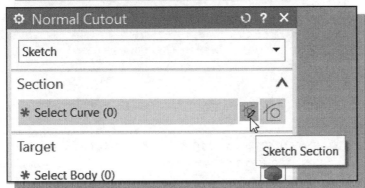

2. In the *Extrude panel*, click on the **Sketch Section** icon to enter the sketch dialog box.

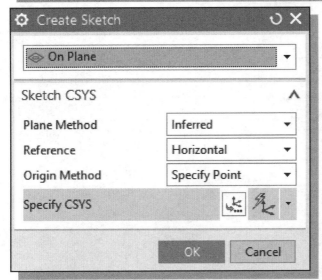

3. Set the *sketch type* to **On Plane** as shown.

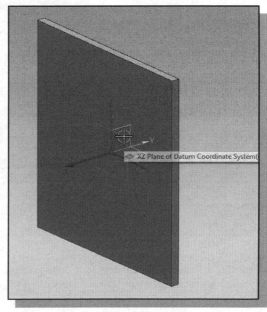

4. Select the **XZ Plane** of the Datum feature to align the sketching plane.

5. Click on the **OK** button to accept the settings.

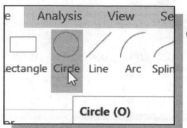

6. Activate the **Circle** command in the *Curve toolbar* as shown.

7. On your own, create a diameter **0.745** circle aligned to the origin as shown.

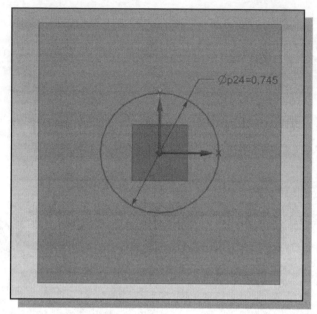

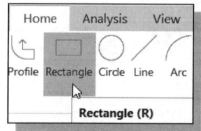

8. On your own, create the four **closed regions** on top of the base feature as shown.

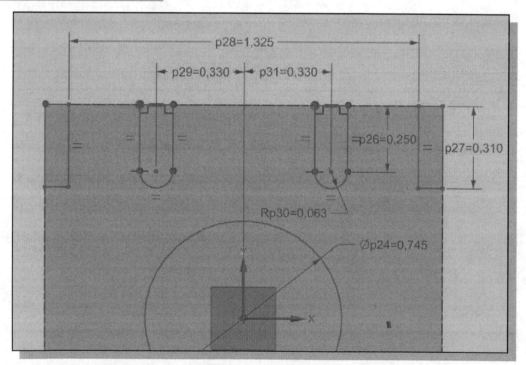

9. On your own, create a set of identical closed regions on the bottom of the base feature as shown. (Hint: Apply the necessary geometric constraints to assure alignments and dimensions.)

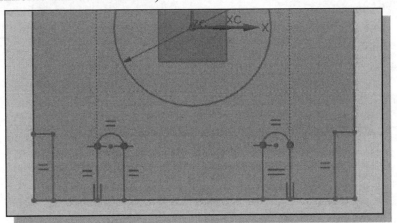

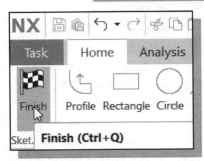

10. Select **Finish**, exiting the 2D Sketch mode and returning to the *Sheet Metal Tab* panel.

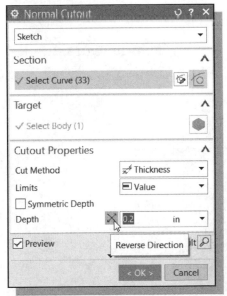

11. In the *Normal Cutout panel*, confirm the 33 curves are selected. Click **Reverse Direction** if necessary.

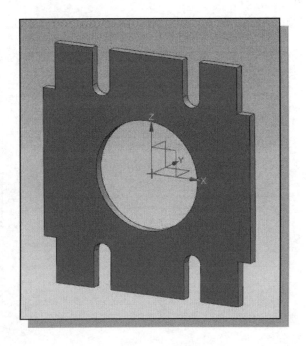

12. Click on the **OK** button to accept the settings and create the cut feature as shown.

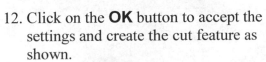

Create a Flange Feature

❖ Sheet metal flange features consist of a flat face, which has a bend that connects the flat face to an existing straight edge of the sheet metal model. Flange features are added by selecting one or more edges and by specifying a set of options which determines the size and position of the material added. The **Flange** command can be used to build additional sections on a selected edge and with extra controls on the bend.

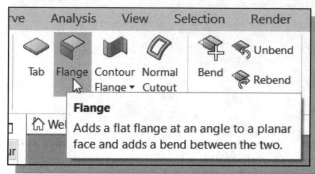

1. In the *Sheet Metal Features* panel, select **Flange** to activate the command.

❖ The *Flange* dialog box contains many settings to create a flange feature.

2. Confirm the *Width Option* is set to **Full** and adjust the length to **0.5** inches as shown in the figure.

3. Confirm the *Flange Angle* to **90** degrees and set the *Length Reference* to **outside** as shown.

4. Set the *Inset* to **Material Outside** and no Offset as shown.

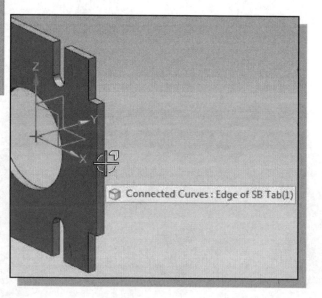

- The **Flange** command requires the selection of an existing straight edge and creates a flat feature.

5. Select the **back edge** on the right side as shown.

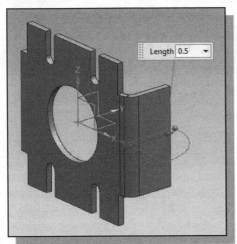

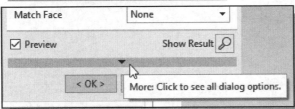

- The **Flange** command creates the flat side feature using the settings in the *Flange* panel.

❖ Note the relief cut is also added when the two flanges are created.

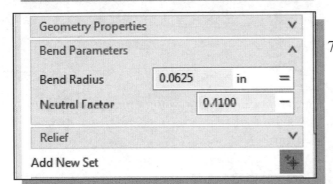

6. Click on the **arrow** icon near the bottom to expand the list as shown.

7. Confirm the *Bend Radius* is set to **0.0625** and *K Factor* to **0.41** as shown.

8. Click on the **OK** button to accept the settings and create the flange feature as shown.

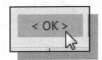

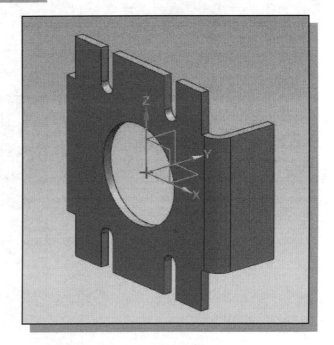

Confirm the Flange Location

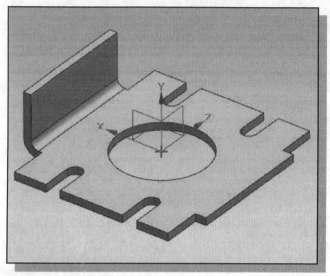

1. On your own, dynamically rotate the display so that you are viewing the back side of the base feature as shown.

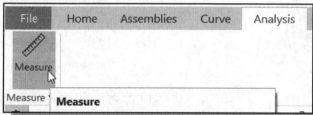

2. Switch to the **Analysis** tab and activate the **Measure Distance** command as shown.

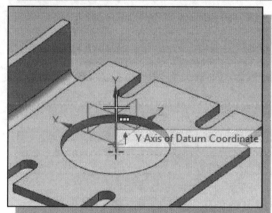

3. Select the **Y axis** of the coordinate system as the first object to measure the distance from.

4. Select the **inside surface** of the flange feature as the second object to measure the distance from.

5. On your own, confirm the distance between the center of the main face to the inside surface of the flange is **0.754**.

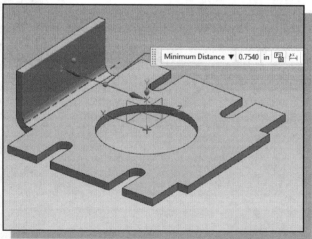

Add another Extruded Feature

❖ The main section of a sheet metal design is generally treated as the stationary portion of the design, to which all the other sections are added to form the final design. Additional extruded features, either adding or cutting features, can be done through the **Extrude** command.

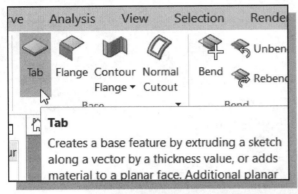

1. In the *Feature* toolbar, select the **Tab** command in the *Ribbon Toolbar* as shown.

2. Rotate the model roughly to the orientation as shown below. Select the **Flange face** by left clicking on the surface.

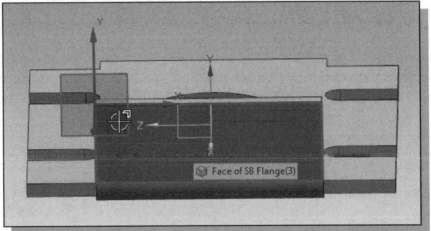

3. On your own, construct the 2D sketch and apply the proper constraints and dimensions as shown. (The sketch is symmetrical with two sets of parallel inclined lines.)

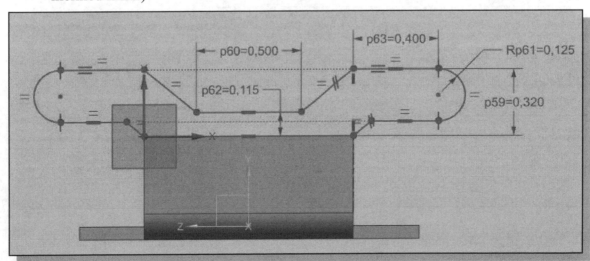

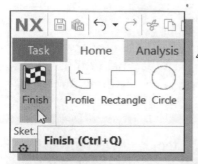

4. Click **Finish** to end the *2D Sketch* mode and return to the *Sheet Metal Features* panel.

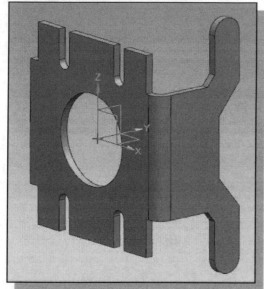

5. Click **OK** to proceed to creating the face feature as shown.

• The *flange feature* is a flat rectangular shape, but it can be easily modified to more complex shapes using the *Extrude* command.

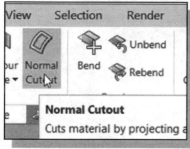

6. In the *Feature* toolbar, select the **Normal Cut** command in the *Ribbon Toolbar* as shown.

7. On your own, create the **three cut slots** as shown.

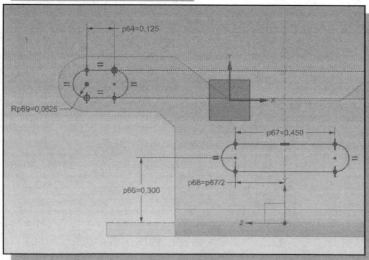

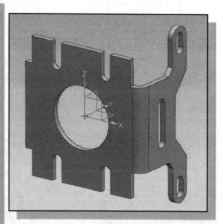

Create a Bend Feature

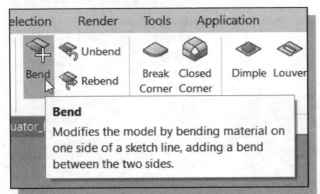

1. In the *Menu list*, select **Insert →
   Bend → Bend** as shown.

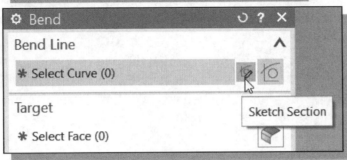

2. In the *Bend dialog* box, click
   Sketch Section to enter the
   NX sketcher mode.

3. Select the **outside surface** of the *Face* feature to align the sketching plane as
   shown.

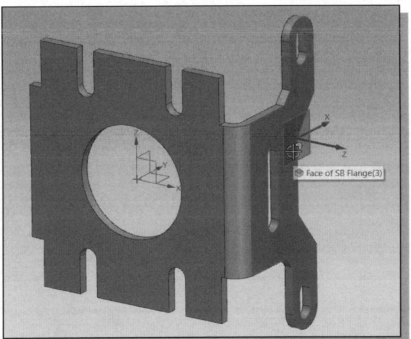

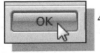

4. Confirm the *Sketch type* is set to **On Plane** and click on the **OK** button
   to accept the settings and enter the 2D sketch mode.

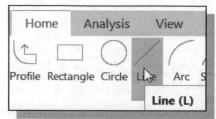

5. In the *2D Sketch* panel, select the **Line** command as shown.

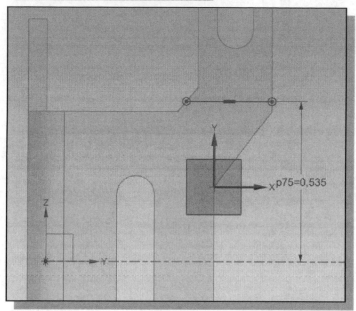

6. On your own, construct a **horizontal line** and apply a vertical dimension, measuring **0.535** to the origin of the world coordinate system, as shown.

❖ Note that the two endpoints of the horizontal line are on the two edges of the existing surface.

7. Click **Finish** to end the *2D Sketch* mode and return to the *Bend Feature* panel.

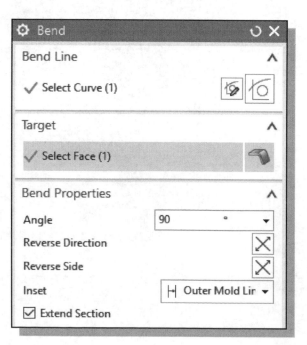

• Note the line we just created is pre-selected as shown in the figure.

8. On your own, examine the effects of the **Reverse direction** and **Reverse side** options.

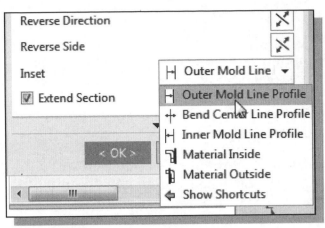

9. On your own, experiment with the different effects with the *Bend Inset* option.

10. Set the *Bend Inset* option to **Outer Mold Line profile** as shown in the figure.

 11. Click **OK** to create the *Bend* feature.

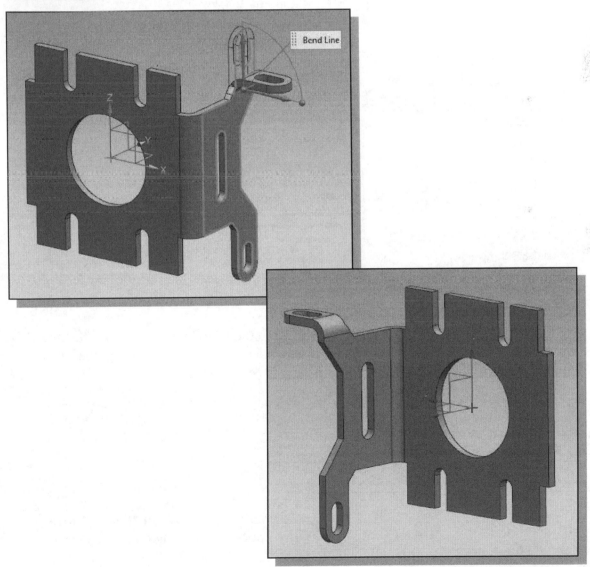

Create Mirrored Features

➢ In *NX*, sheet metal features can also be mirrored just like any regular solid feature. The mirrored feature is parametrically linked to the original parametric definitions.

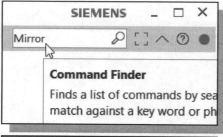

1. Enter **Mirror** in the *Command Finder* box to find the desired command.

   • By default, some of the NX commands are not available in the ribbon toolbars; the Command Finder can be used to locate the desired commands.

2. Select **Mirror Feature** in the Command Finder dialog box as shown.

➢ The **Mirror Feature** command provides several options that can be used to create copies of features.

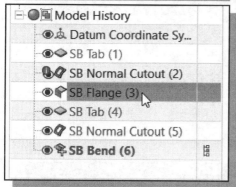

3. Select the **flange feature** in the *Model History Tree* as shown.

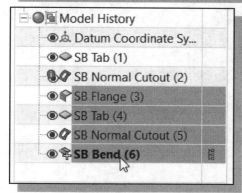

4. Hold down the [**Ctrl**] key and select the three features below the flange feature in the *Model History Tree* as shown.

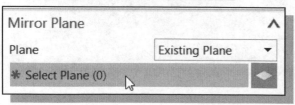

5. Activate the **Select Mirror Plane** option by clicking on the icon as shown.

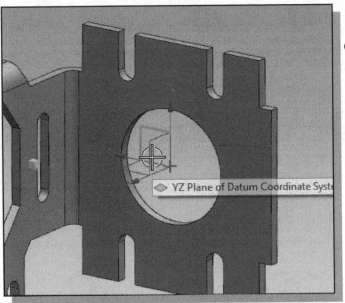

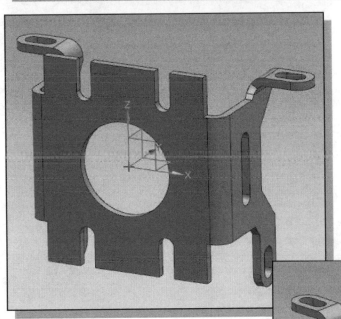

6. Select the **YZ Plane** of the coordinate system in the *graphics area* as shown.

7. Click **OK** to create the *Mirrored* features.

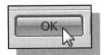

8. On your own, **save** the completed sheet metal design.

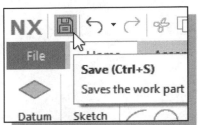

Create a 2D Sheet Metal Drawing

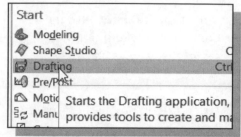

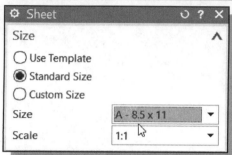

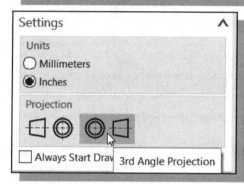

1. Click on the **File** icon in the *Ribbon* toolbar area to display the available options.

2. Select **Drafting** in the *Applications list*.

3. Select **A-8.5X11** from the *sheet size* list.

4. Confirm the **Scale** is set to **1:1**, which is *Full Scale*.

➢ Note that *Sheet 1* is the default drawing sheet name that is displayed in the *Drawing Sheet Name* area.

5. In the settings area, confirm the *Units* is set to **Inches** and the *Projection* type is set to **Third Angle of projection** as shown.

6. **Uncheck** the *Always Start View Creation Command* option.

7. Click **OK** to accept the settings and proceed with the creation of the drawing sheet. Click **Close** to end the create view wizard if it appeared.

Import the Pre-Defined Title Block

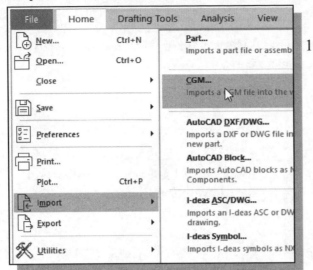

1. Select the **Import CGM** command through the *File* pull-down menu:
   [File] → [Import] → [CGM]

2. Enter or select **Title_A_Master_sheet 1.cgm** in the *Import CGM* window as shown.

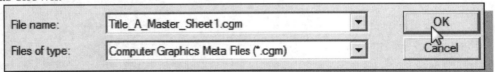

3. Click **OK** to proceed with importing the selected CGM file.

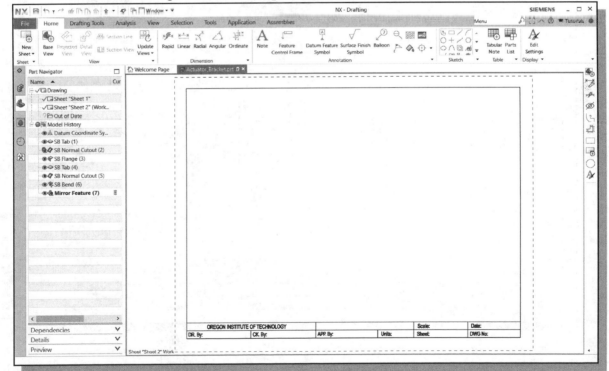

❖ The *Import* and *Export* commands provide a fairly flexible way to reuse *title block* and *borders*.

Create 2D Views

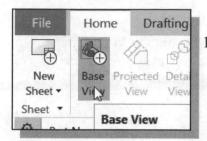

1. Click on the **Base View** icon in the *View toolbar* to create a new base view.

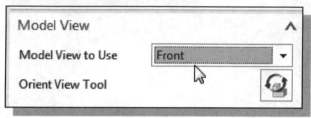

2. In the *Base View* option toolbar, set the view option to **Front** as shown.

3. Place the **Base View**, the *Front view* of the model, in the left side of the drawing sheet as shown.

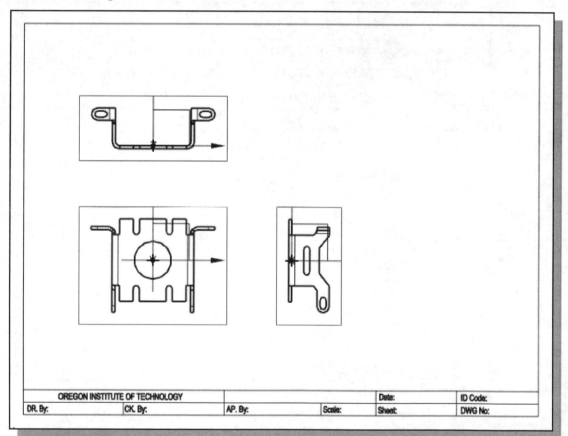

4. On your own, create projected **top** and **side** views as shown in the above figure; click once with the middle-mouse-button to end the create view command.

Create the Associated Flat Pattern View

❖ A sheet metal flat pattern is the shape of the sheet metal part before it is formed. A flat pattern is required to create drawings for manufacturing. The flat pattern shows the shape of the sheet metal part before it is formed showing all the bend lines, bend zones, punch locations, and the shape of the entire part with all bends flattened and bend factors considered.

The *NX* Flat Pattern command calculates the material and layout required to flatten a 3D sheet metal model. Once the associated flat pattern view is created, the part *browser* window displays a *Flat Pattern* item and the flattened state of the model is available in the drafting application. The flat pattern is typically created relative to the sketched base tab feature, but this can be different if necessary.

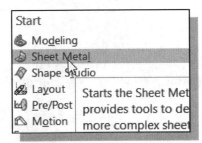

1. Switch back to the *Sheet Metal* application by clicking on the *File* pull-down menu and select **Sheet Metal** as shown.

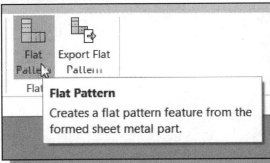

2. In the *Sheet Metal Features* toolbar, select the **Flat Pattern** command by clicking the left-mouse-button on the icon.

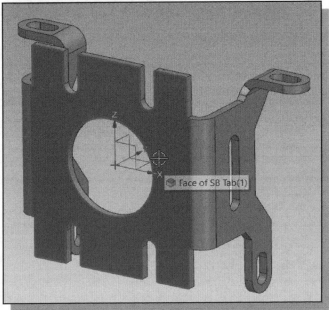

3. Select the **front surface** of the *Tab feature* as the center surface, where the flat pattern will use as the calculation reference.

4. Click **OK** to accept the settings and proceed with the creation of the drawing sheet.

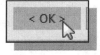

❖   *NX* calculates the material and layout required to flatten the 3D sheet metal model and displays a message indicating a new view has been created.

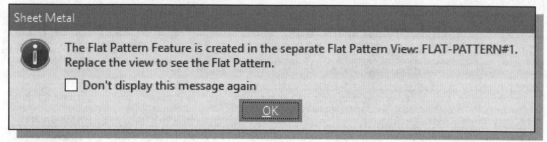

   5.   Click **OK** to close the message window.

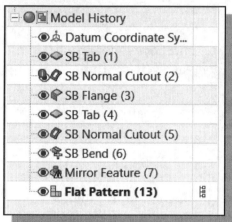

❖   Notice that an additional item is added in the model history tree as shown. The **Flat pattern view** is now available in the *Drafting application.*

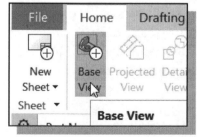

6.   Switch back to the **Drafting** *Application.*

7.   On your own, right-mouse-button on the 2D drawing sheet and choose **Update** to include the new changes we just made.

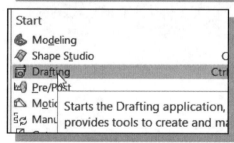

8.   Click on the **Base View** icon in the *View toolbar* to create a new base view.

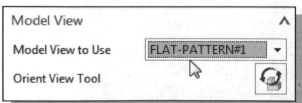

9.   Select the newly created **Flat Pattern view** from the pull-down list as shown.

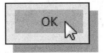

10. If the view appeared in the wrong orientation, use the Orient View tool to adjust the orientation.

11. Select one of the axes to adjust the orientation; set the orientation as shown.

12. Click **OK** to accept the settings.

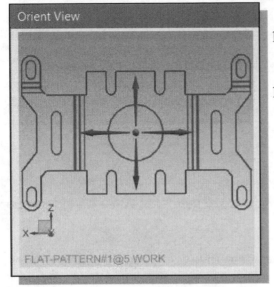

13. On your own, place the **flat pattern view** to the right side of the drawing sheet as shown. Note that *NX* provides information about the bends in the design.

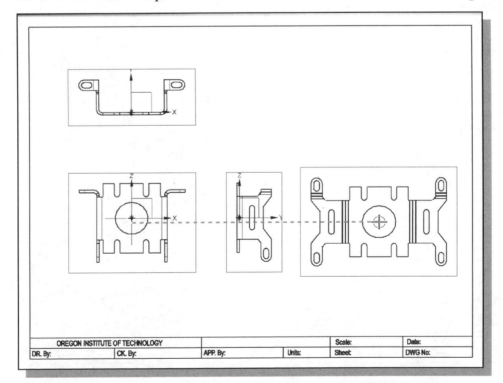

14. Click once with the middle-mouse-button to end the create view command.

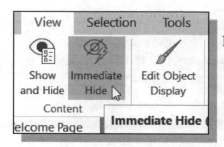

15. Select the **Immediate Hide** command in the *View tab* of the ribbon toolbar as shown.

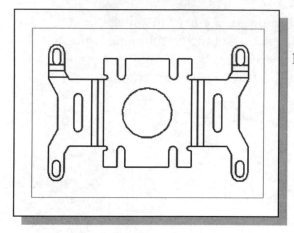

16. On your own, select all of the notes and remove them from the flat pattern view as shown.

17. On your own, add an Isometric and reposition the 2D views roughly as shown in the below figure.

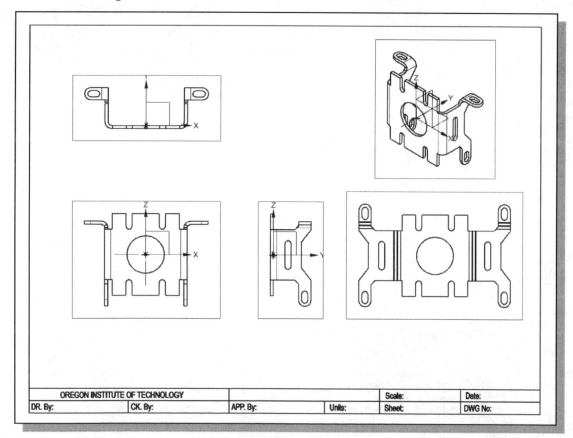

Confirm the Flattened Length

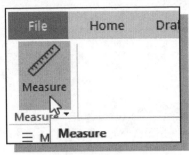

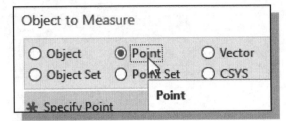

1. In the *Analysis* ribbon tab, select the **Measure** icon as shown.

2. Select **Point** in the *Object to Measure* list.

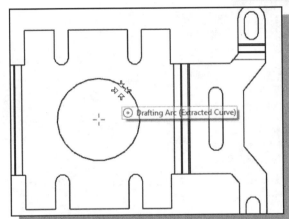

3. In the *Flat Pattern view*, select the large circle; the **center point** is highlighted.

4. Select the mid-point of the **right edge** to measure.

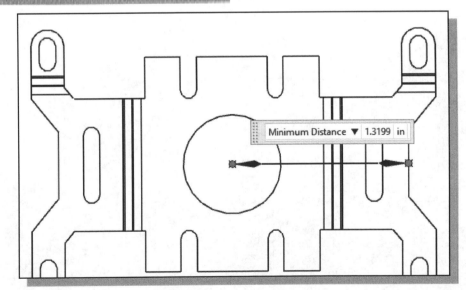

L=A(R+KT)   (Refer to page 5 for the details on this equation.)

$\quad$ = π/2(t +0.41t)

$\quad$ = 0.13843

Flattened Length = W+H+L

$\qquad$ =(1.508/2-t)+(0.615-2t)+0.13843=1.31992

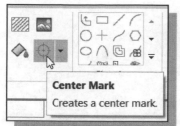

Center Mark
Creates a center mark.

5. Select **Center Mark** in the *Drawing Annotation* panel as shown.

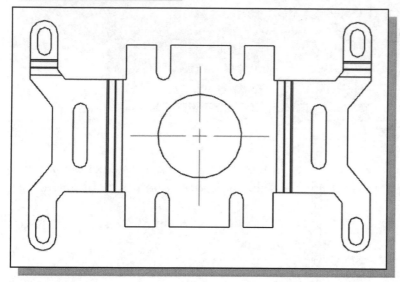

6. Click on the **large circle** of the flat pattern view to place the associated center lines as shown.

7. On your own, create the **center marks** for all of the circular features as shown. Also turn off the display of view borders and datum planes.

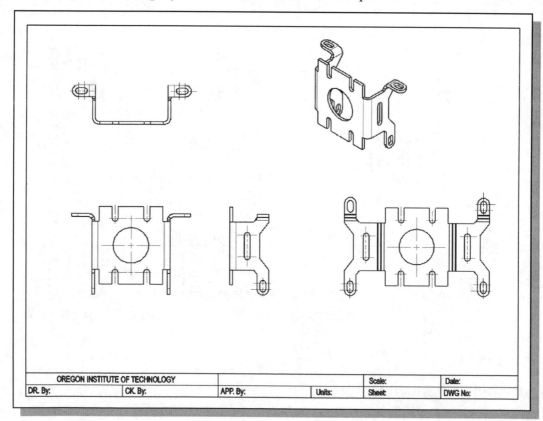

OREGON INSTITUTE OF TECHNOLOGY				Scale:	Date:
DR. By:	CK. By:	APP. By:	Units:	Sheet:	DWG No:

Set up Dimensions Preferences

1. Select **Drafting** in the *Preferences* menu as shown.

2. Turn off the **Line Between Arrows** option under the *Dimension →Dimension Line* as shown.

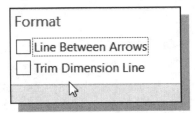

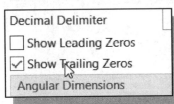

3. Reset the **Leading Zeros** and **Trailing zeros** options under the *Dimension → Text → Units* as shown.

4. On your own, set up the **Extension Line settings** as shown in the below figure.

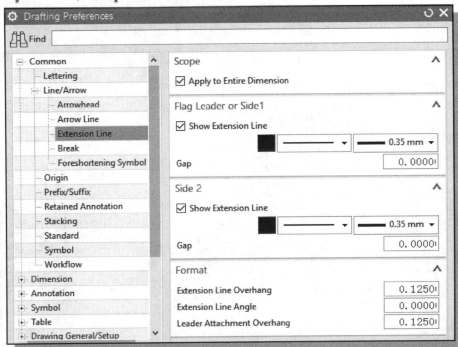

❖ On your own, complete the multiview drawing by adding the necessary dimensions and text as shown on the next page.

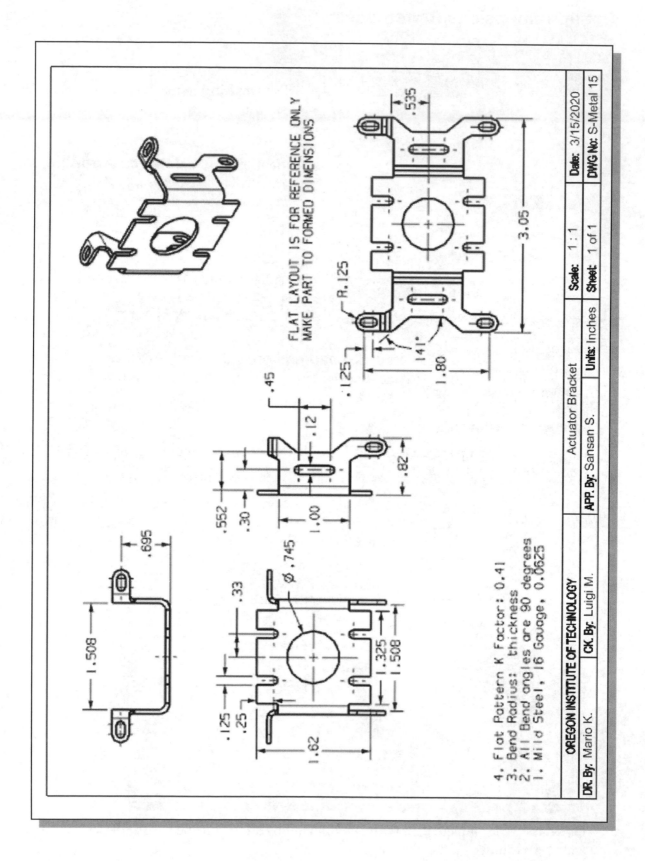

FLAT LAYOUT IS FOR REFERENCE ONLY
MAKE PART TO FORMED DIMENSIONS

.535

R.125

3.05

.125

141°

1.80

.45

.12

.82

.552

.30

1.00

.695

.33

Ø.745

.325

1.508

1.508

.125

.25

1.62

4. Flat Pattern K Factor: 0.41
3. Bend Radius: thickness
2. All Bend angles are 90 degrees
1. Mild Steel, 16 Gauge, 0.0625

OREGON INSTITUTE OF TECHNOLOGY

		Actuator Bracket		Scale: 1 : 1	Date: 3/15/2020
DR. By: Mario K.	CK. By: Luigi M.	APP. By: Sansan S.	Units: Inches	Sheet: 1 of 1	DWG No: S-Metal 15

Review Questions: (Time: 30 minutes)

1. List and describe two of the more commonly used sheet metal processes.

2. Which command do we issue to display the flat pattern of a 3D sheet metal design?

3. What is the **k-factor** used in sheet metal processes?

4. How is the **k-factor** used to calculate the flattened length in sheet metal flat patterns?

5. List and describe two of the factors that can change the k-factor value.

6. List and describe two of the settings available in the **Sheet Metal Preferences** in *NX*.

7. List and describe the differences between the **Flange** and **Tab** commands.

8. How do you create a *flat pattern* view of a sheet metal design?

9. In the *NX* sheet metal module, can the feature-duplicating commands, such as **Mirror**, be used on sheet metal features?

10. In the *Drawing View*, which option allows us to display the *flat pattern* of the sheet metal design?

11. Is the *flat pattern* item always available in the sheet metal part *browser* window?

12. Can we create a sheet metal feature that is at a 30 degree angle to the base tab feature?

Exercises: (Time: 125 minutes)

(Create the 3D model and the associated 2D drawings. All dimensions are in inches.)

1. Cooling Fan Cover

1. No. 16 Gauge (0.0625) Mild Steel
2. Standard Straight Relief
3. Flat Layout K-Factor: 0.44
4. Bend Radius: Thickness
5. All Bend Angles are 90 degrees

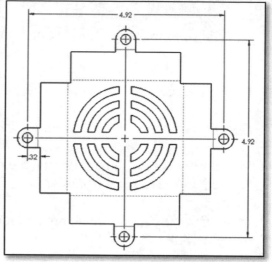

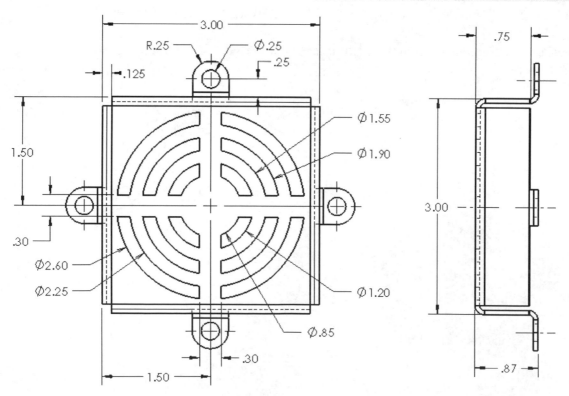

Sheet Metal Base (No. 11 Gauge Mild Steel, K-Factor 0.40.)

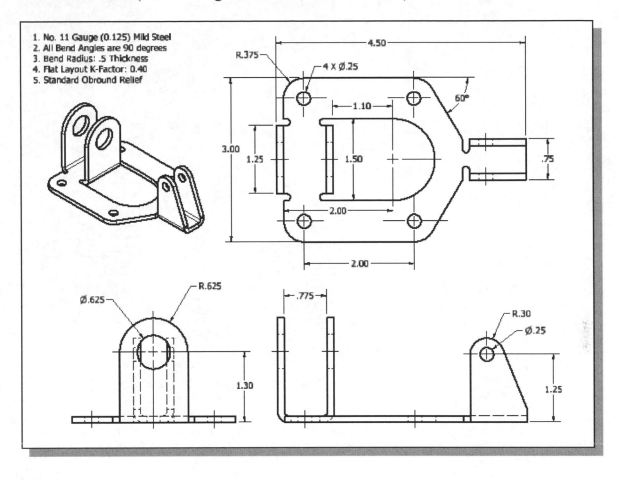

1. No. 11 Gauge (0.125) Mild Steel
2. All Bend Angles are 90 degrees
3. Bend Radius: .5 Thickness
4. Flat Layout K-Factor: 0.40
5. Standard Obround Relief

Notes:

Chapter 14
Assembly Modeling – Putting It All Together

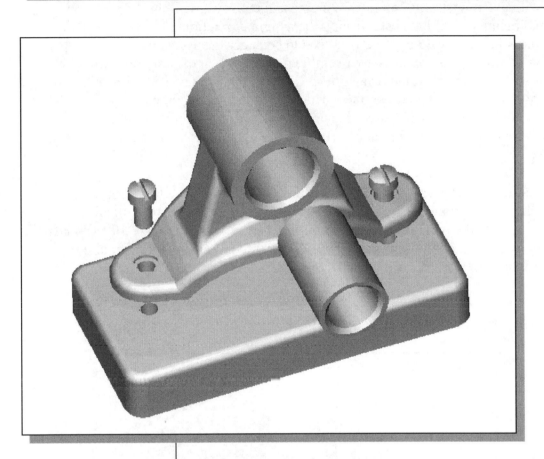

Learning Objectives

- **Understand the Assembly Modeling Methodology**
- **Add Existing Parts in the Assembly Modeler Mode**
- **Understand and Utilize Assembly Constraints**
- **Understand the Degree of Freedom concept**
- **Create Exploded Assemblies**
- **Create an Assembly Drawing from the Solid Model**

Introduction

In the previous lessons, we have gone over the fundamentals of creating basic parts and drawings. In this lesson, we will examine the assembly modeling functionality of *NX*. We will start with a demonstration on how to create and modify assembly models. The main task in creating an assembly is establishing the assembly relationships between parts. To assemble parts into an assembly, we will need to consider the assembly relationships between parts. It is a good practice to assemble parts based on the way they would be assembled in the actual manufacturing process. We should also consider breaking down the assembly into smaller subassemblies, which helps the management of parts. In *NX*, a subassembly is treated the same way as a single part during assembling. Many parallels exist between assembly modeling and part modeling in parametric modeling software such as *NX*.

NX provides full associative functionality in all design modules, including assemblies. When we change a part model, *NX* will automatically reflect the changes in all assemblies that use the part. We can also modify a part in an assembly. **Full associative functionality** is the main feature of parametric solid modeling software that allows us to increase productivity by reducing design cycle time.

The Shaft Support Assembly

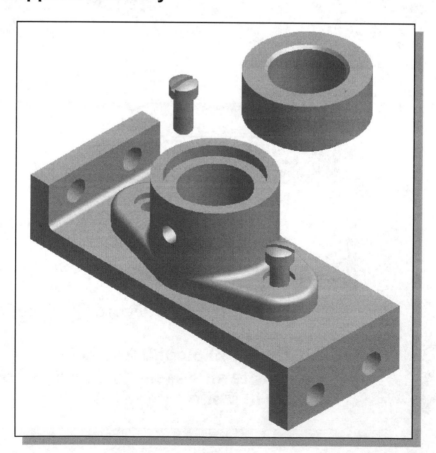

Assembly Modeling Methodology

The *NX* assembly modeler provides tools and functions that allow us to create 3D parametric assembly models. An assembly model is a 3D model with any combination of multiple part models. *Parametric assembly constraints* can be used to control relationships between parts in an assembly model.

NX can work with any of the following assembly modeling methodologies:

The Bottom Up approach
> The first step in the *bottom up* assembly modeling approach is to create the individual parts. The parts are then pulled together into an assembly. This approach is typically used for smaller projects with very few team members.

The Top Down approach
> The first step in the *top down* assembly modeling approach is to create the assembly model of the project. Initially, individual parts are represented by names or symbolically. The details of the individual parts are added as the project gets further along. This approach is typically used for larger projects or during the conceptual design stage. Members of the project team can then concentrate on the particular section of the project to which he/she is assigned.

The Middle Out approach
> The *middle out* assembly modeling approach is a mixture of the bottom-up and top-down methods. This type of assembly model is usually constructed with most of the parts already created and additional parts are designed and created using the assembly for construction information. Some requirements are known and some standard components are used, but new designs must also be produced to meet specific objectives. This combined strategy is a very flexible approach for creating assembly models.

The different assembly modeling approaches described above can be used as guidelines to manage design projects. Keep in mind that we can start modeling our assembly using one approach and then switch to a different approach without any problems.

In this lesson, the *bottom up* assembly modeling approach is illustrated. All of the parts (components) required to form the assembly are created first. *NX's* assembly modeling tools allow us to create complex assemblies by using components that are created in part files or are placed in assembly files. A component can be a subassembly or a single part, where features and parts can be modified at any time. The sketches and profiles used to build part features can be fully or partially constrained. Partially constrained features may be adaptive, which means the size or shape of the associated parts are adjusted in an assembly when the parts are constrained to other parts. The basic concept and procedure of using the adaptive assembly approach is demonstrated in the tutorial.

Additional Parts

- Four parts are required for the assembly: (1) **Collar**, (2) **Bearing**, (3) **Base-Plate** and (4) **Cap-Screw**. Create the four parts as shown below, then save the models as separate part files: *Collar*, *Bearing*, *Base-Plate*, and *Cap-Screw*. (Close all part files or exit *NX* after you have created the parts.)

(1) **Collar**

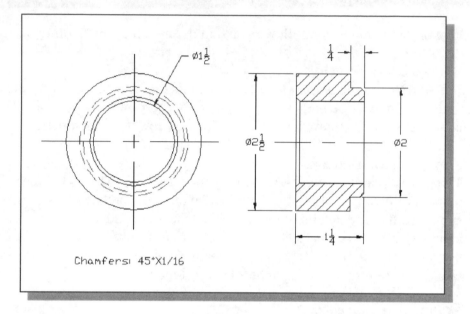

(2) **Bearing** (Rounds & Fillets: *R 1/8*. Construct the part with the datum origin aligned to the bottom center.)

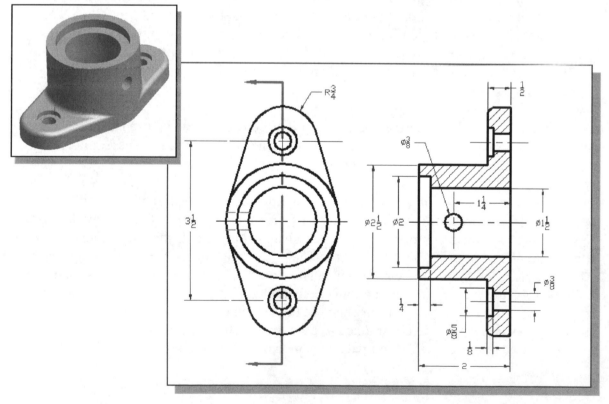

(3) **Base-Plate** (Construct the part with the front view aligned to the XZ plane.)

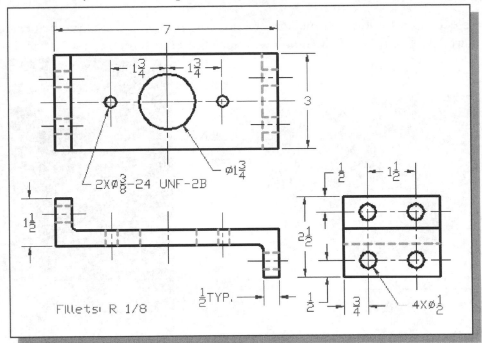

(4) *Cap-Screw*

> We will omit the threads in this model. Threads contain complex three-dimensional curves and surfaces; it will slow down the display considerably. Hint: create a revolved feature using the profile shown below.

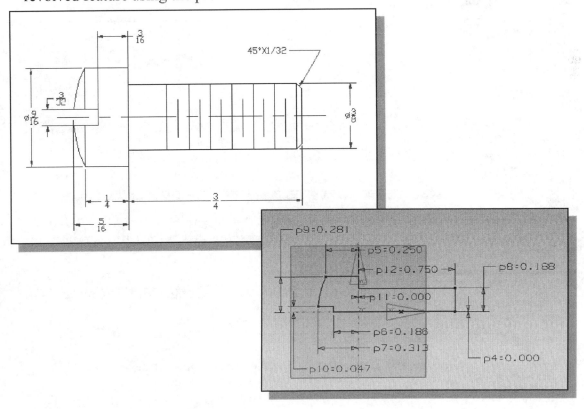

Start NX

1. Select the *NX* option on the *Start* menu or select the *NX* icon on the desktop to start *NX*. The *NX* main window will appear on the screen.

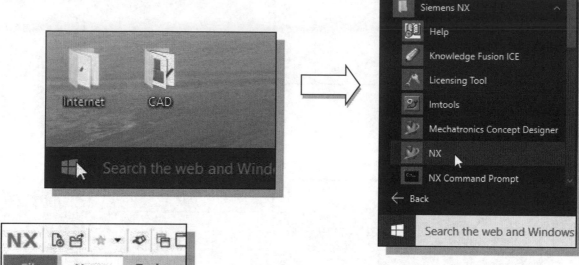

2. Select the **New** icon with a single click of the left-mouse-button (MB1) in the *Standard toolbar area*.

3. Select the **Inches** units as shown in the below figure.

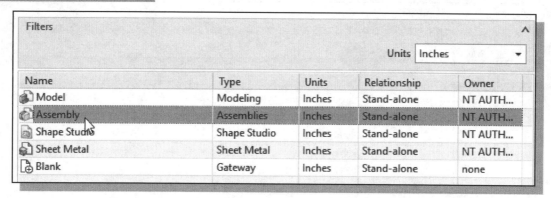

4. Select **Assembly** in the *Template list*. Note that the *Assembly template* will allow us to switch directly into the *Assemblies* task with settings related to creating assembly models.

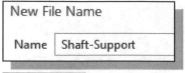

5. In the *New File Name* area, enter **Shaft-Support** as the *File Name*.

6. Click **OK** to proceed with the New File command.

Loading and Placing the First Component

- The first component placed in an assembly should be a fundamental part or subassembly. The first component in an assembly file sets the orientation of all subsequent parts and subassemblies. The origin of the first component is aligned to the origin of the assembly coordinates and the part is grounded (all degrees of freedom are removed). The rest of the assembly is built on the first component, the *base component*. In most cases, this *base component* should be one that is **not likely to be removed** and **preferably a non-moving part** in the design. Note that there is no distinction in an assembly between components; the first component we place is usually considered the *base component* because it is usually a fundamental component to which others are constrained. For our project, we will use the ***Base-Plate*** as the base component in the assembly.

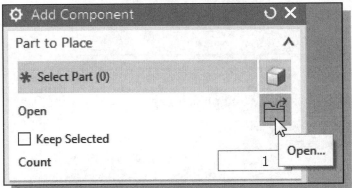

1. The **add component** command is automatically activated. In the **add component** window, click **Open** to load an existing part.

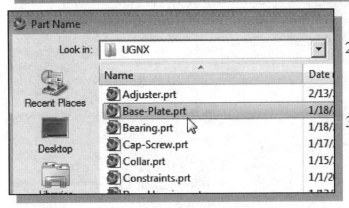

2. Select the ***Base-Plate*** (part file: ***Base-Plate.prt***) in the list window.

3. Click **OK** to accept the selection.

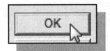

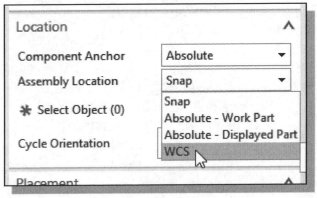

4. Set the *Component Anchor* to **Absolute** and the *Assembly location* to use the **World Coordinate System (WCS)** as shown.

5. Set the placement option to the **Absolute Origin** (set **X, Y** and **Z** to **0.00**) of the *World Coordinate System* as shown.

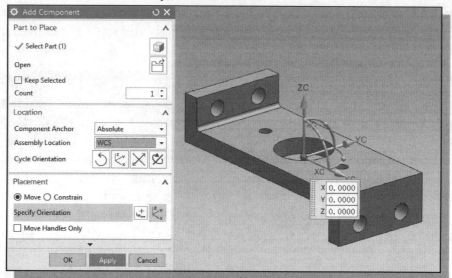

6. Click **Apply** to accept the settings and place the part.

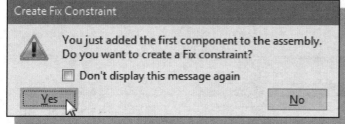
7. Since this is the first component placed in the assembly, by default, NX will add a fixed assembly constraint to it. Click **Yes** to create the constraint.

Place the Second Component

➢   We will retrieve the *Bearing* part as the second component of the assembly model.

1. In the **add component** window, click **Open** to load an existing part.

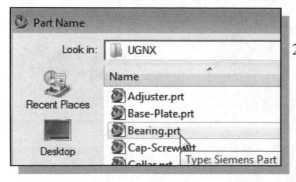
2. Select the **Bearing** design (part file: **Bearing.prt**) in the list window. And click on the **OK** button to retrieve the model.

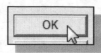

- Note that in NX there are two methods to place the components in the Assembly: (1) Move – position the component in the assembly without adding any assembly constraints. (2) Constrain – add assembly constraints to set the relations in between components. Note that we can add/modify assembly constraints at any time, not just during the load components stage.

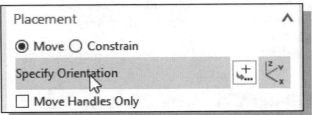

3. Set the placement method to **Move** and select **Specify Orientation** as shown in the figure.

4. On your own, move the part in front of the Base-Plate component by dragging the origin of the movement control.

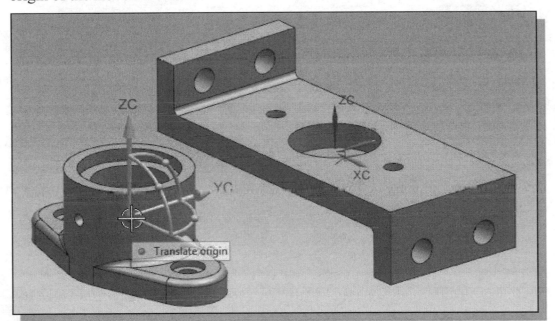

5. Click **OK** to accept the selection and load the *Bearing* file.

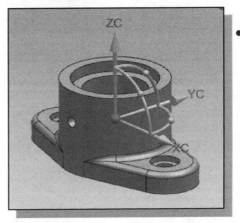

- The component movement control can be used to move components in the 3D environment; use the left mouse-button to *drag and drop*, click on the arrows or grip-points. The straight arrows indicate translation control in the X, Y and Z directions; the arcs are for rotations about the X, Y and Z axes.

Degrees of Freedom

- Each component in an assembly has six **degrees of freedom (DOF)**, or ways in which rigid 3D bodies can move: movement along the X, Y, and Z axes (translational freedom), plus rotation around the X, Y, and Z axes (rotational freedom). *Linear DOFs* allow the part to move in the direction of the specified vector. *Rotational DOFs* allow the part to turn about the specified axis.

- In NX, the **Show Degrees of Freedom** command can be used to quickly examine the remaining DOF of any components in the assembly.

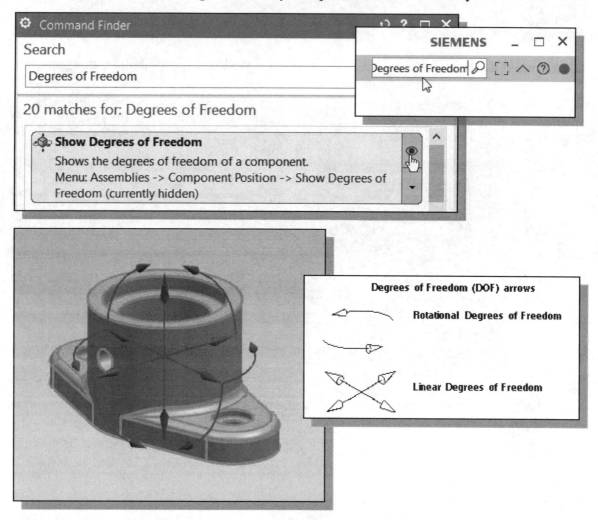

- In parametric modeling, the degrees-of-freedom symbol generally shows the remaining degrees of freedom (both translational and rotational) for one or more components of the active assembly. When a component is fully constrained in an assembly, the component cannot move in any direction. The position of the component is fixed relative to other assembly components. All of its degrees of freedom are removed. When we place an assembly constraint between two selected components, they are positioned relative to one another. Movement is still possible in the unconstrained directions.

Assembly Constraints

To assemble components into an assembly, we need to establish the assembly relationships between components. It is a good practice to assemble components the way they would be assembled in the actual manufacturing process. **Assembly constraints** create a parent/child relationship that allows us to capture the design intent of the assembly. Because the component that we are placing actually becomes a child to the already assembled components, we must use caution when choosing constraint types and references to make sure they reflect the intent.

It is usually a good idea to fully constrain components so that their behavior is predictable as changes are made to the assembly. Although leaving some degrees of freedom open can sometimes help retain design flexibility. As a general rule, we should use only enough constraints to ensure predictable assembly behavior and avoid unnecessary complexity.

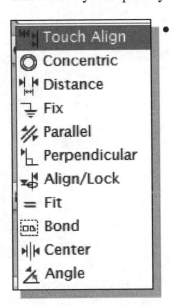

- In *NX*, eleven different assembly constraints are available. Assembly models are created by applying proper *assembly constraints* to the individual components. The constraints are used to restrict the movement between parts. Constraints eliminate rigid body degrees of freedom (**DOF**). A 3D part has *six degrees of freedom* since the part can rotate and translate relative to the three coordinate axes. Each time we add a constraint between two components, one or more DOF is eliminated. The movement of a fully constrained part is restricted in all directions. Six basic types of assembly constraints are available in *NX*: Align, Touch, Angle, Parallel, Perpendicular and Align/Lock. Each type of constraint removes different combinations of rigid body degrees of freedom. Note that it is possible to apply different constraints and achieve the same results.

> **Touch (Mate)** – This constraint positions components face-to-face, or adjacent to one another, with faces flush. Removes one degree of linear translation and two degrees of angular rotation between planar surfaces. Selected surfaces point in opposite directions. Touch constraint positions selected faces normal to one another, with faces coincident.

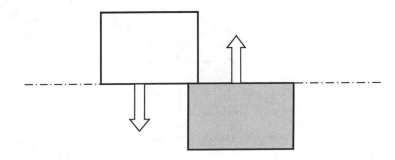

➢ **Touch/Align** – This constraint makes two planes coplanar with their faces aligned in the same direction. Selected surfaces point in the same direction. *Align* constraint aligns components adjacent to one another with faces aligned. For axisymmetric objects, *Align* can be used to align two circles or cylindrical surfaces, including their center axes and planes. Selected circular surfaces can become co-axial or tangent to each other.

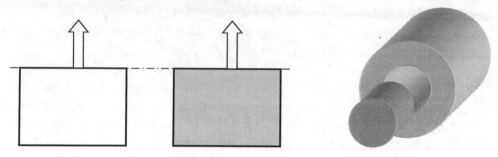

➢ **Angle** – Creates an angular assembly constraint between parts, subassemblies, or assemblies. Selected surfaces point in the direction specified by the angle.

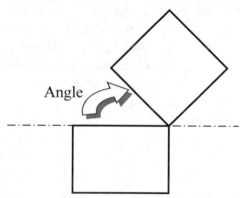

➢ **Parallel**– Constrain the surfaces or direction vectors of two selected objects as parallel to each other.

➢ **Perpendicular**– Constrain the surfaces or direction vectors of two objects as perpendicular to each other.

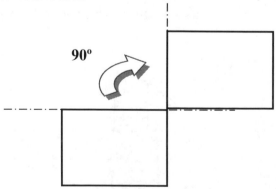

➢ **Concentric** – Position the center of one component to the center of the other center of component.

➢ **Center** – Position the center of one component everywhere along the center of the other, or center one or two components between a pair of components. The components typically involve one circular object constraining to flat faces. For example, putting a pin in a slot, use the Center 2-1 option, pick two sides to the slot, then 1 center line of the pin. For a flat sided pin, use the Center 2-2, select each side of the slot, then each side of the pin.

➢ **Distance**– Constraint positions components face-to-face, and can be offset by a specified **distance**. Specifies the minimum 3D distance between two objects. You can control which side of the surface the solution should be by using positive or negative values. Distance constraint is similar to the Mate constraint, which also positions selected faces normal to one another, with faces coincident.

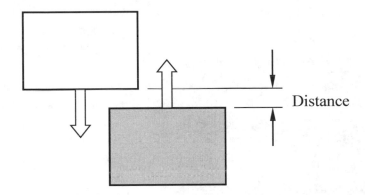

➢ **Fit** – This constraint brings two cylindrical faces together. Note that they should have the same radius. This constraint is useful for locating pins or bolts in holes. If the radii later become non-equal, the constraint is invalid.

➢ **Align/Lock** – Aligns position and fully constrains rotationally-symmetric components, selected faces, planes, cylinders, spheres, and cones to contact at the point of tangency. A component is constrained using an axis, and simultaneously prevents rotation about that axis. For example, an Align/Lock constraint can be used to position and lock a bolt in a hole. An Align/Lock constraint leaves only one degree of translational freedom open.

➢ **Fix** – A *Fix* constraint fixes a component at its current position in the 3D space; only one component is selected for this constraint. Generally, we should apply a Fix constraint on at least one component in an assembly, preferably the first component.

➢ **Bond** – A *Bond* constraint is between two or more components, and what it does is 'fix' the relationship between the referenced components but not relative to the space in which these components are found.

Apply the First Assembly Constraint

- We are now ready to assemble the components together. We will start by placing assembly constraints on the *Bearing* and the *Base-Plate*.

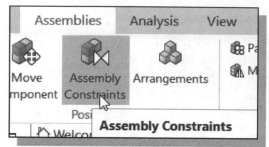

1. In the *Component Position* toolbar, pick the **Assembly Constraints** command as shown.

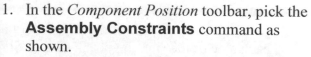

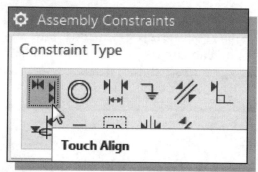

2. Choose the **Touch Align** constraint.

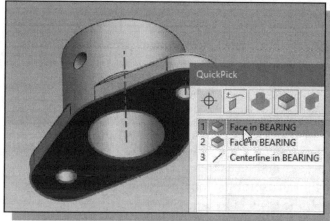

3. On your own, use the **middle-mouse-button (MB2)** to dynamically rotate the model to view the bottom of the *Bearing* part, as shown in the figure.

4. Click on the **bottom face** of the *Bearing* part as the first selection to apply the constraint to, as shown in the above figure.

5. Select the **top horizontal surface** of the base part as the second selection for the Touch-Mate alignment command. Note the Touch-Mate constraint requires the selection of opposite direction of surface normal.

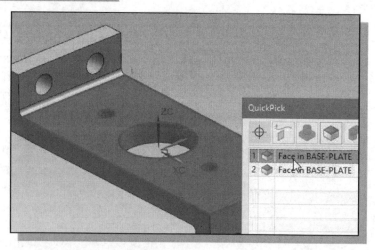

> **Do not** click on the **OK** button yet; we will continue to apply additional constraints to place the Bearing part.

Apply another Align Constraint

❖ In NX, the *Touch Align* constraint can be viewed as the general all-purpose constraining tool. Besides aligning flat surfaces, the Align constraint can also be used to align axes of cylindrical features.

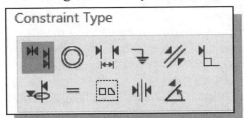

1. Confirm the constraint type is set to **Touch Align** as shown.

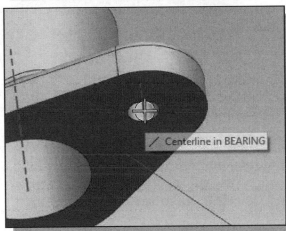

2. Select the **center axis** of the right counter bore hole of the *Bearing* part as shown. (Hint: Use the *Dynamic Viewing* option to assist the selection.)

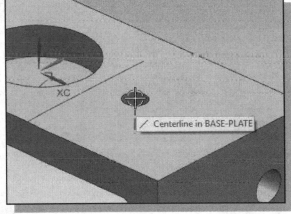

3. Select the **center axis** of the right small hole on the *Base-Plate* part as shown.

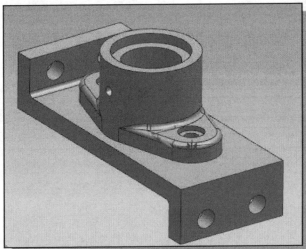

4. Click **Apply** to accept the established constraints.

5. Click **Cancel** to exit the *Assembly Constraints* command.

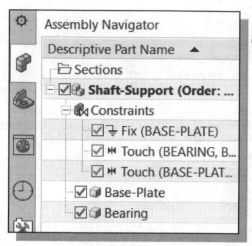

6. Inside the *Assembly navigator*, click on the plus sign in front of the **Constraints** item to expand the list, and note all of the applied constraints are displayed.

➢ The *Assembly navigator* can be used to manage the existing constraints.

➢ The *Bearing* part may appear to be placed and constrained correctly. But it can still rotate about the aligned vertical axis.

Constrained Move

❖ To see how well a component is constrained, we can also perform a constrained move. A constrained move is done by using the **Move Component** command, which allows us to drag the component in the graphics window with the left-mouse-button. A constrained move will honor previously applied assembly constraints. That is, the selected component and parts constrained to the component move together in their constrained positions.

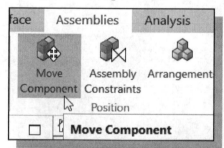

1. In the *Assembly Panel*, select the **Move Component** command by left-mouse-clicking the icon.

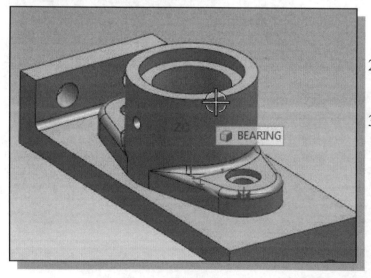

2. Inside the *graphics window*, select the *Bearing* part.

3. Click once with the **middle-mouse-button (MB2)** to accept the selection.

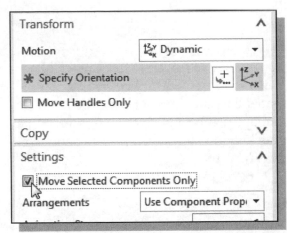

4. Confirm the *Motion* option to **Dynamic**, as shown in the figure.

5. Expand the additional options in the Transform dialog box.

6. Switch **ON** the *Move Selected Components Only* option as shown.

7. Move the cursor near the horizontal ring, the **Rotate about ZC-axis ring,** in the displayed coordinate system as shown. Press and hold down the left-mouse-button and drag the *Bearing* part.

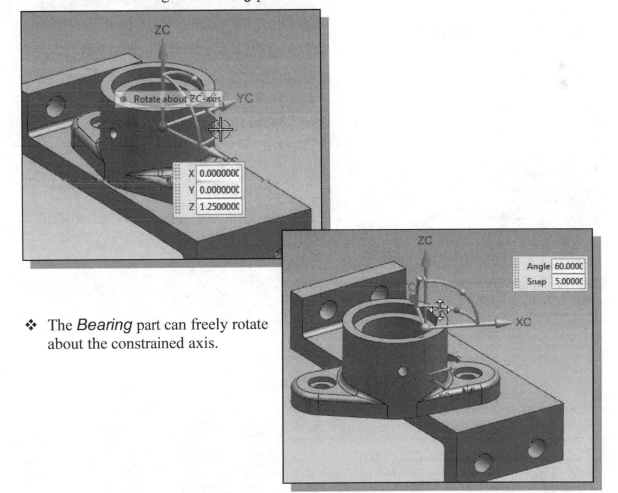

❖ The *Bearing* part can freely rotate about the constrained axis.

8. On your own, dynamically rotate and view the alignment of the *Bearing* part.

9. On your own, rotate the *Bearing* part roughly to the position as shown in the above figure and click **OK** to exit the **Move Component** command.

Show Degrees of Freedom

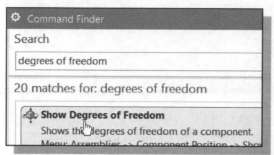

1. In the *command finder*, search and activate the **Show Degrees of Freedom** command as shown. (Note that the command can be added to the ribbon toolbar by clicking on the eyeball icon as shown on page 14-10.)

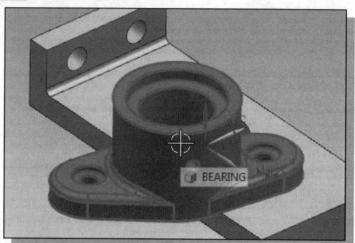

2. Select the **Bearing** part in the graphics window.

3. Click once with the **middle-mouse-button (MB2)** to accept the selection.

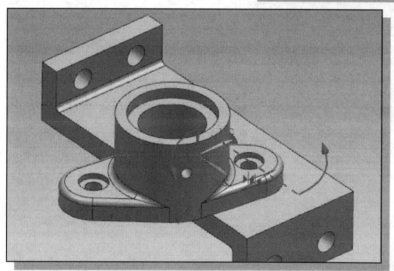

➢ Note that two rotational arrows are displayed, indicating the component can still pivot about the center axis of the right hole.

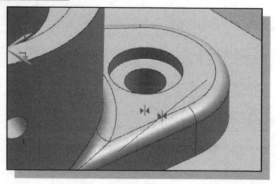

➢ Also note the applied assembly constraint symbols indicating the type and locations of the constraints.

Apply another Assembly Constraint

For the *Bearing* part, we will apply another Align constraint to constrain the left hole and eliminate the last rotational DOF.

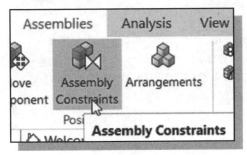

1. In the *Assembly Panel*, select the **Assembly Constraints** command by left-mouse-clicking the icon.

2. In the *constraint Type* option panel, set the constraint type to **Touch Align** constraint as shown.

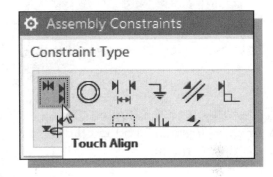

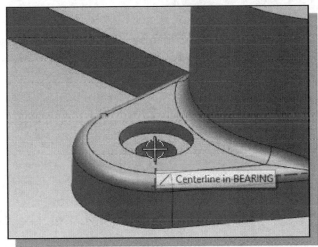

3. Zoom in and select the **center axis** of the left small hole of the *Bearing* part as the first part for the Align constraint.

4. Select the **center line** of the drill hole on the *Base-Plate* part as the second item for the Align constraint.

5. Click on the **OK** button to apply the constraint and exit the *Assembly constraints* command.

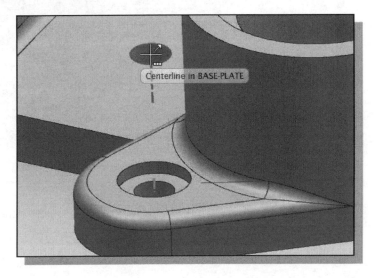

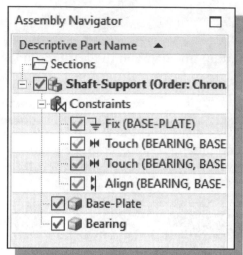

➢ Notice the new constraint is also added in the *Assembly Navigator*.

6. Inside the graphics window, **right-mouse-click** on the *Bearing* part to bring up the option menu and select **Update Display** to reset the display of the model.

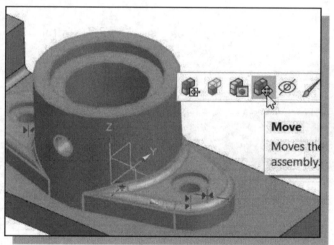

7. Inside the graphics window, **left-mouse-click** on the *Bearing* part to bring up the option menu.

8. Select the **Move** command as shown.

9. On your own, try to perform a constrained move.

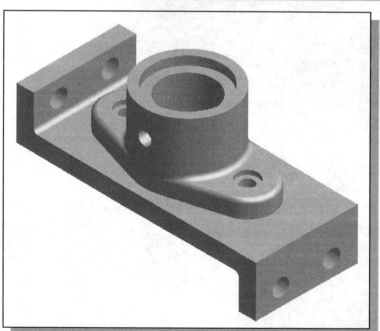

❖ Note the *Bearing part* is now **fully constrained,** and therefore a *constrained move* cannot be done.

Placing the Third Component

> ➢ We will retrieve the *Collar* part as the third component of the assembly model.

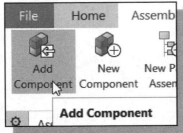

1. In the *Assembly Panel* select the **Add Component** command by left-mouse-clicking the icon.

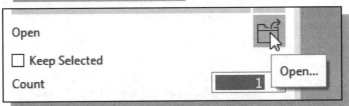

2. In the **add component** window, click **Open** to load an existing part.

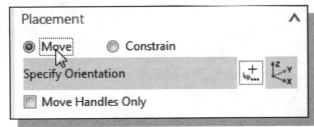

3. Select the ***Collar*** design (part file: ***Collar.prt***) in the list window. And click on the **OK** button to retrieve the model.

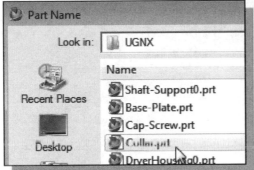

4. Set the positioning method to ***Move*** as shown in the figure.

5. On your own, reposition the Collar model so that it does not overlap with the current assembly mode.

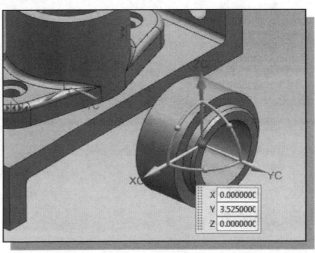

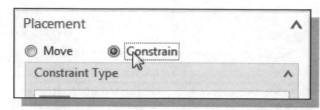

6. Set the placement method to *Constrain* as shown in the figure.

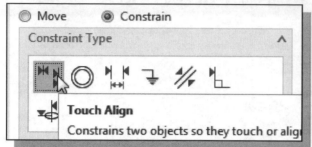

➢ Note the assembly constraints become available.

7. In the *Mating Type* option panel, confirm the constraint type is set to **Touch Align** constraint as shown.

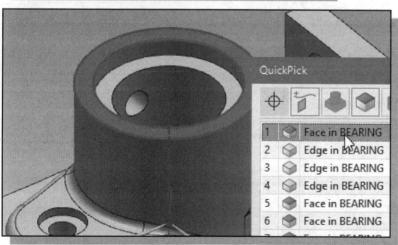

8. Click on the **outer circular face** at the bottom of the *Collar* part as the first selection to apply the mate constraint, as shown in the figure.

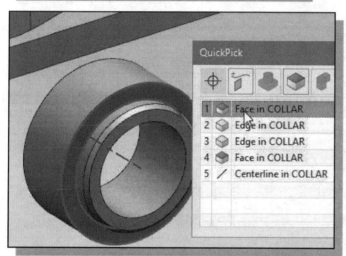

9. Select the top circular surface of the bearing part as the second selection for the **Touch** alignment command. Note the *Touch* constraint requires the selection of opposite direction of surface normals.

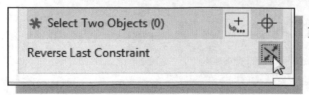

10. If the collar is upside-down, click on the **reverse direction** icon once to flip the alignment of the surfaces.

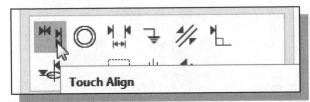

11. Confirm the positioning method is still set to *Touch Align* as shown in the figure.

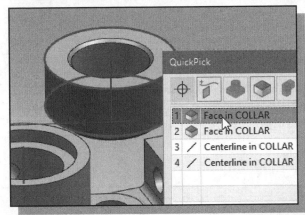

12. Select the **outer cylindrical surface** of the *Collar* part as shown.

13. Select the **outer cylindrical surface** of the *Bearing* part as shown. If necessary, select the corresponding surface in the QuickPick window.

14. Click on the **OK** button to exit the assembly constraints command.

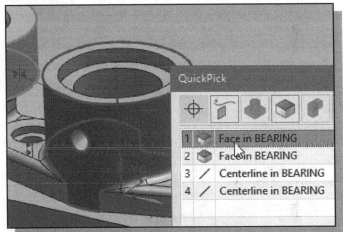

➢ Notice the resulting assembly is not quite what we have intended.

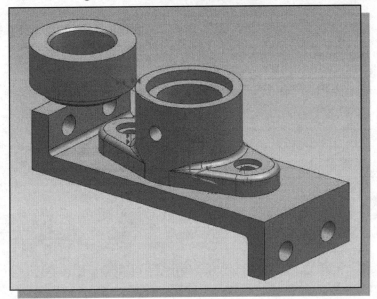

Redefine the Applied Constraints

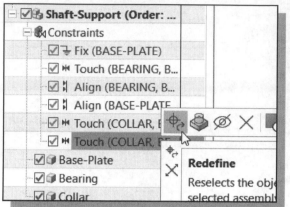

1. Inside the *Assembly Navigator* window, right-mouse-click on the fourth constraint to bring up the option menu and select **Redefine** as shown.

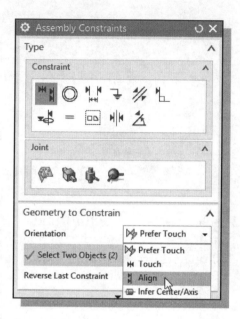

➢ Note the Assembly Constraints command is activated and the two constrained surfaces are highlighted.

2. Click on **Align** to adjust the alignment of the components.

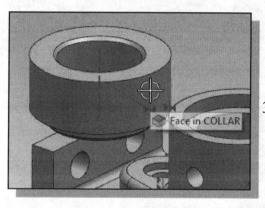

3. On your own, hold down the shift key and select the two highlighted surfaces to de-select them.

4. Select the **Centerlines** of the *Collar* and *Bearing* components, which will assure the proper alignment of the two pieces.

5. Click **OK** to exit the *Assembly Constraints* command.

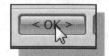

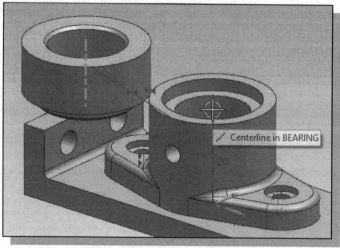

➢ Note that the *Collar* part is now correctly aligned to the vertical axis of the Bearing component. Can the Collar part still move in the assembly?

Apply Assembly Constraints on Datum Planes

- The *datum planes* of parts can also be used to constrain the assembly model. We will first switch on the datum planes in the assembly model.

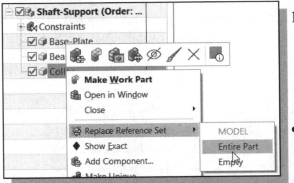

1. Inside the *Assembly Navigator* window, right-mouse-click on the *Collar part* to bring up the option menu and select **Replace Reference Set → Entire Part** as shown.

- Note the datum planes of the **Collar** part now are visible in the assembly.

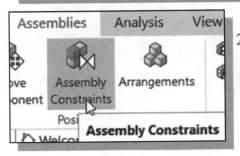

2. In the *Assembly Panel*, select the Assembly Constraints command by left-mouse-clicking the icon.

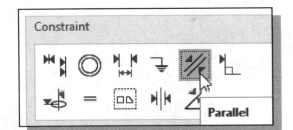

3. Set the constraint type to **Parallel** as shown.

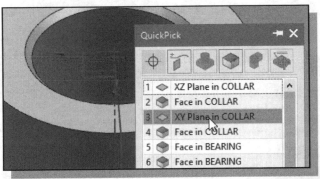

4. Select the **XY Plane** of the *Collar* part as shown.

5. Select the front face of the Base Plate part and note the selected datum plane is now parallel to the front face of the Base plate part.

6. Click **OK** to accept the settings.

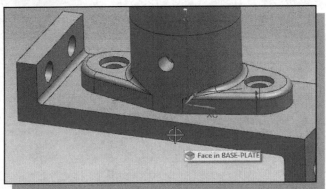

Assemble the First Cap-Screw

- Two *Cap-Screw* parts are needed to complete the assembly model.

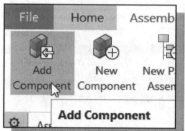

1. In the *Assembly Panel* (the toolbar that is located to the bottom of the graphics window), select the **Add Component** command by left-mouse-clicking the icon.

2. In the **add component** window, click **Open** to load an existing part.

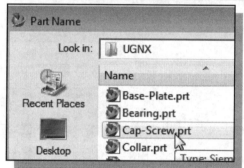

3. Select the ***Cap-Screw*** design (part file: ***Cap-Screw.prt***) in the list window. And click on the **OK** button to retrieve the model.

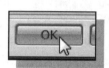

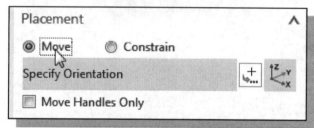

4. Confirm the *positioning* method is set to ***Move*** as shown in the figure.

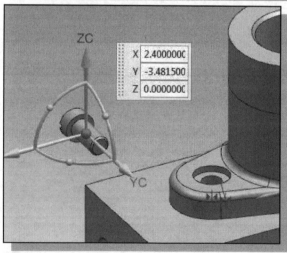

5. On your own, reposition the Collar model so that it does not overlap with the current assembly mode.

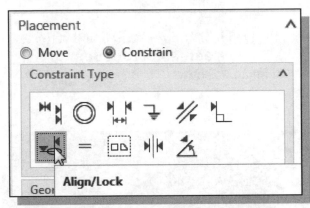

6. Set the placement method to *Constrain* as shown in the figure.

7. In the *Type* option panel, choose **Align/Lock** constraint as shown.

- Note that *NX* expects us to select two centerlines of two components.

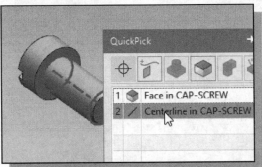

8. Select the **center axis** of the shaft of the *Cap-Screw* part as shown.

- Select **Centerline** in Cap-Screw in the QuickPick dialog box if necessary.

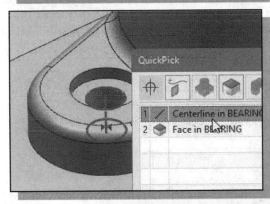

9. Select the **center axis** of the counter bore surface on the *Bearing* part as shown. (Click the flip direction button if the screw is upside down.)

10. Click on the **OK** button to apply the constraint.

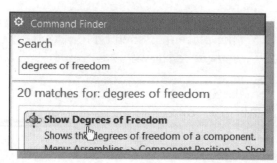

11. In the *command finder*, search and activate the **Show Degrees of Freedom** command as shown.

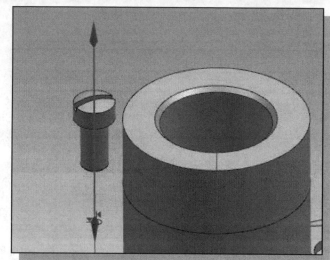

12. Select the **Cap-Screw** component.

13. Click once with the **middle-mouse-button (MB2)** to accept the selection

➢ Note the **Cap-Screw** component has one DOF open, which is along the aligned axis direction.

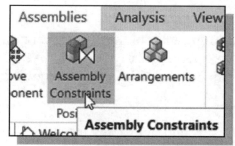

14. In the *Assembly Panel*, select the **Assembly Constraints** command by left-mouse-clicking the icon.

15. On your own, add a **Prefer Touch** constraint to properly assemble the *Cap-Screw* part.

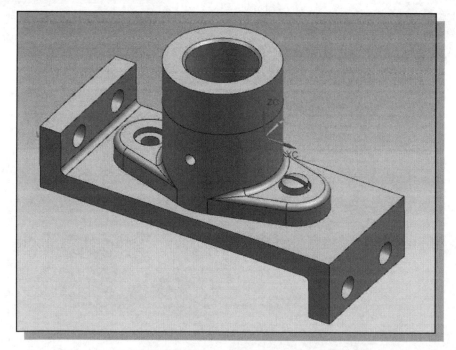

Placing the Second Cap-Screw Part

- For the **Shaft-Support** assembly, we need two copies of the **Cap-screw** part. The second cap-screw will be a separate copy of the original part, each with its own DOF and assembly constraints.

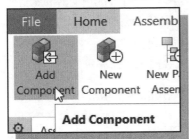

1. In the *Assembly Panel*, select the **Add Component** command by left-mouse-clicking the icon.

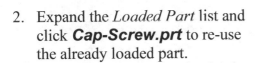

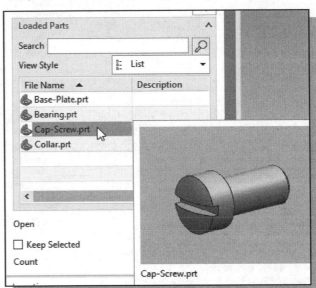

2. Expand the *Loaded Part* list and click **Cap-Screw.prt** to re-use the already loaded part.

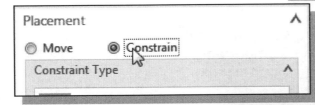

3. Confirm the positioning method is set to **Constrain** as shown in the figure.

> On your own, use the proper constraints and assemble the second *Cap-Screw* in place as shown in the figure.

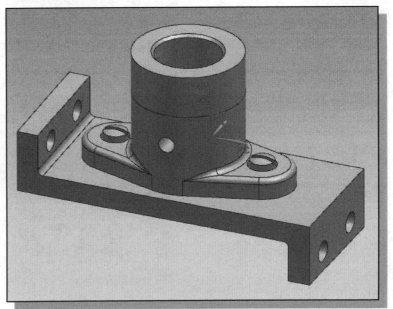

Exploded View of the Assembly

Exploded assemblies are often used in design presentations, catalogs, sales literature, and in the shop to show all of the parts of an assembly and how they fit together. In *NX*, an exploded assembly can be created by two methods: (1) using the **Auto-Explode** command, which provides a very simple and quick method to create exploded view; (2) using the **Edit Explosion** to arrange the individual components. For our example, we will create an exploded assembly by using the Auto-Explode command.

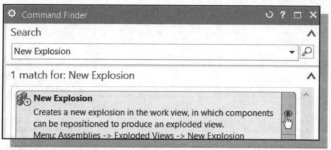

1. In the *Command Finder,* search for the **New Explosion** command as shown. Click on the eyeball icon to add the command group to the ribbon toolbar.

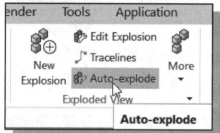

2. Activate the **New Explosion** command as shown

3. Click **OK** to accept the default explosion name.

    •   Note that in *NX*, multiple explosion views can be set up within the same assembly model.

4. In the *Exploded View Toolbar,* select the **Auto-Explode Components** command by left-mouse-clicking once on the icon.

5. Select the **Collar** part as shown.

6. Click once with the **middle-mouse-button (MB2)** to accept the selection.

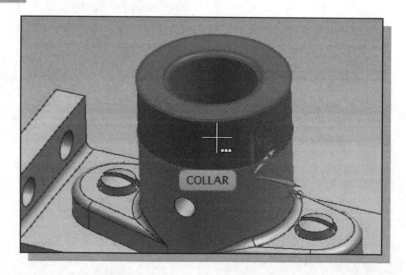

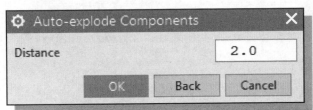

7. Enter **2.0** in the *Distance* option box.

8. Click **OK** to accept the setting.

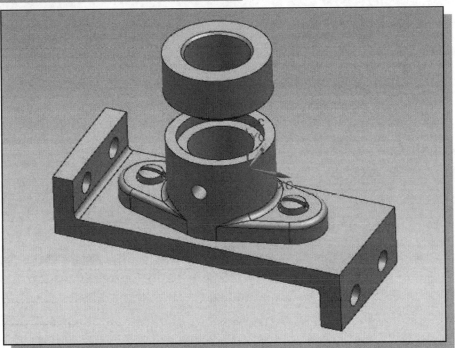

- The **Auto-Explode Components** command will move parts, based on the applied constraints, in a more commonly accepted fashion.

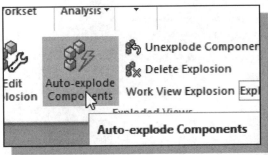

9. On your own, activate **Auto-Explode Components** command again.

10. Inside the graphics window, select the **Cap-Screw** part as shown.

11. Click **OK** to accept the selection.

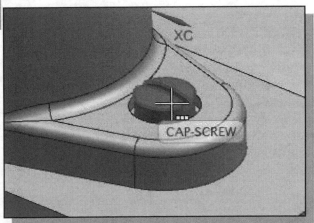

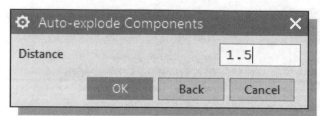

12. Enter **1.5** in the *Distance* option box.

13. Click **OK** to accept the setting.

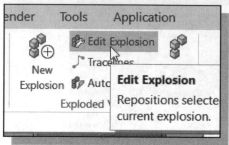

14. In the *Exploded View Toolbar,* select the **Edit Explosion** command by left-mouse-clicking once on the icon.

➢ Note the Edit Explosion dialog box shows the options to manually move the components.

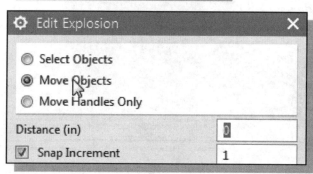

15. Select the **Bearing** part.

16. In the *Edit Explosion* dialog box, activate the **Move Objects** option as shown.

17. On your own, use the movement control handles to reposition the Bearing part.

18. On your own, repeat the above steps and complete the exploded assembly by repositioning the components as shown in the figure.

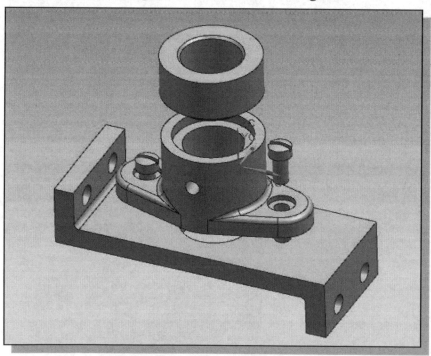

Switch between the Exploded/Unexploded Views

1. In the *Command Finder,* search and activate the **Hide Explosion** command as shown in the below figure.

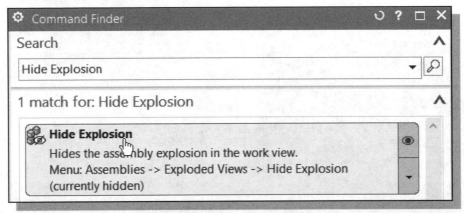

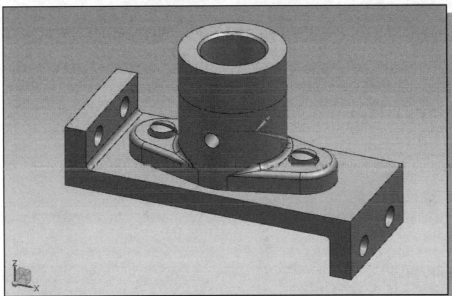

- Note that we can quickly switch the exploded view back on by using the **Show Explosion command** as shown.

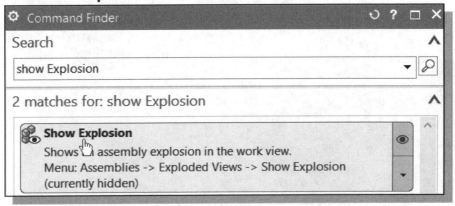

Edit the Components

- The *associative functionality* of *NX* allows us to change the design at any level, and the system reflects the changes at all levels automatically.

    1. Right-mouse-click once on the **Bearing** part to bring up the option menu and select **Open in Window** in the option list as shown.

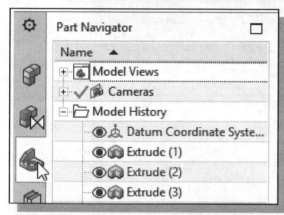

❖ Note that we are opening up the *Bearing* part in the *NX Assembly Navigator*. We will need to switch to the *Part Navigator* to edit the design.

2. Activate the ***Part Navigator*** by clicking on the tab as shown.

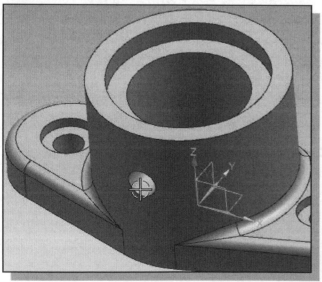

3. Double click, with the *left-mouse-button,* on the **small hole feature** to enter the editing mode.

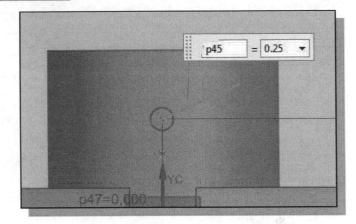

4. Click on the **Sketch Section** button to enter the 2D sketcher mode.

5. On your own, adjust the **diameter** of the small *Drill Hole* to **0.25** as shown.

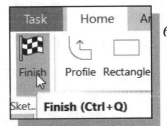

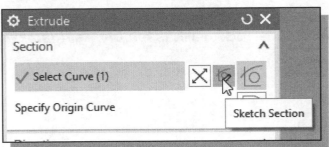

6. Click **Finish Sketch** to exit the NX sketcher mode and return to the *Feature option dialog window*.

7. Click **OK** to proceed with updating the model.

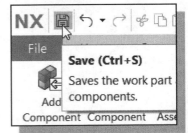

8. Click the **Save** button to save the modification of the bearing part.

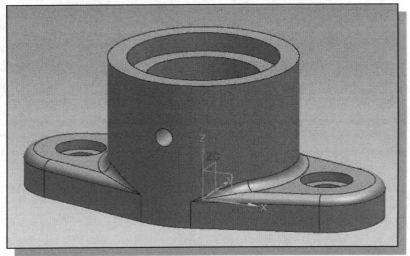

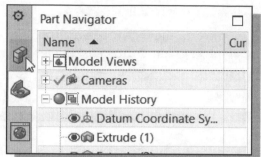

9. Switch back to the ***Assembly Navigator***.

10. Right-mouse-click once on the **Bearing** part to bring up the option menu and select **Open Parent in Window – Shaft-Support** in the option list as shown.

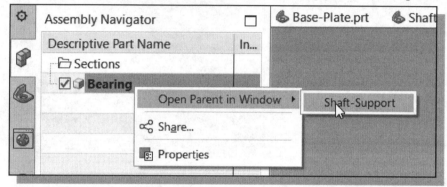

➤ *NX* has updated the bearing part in all levels, including the current *Assembly Mode*.

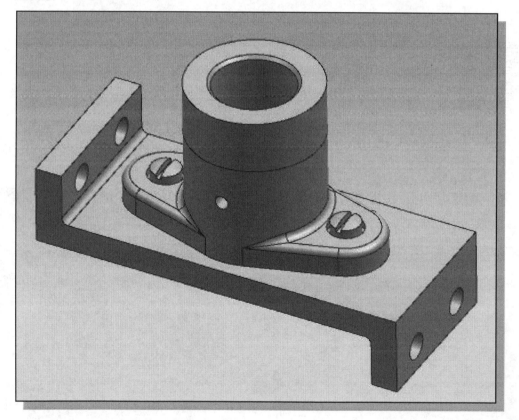

Set up a Drawing of the Assembly Model

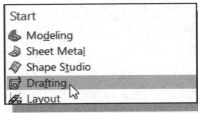

1. Click on the **File** tab in the *Ribbon* toolbar area to display the available options.

2. Select **Drafting** from the Applications list.

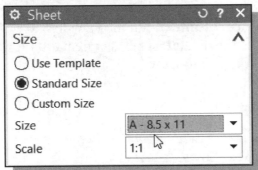

3. Select **A-8.5X11** from the *sheet size* list. Confirm the Scale is set to 1:1, which is Full Scale.

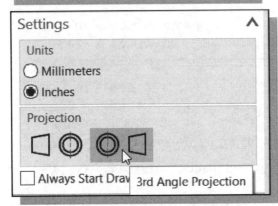

4. Confirm the *Projection* type is set to **Third Angle of projection** and uncheck the **Start Base View** option.

5. Click **OK** to accept the settings and proceed with the creation of the drawing sheet.

6. Click **Close** to exit the view creation wizard.

Importing the Title Block

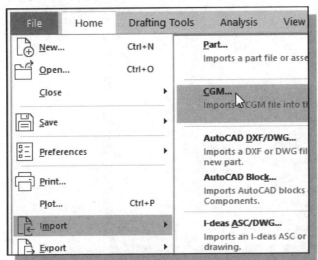

1. Select the **Import CGM** command through the *File* pull-down menu:
   [File] → [Import] → [CGM]

2. Enter or select **Title_A_Master_sheet 1.cgm** in the *Import CGM* window and click **OK** to proceed with importing the selected CGM file.

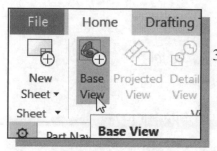

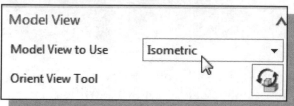

3. Click on the **Base View** in the *Drawing Layout toolbar* to create a new base view.

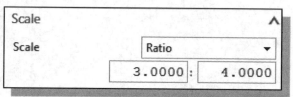

4. Select **Isometric** in the view list, to use the default isometric orientation for the drawing view.

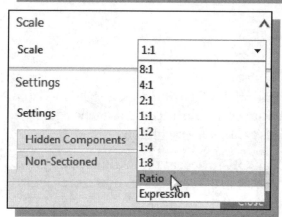

5. Click on the *Scale option* and choose **Ratio** as shown.

6. Set the *Scale factor* to **3.0:4.0** in the input box as shown (0.75 scale).

7. Move the cursor inside the graphics window and place the ***Base*** view near the left side of the *Border* as shown.

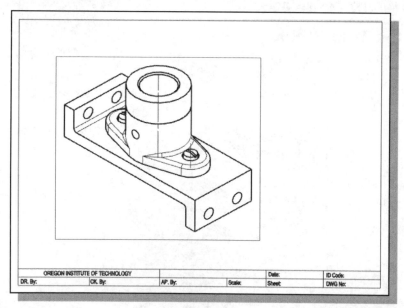

8. Click **Close** to exit the *projected view* option.

Create a Parts List

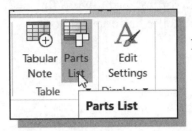

1. Select the **Parts List** command through the *Table* toolbar.

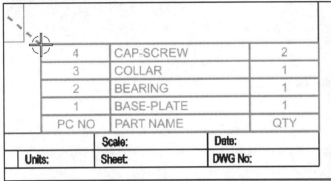

2. A rectangle representing a *Parts List* appears on the screen. **Move** the box so that it aligns with the upper right corner of the title block as shown.

3. Click once with the left-mouse-button to place the **Parts List** as shown.

*   Note that NX automatically fills the content of the parts list.

Complete the Assembly Drawing

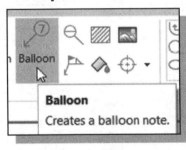

1. Select the **Balloon** command through the *table toolbar* as shown.

2. Enter **1** in the *Text box*.

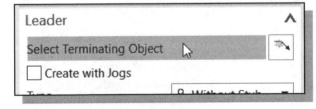

3. Activate the **Select Terminating Object** command as shown.

4. Select the **Base Plate part** by clicking on the bottom right edge as shown.

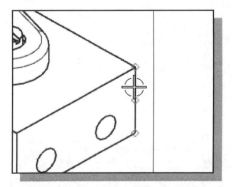

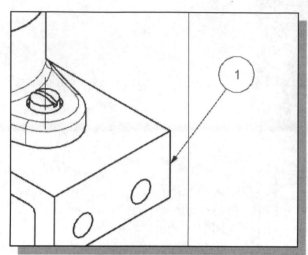

5. Move the cursor upward toward the right and place the balloon as shown.

6. On your own, repeat the above steps and create balloons for all the parts.

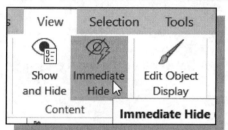

7. On your own, remove the **center lines** and other unwanted objects using the *Immediate Hide* command.

8. On your own, fill in the title block and complete the drawing as shown.

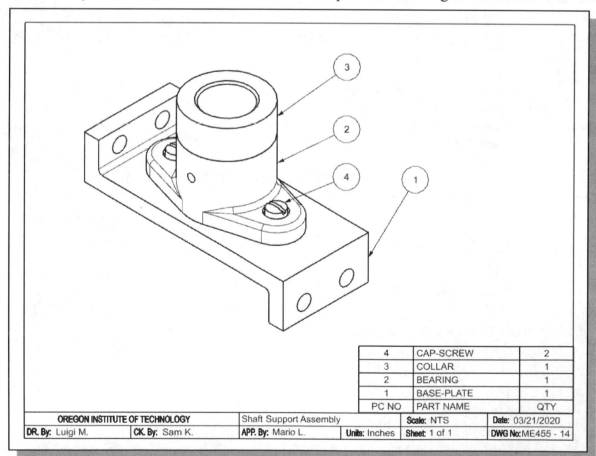

PC NO	PART NAME	QTY
4	CAP-SCREW	2
3	COLLAR	1
2	BEARING	1
1	BASE-PLATE	1

OREGON INSTITUTE OF TECHNOLOGY		Shaft Support Assembly		Scale: NTS	Date: 03/21/2020
DR. By: Luigi M.	CK. By: Sam K.	APP. By: Mario L.	Units: Inches	Sheet: 1 of 1	DWG No: ME455 - 14

Summary of Modeling Considerations

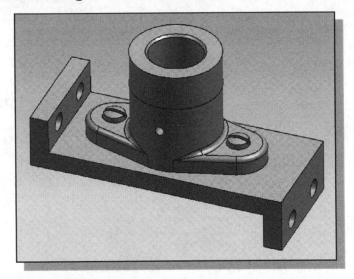

- **Design Intent** – determine the functionality of the design; select features that are central to the design.

- **Order of Features** – consider the parent/child relationships necessary for all features.

- **Dimensional and Geometric Constraints** – the way in which the constraints are applied determines how the components are updated.

- **Relations** – consider the orientation and parametric relationships required between features and in an assembly.

Review Questions:

1. What is the purpose of using *assembly constraints*?

2. List three of the commonly used *assembly constraints*.

3. Describe the difference between the **Touch** constraint and the **Align** constraint.

4. In an assembly, can we place more than one copy of a part? How is it done?

5. How should we determine the assembly order of different parts in an assembly model?

6. How do we adjust the information listed in the **parts list** of an assembly drawing?

7. In *NX*, describe the procedure to create **balloon callouts**.

8. Create sketches showing the steps you plan to use to create the four parts required for the assembly shown on the next page:

Ex.1)

Ex.2)

Ex.3)

Ex.4)

Exercises:

1. **Wheel Assembly** (Create a set of detail and assembly drawings. All dimensions are in mm.)

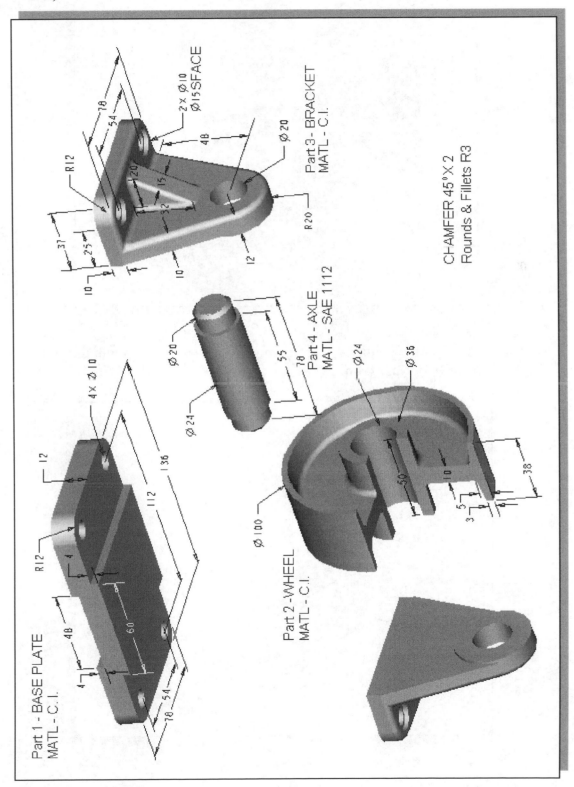

2. **Vise Assembly** (Create a set of detail and assembly drawings. All dimensions are in inches.)

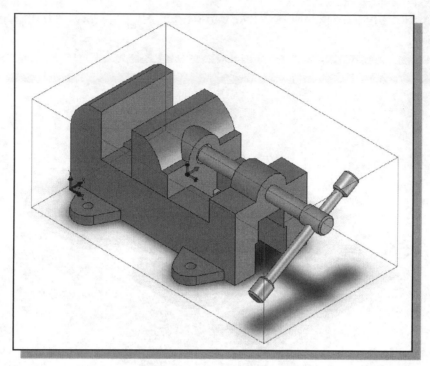

(a) **Base:** The 1.5 inch wide and 1.25 inch wide slots are cut through the entire base. Material: **Gray Cast Iron**.

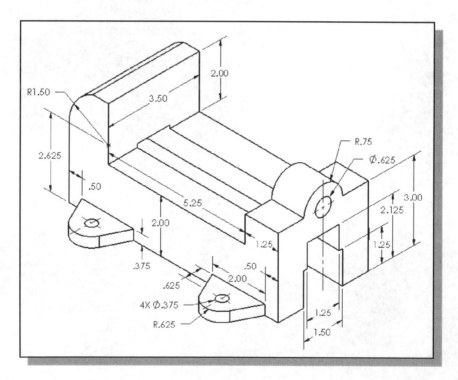

(b) **Jaw:** The shoulder of the jaw rests on the flat surface of the base and the jaw opening is set to 1.5 inches. Material: **Gray Cast Iron**.

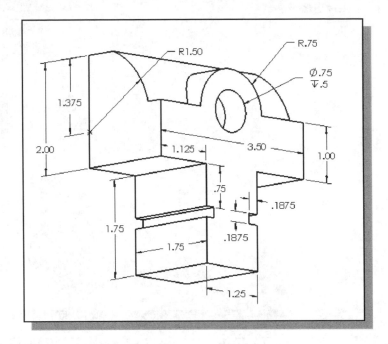

(c) **Key: 0.1875 inch H x 0.375 inch W x 1.75 inch L**. The keys fit into the slots on the jaw with the edge faces flush as shown in the sub-assembly to the right. Material: **Alloy Steel**.

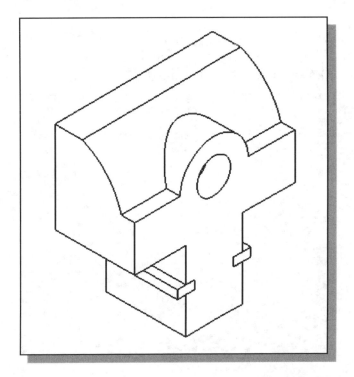

(d) **Screw:** There is one chamfered edge (0.0625 inch x 45°). The flat $\varnothing$ 0.75″ edge of the screw is flush with the corresponding recessed $\varnothing$ 0.75 face on the jaw. Material: **Alloy Steel**.

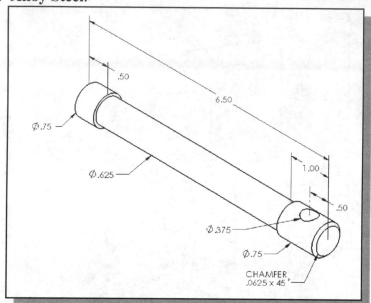

(e) **Handle Rod:** $\varnothing$ **0.375″ x 5.0″ L**. The handle rod passes through the hole in the screw and is rotated to an angle of 30° with the horizontal as shown in the assembly view. The flat $\varnothing$ 0.375″ edges of the handle rod are flush with the corresponding recessed $\varnothing$ 0.735 faces on the handle knobs. Material: **Alloy Steel**.

(f) **Handle Knob:** There are two chamfered edges (0.0625 inch x 45°). The handle knobs are attached to each end of the handle rod. The resulting overall length of the handle with knobs is 5.50″. The handle is aligned with the screw so that the outer edge of the upper knob is 2.0″ from the central axis of the screw. Material: **Alloy Steel**.

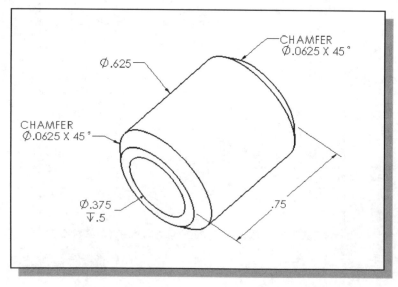

Chapter 15
Advanced Assembly Modeling and Animation

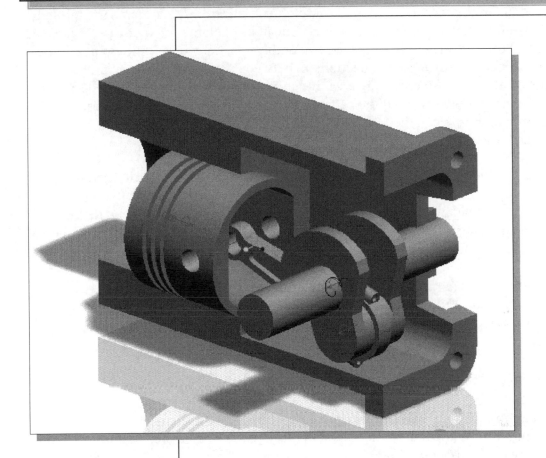

Learning Objectives

- ♦ **Understand the Basic Concepts of performing Basic Motion Analysis**
- ♦ **Using Bodies and Joints**
- ♦ **Using the Joint Connections for specific types of motions**
- ♦ **Create Animation using the Simulation Module**
- ♦ **Output the associated Simulation Video file**

Introduction

In the previous chapter, we went over the fundamentals of creating assembly models. In this chapter, we will examine assembly models that contain moving parts. The main task in creating an assembly is establishing the assembly relationships between parts through the use of assembly constraints. In *Siemens NX*, designs containing moving parts can also be constrained through specially packaged constraint sets, known as **joint connections**. Placing the proper type of **joint connections** in between parts will allow the proper movements of the parts. To create an assembly motion simulation, we will need to first create an assembly model with all the parts placed in the proper locations as described in the previous chapter; we can then transfer the assembly model into the *NX Simulation* module to perform basic **motion analysis**.

In designing machines with moving parts, ***motion analysis*** is usually performed to confirm the proper assembly of the designs, also to check for any other potential problems. One main advantage of using *NX Simulation* is the ability to validate the initial design concepts, check for potential problems and to explore design alternatives without actually creating a physical prototype.

In this chapter, a crank and slider mechanism, commonly seen in engines and compressors, is used to illustrate the basic concepts and procedures of creating assembly models with connections and animations in *NX Simulation*.

NX Motion Simulation

To perform a motion simulation, first create an assembly model then launch the motion simulation module. When a new simulation is created, the system creates a folder with the same name as the assembly model. This simulation folder contains the mechanism data for the assembly model such as simulation files, analysis results files, and so on. The simulation file contains all the simulation data, such as solutions, solution setup, and all motion objects such as bodies, joints, and connectors.

Motion simulation will use the parts in the assembly model as the base and additional simulation components can be added. Note that the applied assembly constraints in the assembly model are **not** used in the motion simulation module. A set of simulation components, such as bodies and links, will be applied on the parts in the assembly model, and these components only exist in the simulation file. This approach allows the edit, change, adjust, and analyze multiple motion design without altering the original assembly model.

For a basic motion simulation, the following steps are required: 1. Start a new simulation. 2. Define motion bodies, which can be a single part or multiple parts. 3. Apply joint connections to define the motion in between motion bodies. 4. Using the NX simulation solver to perform the analysis and calculations. 5. View the analysis results, which can include the animation of the assembly model.

Motion Bodies

In *Siemens NX*, a body represents a single rigid component that moves relative to the other bodies within the assembly. A body may consist of a single *Siemens NX* part or several *Siemens NX* parts fully constrained using joint connections. Each component in an assembly has six **degrees of freedom (DOF)**, or ways in which rigid 3D bodies can move: movement *along* the X, Y, and Z axes (translational freedom), plus rotation *around* the X, Y, and Z axes (rotational freedom). A **ground body** (or **frame**) represents a fixed location in the three-dimensional space where the assembly is referencing its motions. The first object placed in an assembly should typically be the fixed reference of the assembly. It is generally a good habit to assemble the ground piece first when building the assembly.

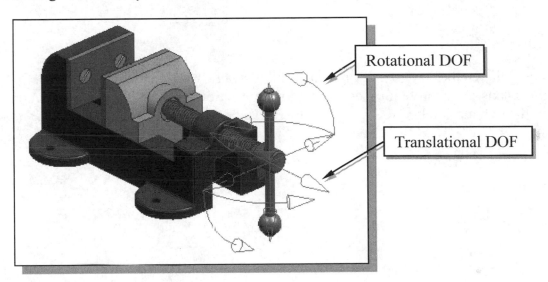

Joint Connections

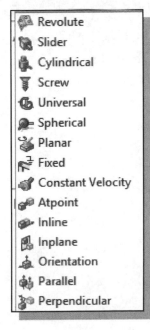

In *Siemens NX*, **joint connections** are special types of packaged constraints that can be used to connect moving components. The applied joint connections will constrain the relative motion between the selected components. The degrees of freedom that a constraint allows can be translation and rotation about three perpendicular axes, as shown in the above figure.

In *Siemens NX*, the available joint connections include Fixed, Revolute, Slider, Cylindrical, Spherical, Planar, Screw, Universal, Constant Velocity, Atpoint, Inline, Inplane, Orientation, Parallel, and Perpendicular.

Joint connections are generally created to prescribe specific types of motions in an assembly. For some connection types, the selected references must be in the same two components of the assembly.

In *Siemens NX motion simulation*, for an assembly with moving components, there are five commonly used joint connections: **Revolute**, **Slider**, **Cylindrical**, **Planar** and **Fixed**.

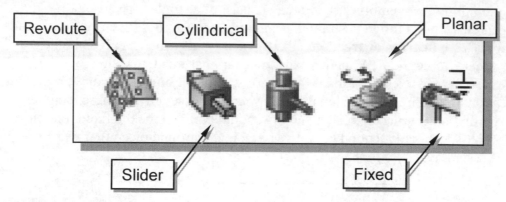

Revolute

A revolute joint connects two motion bodies, allowing one rotational degree of freedom about the Z-axis. A revolute joint does not allow translational movement in any direction between the two motion bodies. A motion driver can be assigned to a slider joint.

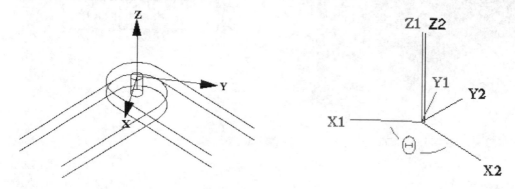

Slider

A slider joint connects two motion bodies, allowing one translational degree of freedom between them. Slider joints do not allow rotational movement between the two motion bodies. A motion driver can be assigned to a slider joint.

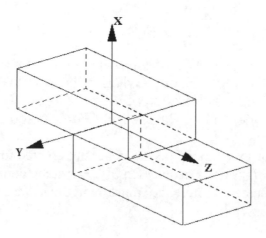

Cylindrical

A cylindrical joint connects two motion bodies, allowing two degrees of freedom: one translational and one rotational about the Z-axes of the links associated with the joint. With a cylindrical joint, the two motion bodies are free to rotate and translate relative to each other about and along the Z-axis. A motion driver can be assigned to a cylindrical joint.

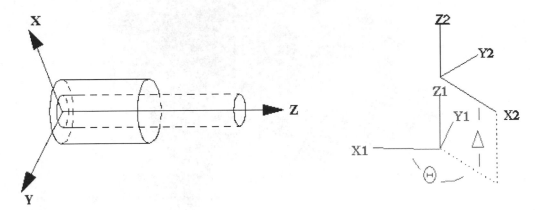

Planar

A planar joint connects two motion bodies, allowing three degrees of freedom between them: two translational and one rotational. In a planar joint, the two links are free to slide and rotate relative to each other while remaining in planar contact.

Two surfaces are in contact but allowed to translate in plane and rotate about the plane normal. The XY planes of the joint on each motion body are parallel

Fixed

A fixed joint connects a body to a fixed position (such as ground), or to another joint. The default position is the center of gravity. Two joints that are connected as fixed move together as one body. A fixed joint allows zero degrees of freedom.

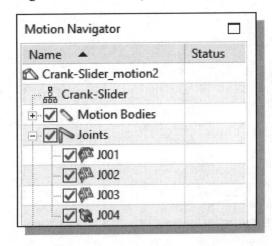

Note that the **Fix the body** option is also available when a body is first defined. This will automatically create a corresponding fixed joint. In the **Motion Navigator**, fixed motion bodies and joints are visually distinct from free bodies and joints. For a fixed body, the rotational and translational axes allow no movement. Also note that the origins of the fixed bodies become coincident.

The Crank-Slider Assembly

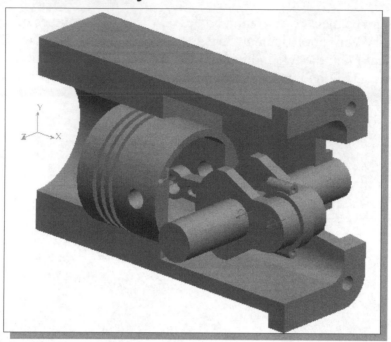

Creating the Required Parts

Five parts are required for this assembly: (1) **End Cap**, (2) **Connecting Rod**, (3) **Base Block**, (4) **Crank Shaft**, and (5) **Piston**. On your own, create the five parts shown below. Save the models as separate part files. (Note the location of the parts relative to the datum planes and close all part files after you have created the parts.)

(1) *End Cap*

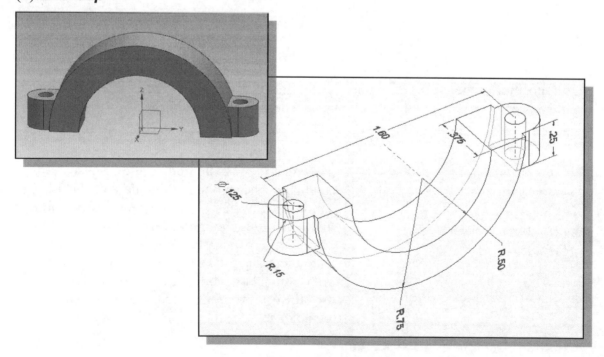

(2) *Connecting Rod*

(Construct the part with the datum planes passing through the center axes of the
cylindrical surfaces.)

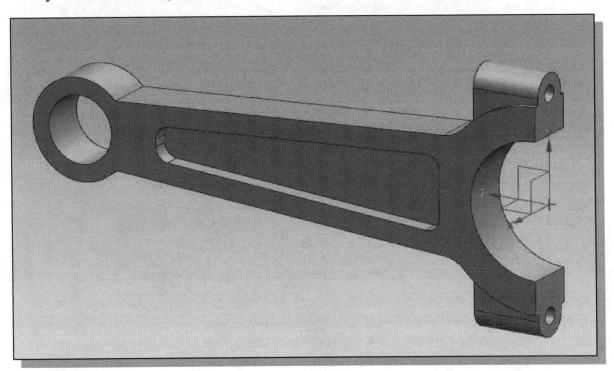

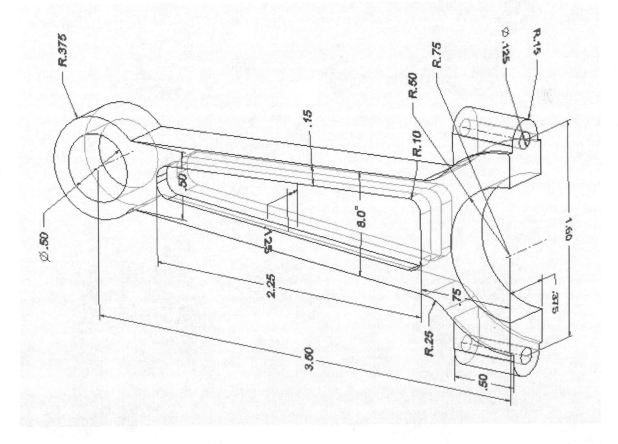

(3) *Base Block*

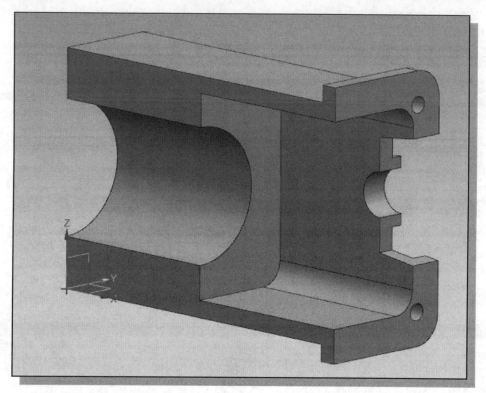

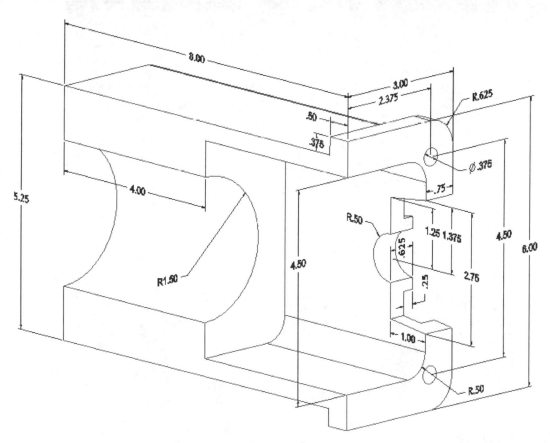

(4) *Crank Shaft*

(Construct the part with the datum planes passing through the center of the main shaft. Create additional datum axes for the cylindrical surfaces if necessary.)

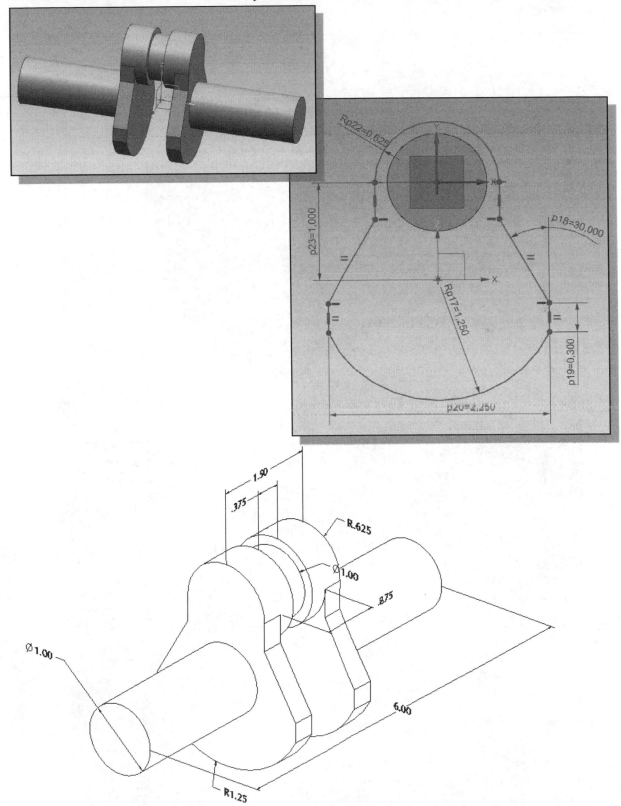

(5) *Piston*

(Construct the part with two of the datum planes passing through the center axis of the main cylinder body. Create additional datum axes for the cylindrical surfaces if necessary.)

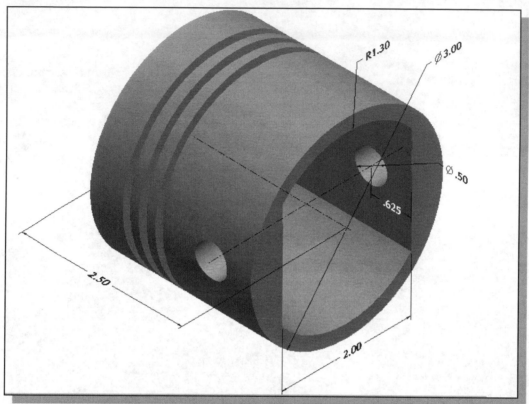

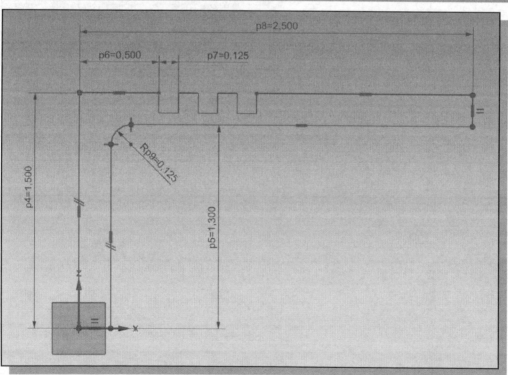

Create the Crank Slider Assembly Model

1. Select the *NX* option on the *Start* menu or select the *NX* icon on the desktop to start *NX*. The *NX* main window will appear on the screen.

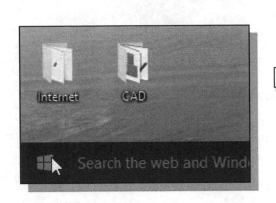

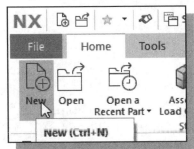

2. Select the **New** icon with a single click of the left-mouse-button (MB1) in the *Standard toolbar area*.

3. Select the **Inches** units as shown in the below figure.

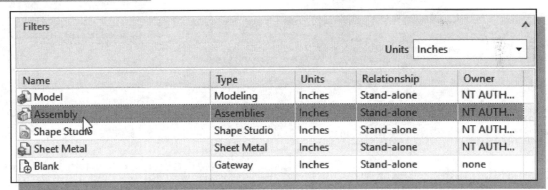

4. Select **Assembly** in the *Template list*. Note that the *Assembly template* will allow us to switch directly into the *Assemblies* task with settings related to creating assembly models.

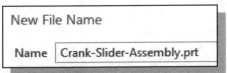

5. In the *New File Name* area, enter **Crank-Slider-Assembly** as the *File Name*.

6. Click **OK** to proceed with the New File command.

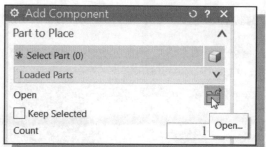

7. The **add component** command is automatically activated. In the **add component** window, click **Open** to load an existing part.

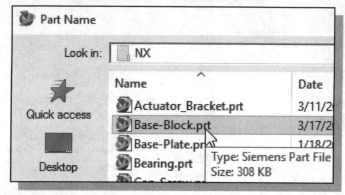

8. Select the ***Base-Block*** (part file: ***Base-Block.prt***) in the list window.

9. Click **OK** to accept the selection.

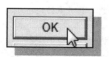

10. Click **Apply** to accept the settings and load the part.

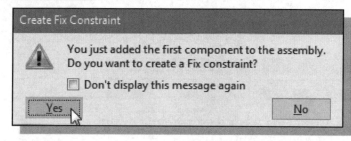

11. Since this is the first component placed in the assembly, by default, NX will add a fixed assembly constraint to it. Click **Yes** to create the constraint.

Assembling the Crank Shaft

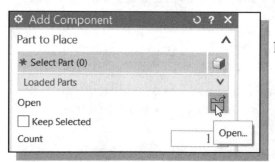

1. In the **add component** window, click **Open** to load an existing part into the assembly model.

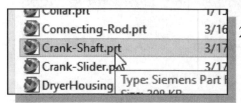

2. Select the ***Crank Shaft*** part (part file: *crank_shaft.prt*) in the list window and pick **OK** to retrieve the model.

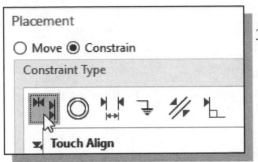

3. Choose the **Touch Align** constraint from the placement constraint type list as shown.

4. Pick and align the two *axes* of the *Crank Shaft* and the *Base Block* as shown.

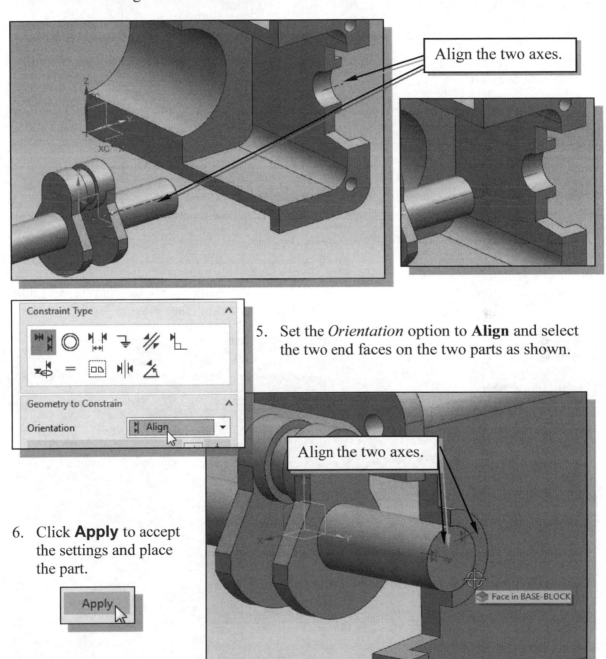

Align the two axes.

5. Set the *Orientation* option to **Align** and select the two end faces on the two parts as shown.

Align the two axes.

6. Click **Apply** to accept the settings and place the part.

Face in BASE-BLOCK

Assembling the Connecting Rod

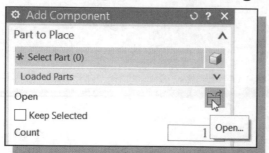

1. In the **add component** window, click **Open** to load another part into the assembly model.

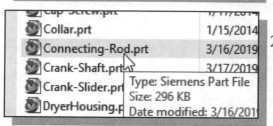

2. Select the **Connecting Rod** part in the list window and pick **OK** to retrieve the model.

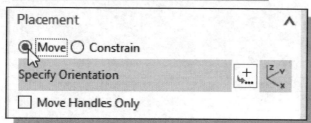

3. On your own, use the **Move** option to reposition the part if necessary.

4. Use the **Constrain** option to align the part into the assembly model as shown.

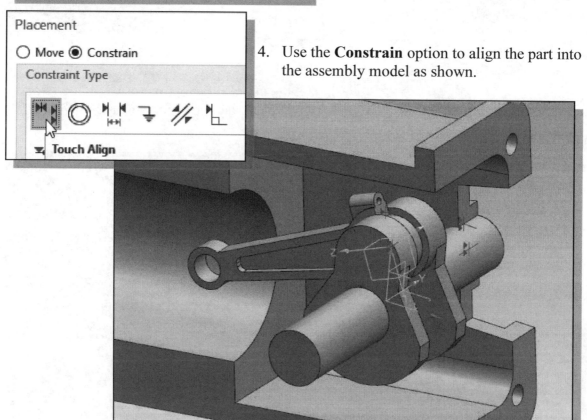

Complete the Assembly

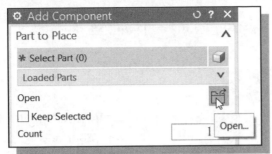

1. In the **add component** window, click **Open** to load another part into the assembly model.

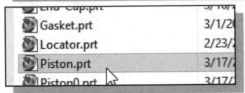

2. Select the *Piston* part in the list window and pick **OK** to retrieve the model.

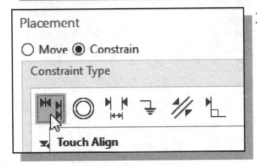

3. Align the two *corresponding* center axes of the *Piston* and the *Base Block* parts as shown.

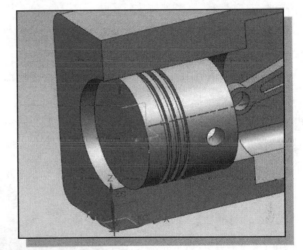

4. On your own, complete the assembly model by adding the **End-Cap** part and apply proper assembly constraints.

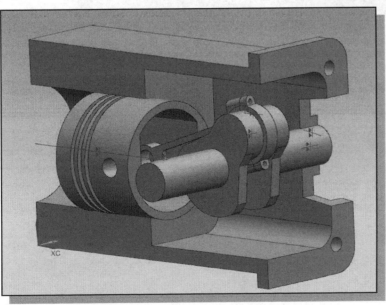

❖ Note that the assembly constraints are applied to assure the proper alignments of all the parts in the assembly model.

Start the NX Motion Simulation Module

The *NX Motion simulation* module can be used to perform a motion analysis that can be done in a relatively short time.

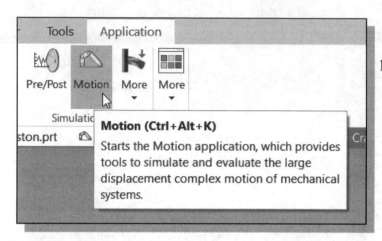

1. Select the **Applications** tab in the *Ribbon* toolbar and select **Motion**.

❖ Note the *NX Motion simulation* package is an integrated *NX* module; the *NX Motion simulation* toolbar appears in the **Ribbon** toolbar.

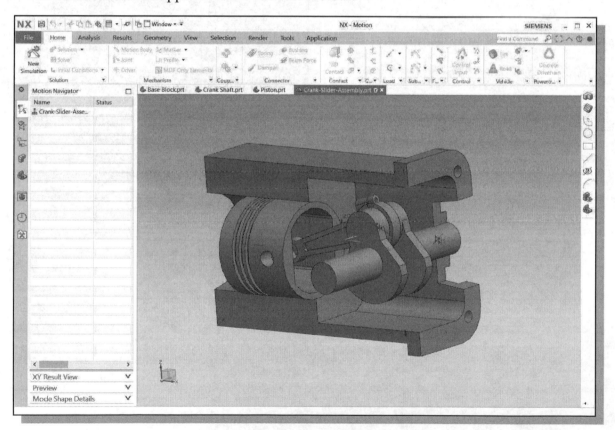

Defining a New Simulation

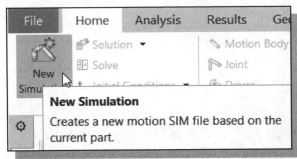

1. In the *Ribbon* toolbar, click on the **New Simulation** icon to launch the command.

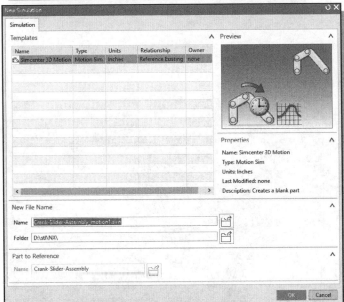

- Note a new simulation file will be created using the assembly model we just assembled together.

2. Select **OK** to accept the default settings and start a new simulation.

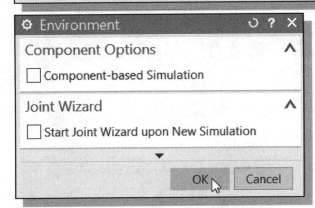

3. In the *Environment* dialog box, set the analysis type to **Dynamics** as shown.

4. Uncheck the Start Joint Wizard upon New Simulation option and click **OK** to accept the settings.

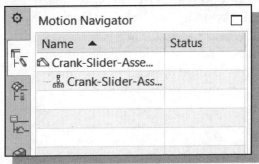

- Inside the *Motion Navigator*, two items are listed: the current simulation and the **Crank-Slider-Assembly** model

Define the Motion Bodies

In *NX Motion simulation*, a **body** represents a single rigid component that moves relative to the other bodies within the assembly. A body may consist of a single *Siemens NX* part or several *Siemens NX* parts in the assembly.

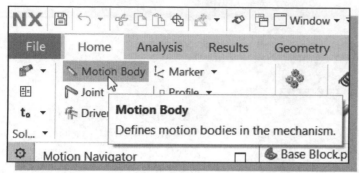

1. Select **Motion Body** in the *Mechanism* toolbar, which is the first icon in the toolbar.

2. Select the **Base-Block** part in the *graphics window* as shown.

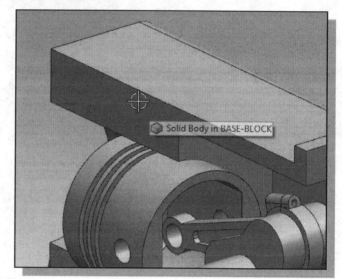

3. In the *Motion Body* dialog box, switch **on** the *Fix the Motion Body without Joint* option, this setting allows the set up of the selected part as a **ground body**.

A **ground body** (or **frame**) represents a fixed location in the three-dimensional space where the assembly is referencing its motions. The first object placed in an assembly should typically be used as the fixed reference of the assembly.

4. Click **Apply** to accept the settings and create the first motion body.

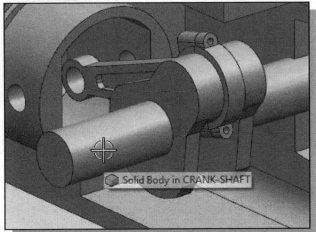

5. Select the **Crank-Shaft** part in the *graphics window* as shown.

6. In the *Motion Body* dialog box, switch **off** the *Fix the Motion Body without Joint* option, as the part is a moving part in the assembly.

7. Click **Apply** to accept the settings and create the second motion body.

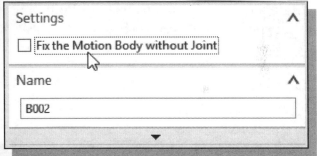

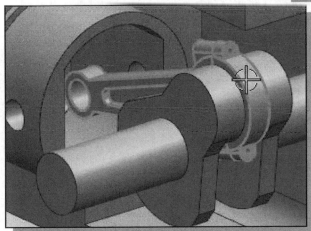

8. Select the **Connecting Rod** and the **End Cap** part in the *graphics window* as shown. These two parts will be defined as one body and therefore move together.

9. Click **Apply** to accept the settings and create the next motion body.

10. On your own, select the **Piston** part and create the last motion body.

- In the *Motion Navigator*, the defined bodies are listed under **Motion Bodies** as shown.

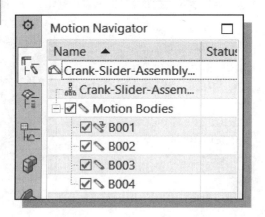

Define the Joint Connections

In *NX Motion simulation*, **joint connections** are special types of packaged constraints that can be used to connect moving components. The applied joint connections will constrain the relative motion between the selected components.

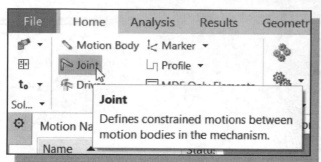

1. Select **Joint** in the *Mechanism* toolbar, which is the second icon in the toolbar.

2. In the *Joint* dialog box, set the **Joint Type** to **Revolute** as shown.

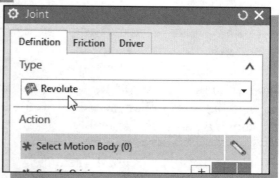

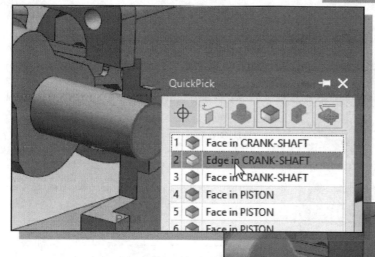

3. Select the **Circular edge of the Crank shaft** part as shown.

- Note that the center of the circular face is now set as the rotational origin and the rotational direction vector is set to be perpendicular to the selected face.

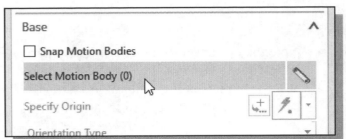

4. Click on the **Select Motion Body** option under the *Base* list as shown.

5. Select a **surface of the Base Block** part as the relative motion reference part as shown. Click **Apply** to accept the settings.

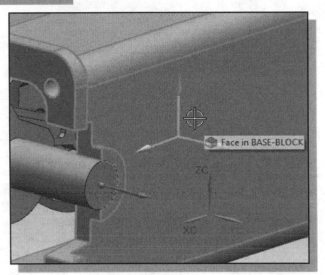

- We have set up the rotational relationship of the crank-shaft relative to the base-block part.

- We will next set up the rotational relationship of the Connecting rod and the Crank Shaft part. The cylindrical joint will maintain the alignment of the axes of the two selected components.

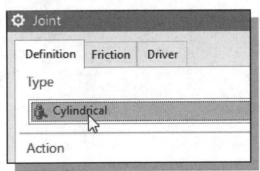

6. In the *Joint* dialog box, set the **Joint Type** to **Cylindrical** as shown.

7. Select the **Circular face of the Connecting Rod** part as shown.

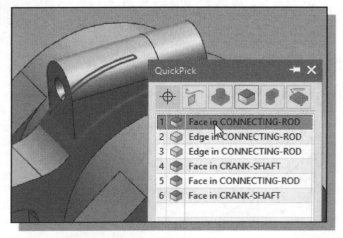

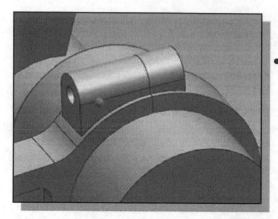

• Note the current origin of the selected part, appearing as an orange colored dot, is not aligned to the desired center axis. We will need to re-define the origin location.

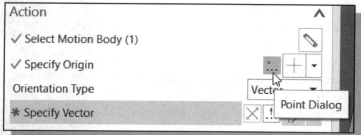

8. Click on the **Point Dialog** button to enter the Point selection option dialog box.

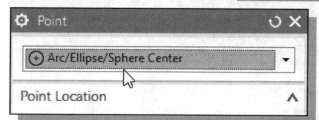

9. Select the **Arc Center** option as shown.

10. Select one of the larger arcs to align the origin to the center point.

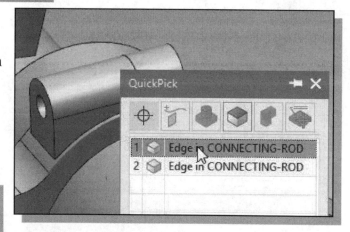

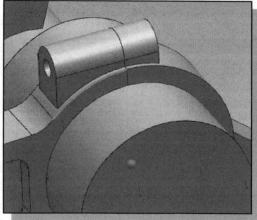

• Note the current origin of the selected part, appearing as an orange colored dot, is now aligned to the desired center axis.

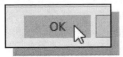

11. Click **OK** to accept the selection.

- Note the selected surface does not automatically set the rotational direction vector.

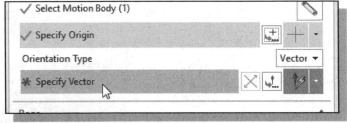

12. Click on the **Specify Vector** icon as shown.

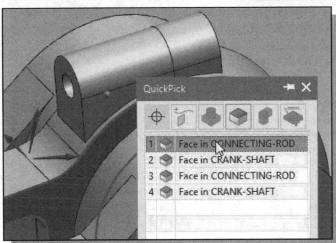

13. Select the **side face of the Connecting Rod** part to set the rotational direction vector as shown.

14. Click on the **Select Motion Body** option under the *Base* list as shown.

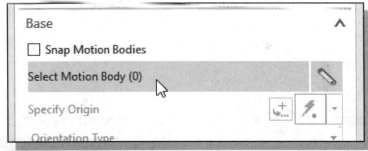

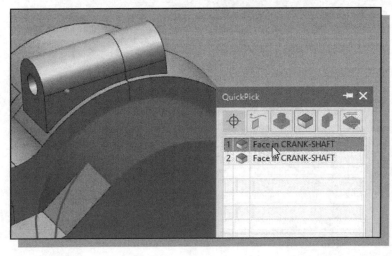

15. Select the **outer Circular face of the Crank shaft** part as the relative motion reference part as shown. Click **Apply** to accept the settings.

- We have set up the rotational relationship of the connecting rod part relative to the crank-shaft part.

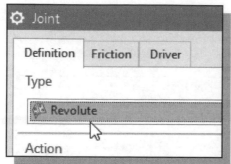

- We will next set up the rotational relationship of the Piston and the connecting rod part.

16. In the *Joint* dialog box, set the **Joint Type** to **Revolute** as shown.

17. Select the **inside circular edge of the Piston** part as shown.

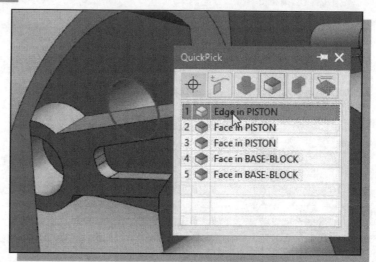

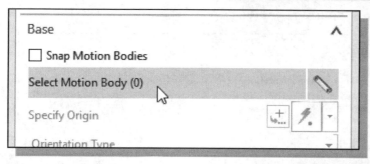

18. Click on the **Select Motion Body** option under the *Base* list as shown.

19. On your own, select the *cylindrical surface* of the **Connecting Rod** part as shown.

20. Click **Apply** to accept the settings.

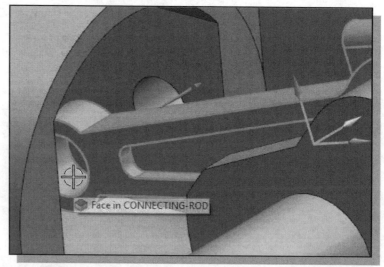

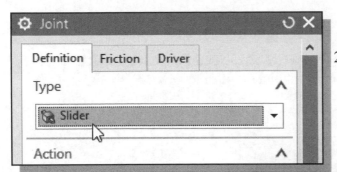

21. In the *Joint* dialog box, set the **Joint Type** to **Slider** as shown.

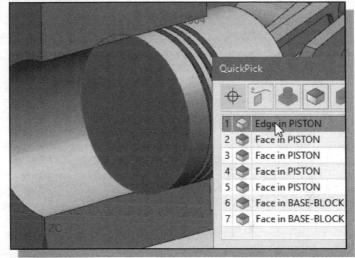

22. Select the **Circular edge of the Crank shaft** part as shown.

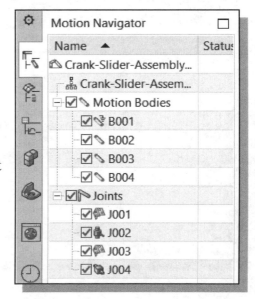

23. On your own, select the *cylindrical surface* of the **Base-Block** part as the Motion Body as shown.

24. Click **OK** to accept the settings and complete the joint motion definitions.

• In the *Motion Navigator*, the four defined Joint connections are listed under **Joints** as shown.

Set up a Motion Driver for the Animation

In *NX Motion simulation*, a *driver* can be used to control a component by specifying a desired position, velocity, or acceleration as a function of time.

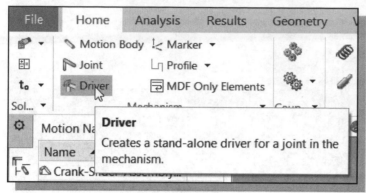

1. Select **Driver** in the *Solution* toolbar as shown.

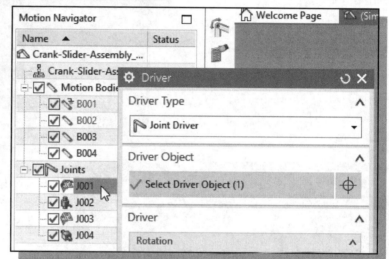

2. In the *Driver* dialog box, set the **Driver Type** to **Joint Driver** as shown.

3. Select the first Joint, **J001**, in the *Motion Navigator* as shown.

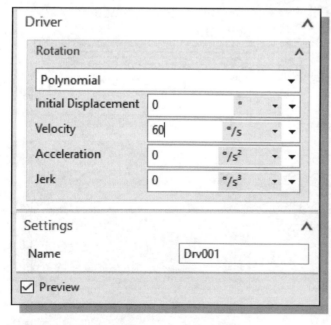

4. For the *Driver Rotation*, choose **Polynomial**.

5. Set the **Velocity** to **60** degrees per second as shown.

6. Click **OK** to accept the settings.

Set up an Animation Analysis

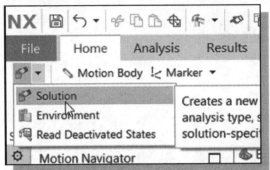

1. Select **Solution** in the *Solution* toolbar as shown.

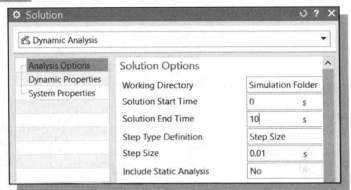

2. In the *Solution* dialog box, set the **Solution Type** to **Normal run** and **Analysis type** to **Kinematics-Dynamics** as shown.

3. Enter **10s** in the *Time box* and **0.01** in the *Steps box* as shown.

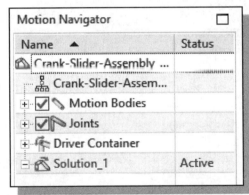

4. Click **OK** to accept the settings and complete the solution definitions.

❖ In the *Motion Navigator,* **Solution_1** is set up and active. Note that multiple solutions can co-exist with different analysis settings.

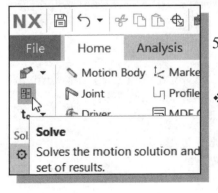

5. Select **Solve** in the *Solution* toolbar to start the Simulation Solver.

❖ Note the information window will appear and display the progress of the solution.

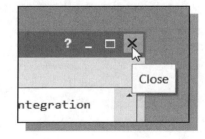

6. Near the upper right corner, click Close to close the information window.

View the Animation

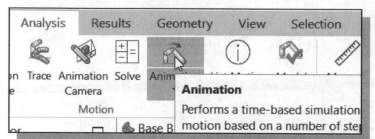

1. Select Animation in the Analysis tab as shown.

2. Click on the **Play** button to watch the animation.

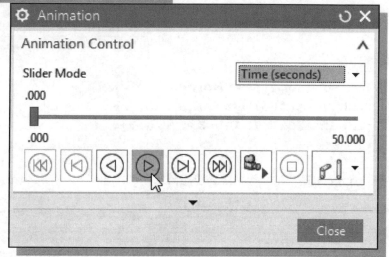

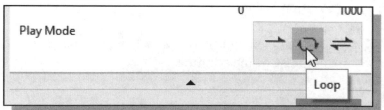

3. On your own, experiment with the **Animation Delay** control.

4. Click the **Play** button to watch the effects of the adjustment on the animation.

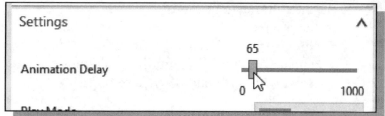

5. On your own, experiment with the **PlayMode** control.

6. Click on the **Reverse** button to run the animation backward.

Output the Animation as a Video file

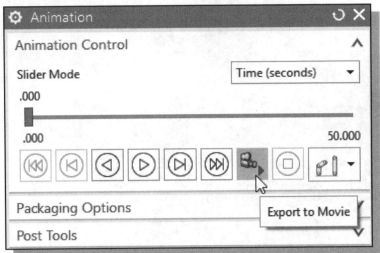

1. Click on the **Record** button in the *dashboard* as shown.

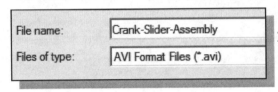

2. Enter Crank-Slider-Assembly as the file name as shown.

- Note that the default **File type** is set to AVI video format.

3. Accept the default settings and click **OK** to save the animation as an **AVI** video file.

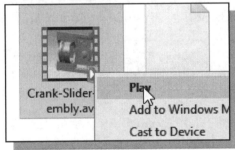

4. On your own, view the captured movie file.

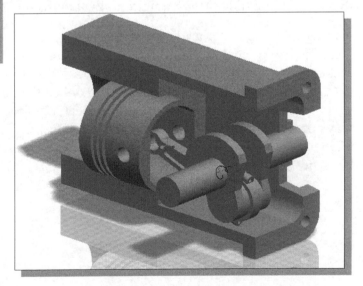

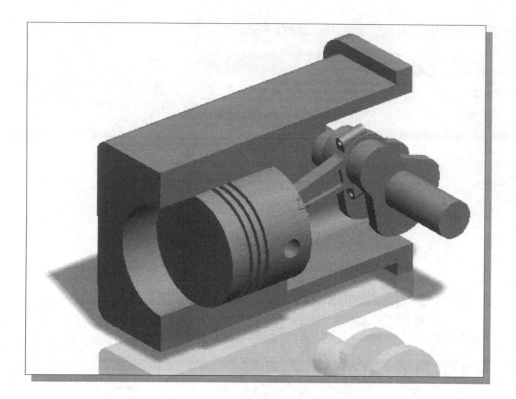

Conclusion

Design includes all activities involved from the original concept to the finished product. Design is the process by which products are created and modified. For many years, designers sought ways to describe and analyze three-dimensional designs without building physical models. With the advancements in computer technology, the creation of parametric models on computers offers a wide range of benefits. Parametric models are easier to interpret and can be easily altered. Parametric models can be analyzed using finite element analysis software, and simulation of real-life loads can be applied to the models and the results graphically displayed. The finalized solid models can also be used directly by manufacturing equipment to manufacture the product.

Throughout this text, various modeling techniques have been presented. Mastering these techniques will enable you to create intelligent and flexible solid models. The goal is to make use of the tools provided by *Siemens NX* and to successfully capture the ***design intent*** of the product. In many instances, only a single approach to the modeling tasks was presented; you are encouraged to repeat all of the tutorials and develop different ways of thinking in accomplishing the same tasks. We have only scratched the surface of *Siemens NX's* functionality. The more time you spend using the system, the easier it will be to perform parametric modeling with *Siemens NX*.

Review Questions:

1. When and why should we use *joint connections* in an assembly?

2. List and describe four of the commonly used *joint connections*.

3. Describe the procedure to apply a **Revolute** connection in an assembly.

4. What is the default video file format to save an animation generated in *NX Motion simulation*?

5. List and describe the procedure to attach a motion driver in *NX Motion simulation*.

6. Can we adjust the playback speed of the animation in *NX Motion simulation*?

7. What is the difference between a *body* and a *ground body* in *NX Motion simulation*?

8. How did we constrain the **Crank Shaft** part in the **Crank-Slider** assembly? Which joint connections did we use in the Motion simulation?

Exercises: (Create Assembly models and animations in *NX Motion simulation.*)

1. **Leveling Assembly** (Create a set of detail and assembly drawings. All Dimensions are in mm.)

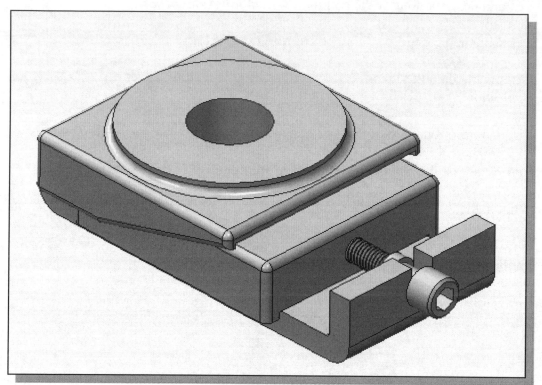

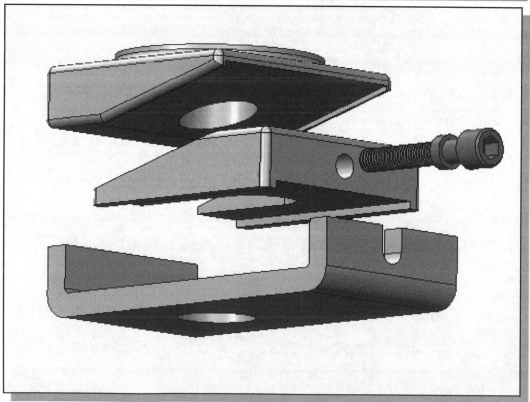

(a) **Base Plate**

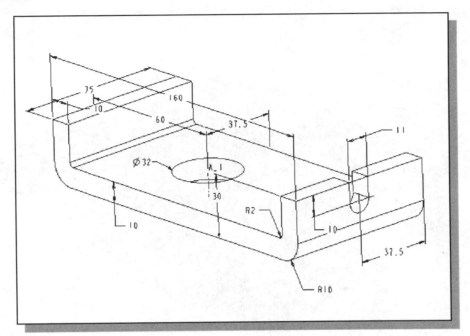

(b) **Sliding Block** (Rounds & Fillets: R3)

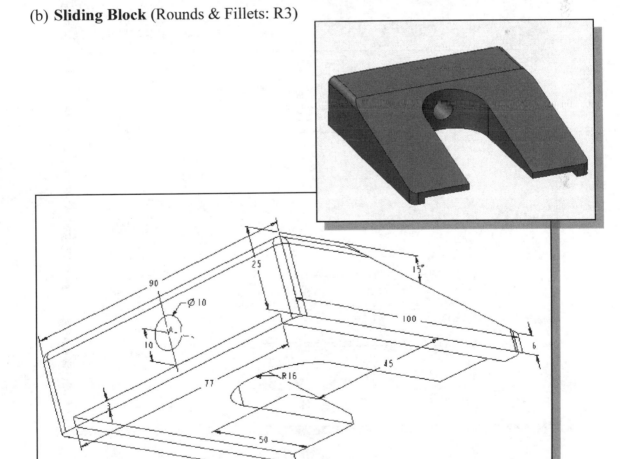

(c) **Lifting Block** (Rounds & Fillets: R3)

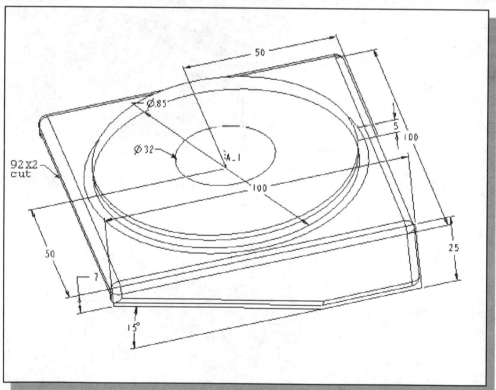

(d) Adjusting Screw (M10 × 1.5)

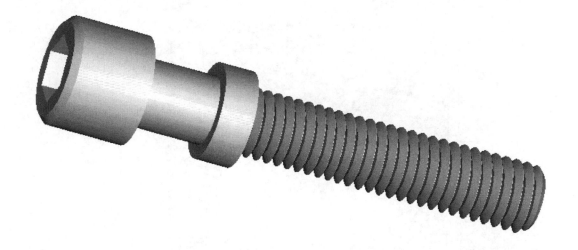

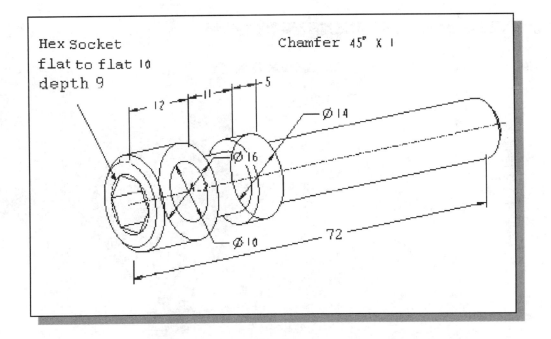

2. **Quick Return Mechanism** (Design and create the necessary parts.)

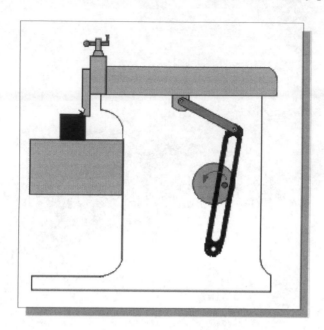

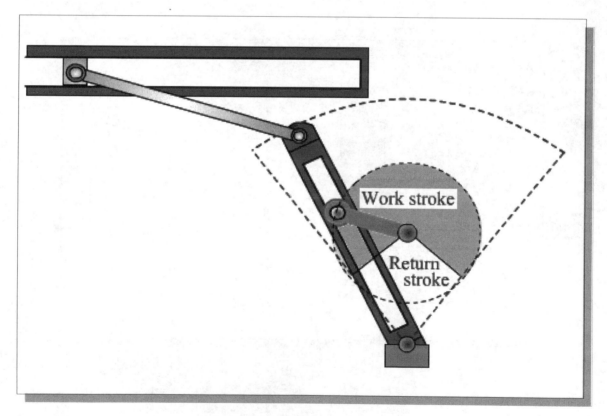

Appendix A

Sec. 206. Standard gauge for sheet and plate iron and steel

For the purpose of securing uniformity, the following is established as the only standard gauge for sheet and plate iron and steel in the United States of America, namely:

Gauge	Frac. Inch	Dec. Inch	mm	oz/ft²	lb/ft²	kg/ft²	kg/m²	lb/m²
0000000	1/2	.5	12.7	320	20.00	9.072	97.65	215.28
000000	15/32	.46875	11.90625	300	18.75	8.505	91.55	201.82
00000	7/16	.4375	11.1125	280	17.50	7.983	85.44	188.37
0000	13/32	.40625	10.31875	260	16.25	7.371	79.33	174.91
000	3/8	.375	9.525	240	15	6.804	73.24	161.46
00	11/32	.34375	8.73125	220	13.75	6.237	67.13	148.00
0	5/16	.3125	7.9375	200	12.50	5.67	61.03	134.55
1	9/32	.28125	7.14375	180	11.25	5.103	54.93	121.09
2	17/64	.265625	6.746875	170	10.625	4.819	51.88	114.37
3	1/4	.25	6.35	160	10	4.536	48.82	107.64
4	15/64	.234375	5.953125	150	9.375	4.252	45.77	100.91
5	7/32	.21875	5.55625	140	8.75	3.969	42.72	94.18
6	13/64	.203125	5.159375	130	8.125	3.685	39.67	87.45
7	3/16	.1875	4.7625	120	7.5	3.402	36.62	80.72
8	11/64	.171875	4.365625	110	6.875	3.118	33.57	74.00
9	5/32	.15625	3.96875	100	6.25	2.835	30.52	67.27
10	9/64	.140625	3.571875	90	5.625	2.552	27.46	60.55
11	1/8	.125	3.175	80	5	2.268	24.41	53.82
12	7/64	.109375	2.778125	70	4.375	1.984	21.36	47.09
13	3/32	.09375	2.38125	60	3.75	1.701	18.31	40.36
14	5/64	.078125	1.984375	50	3.125	1.417	15.26	33.64
15	9/128	.0703125	1.7859375	45	2.8125	1.276	13.73	30.27
16	1/16	.0625	1.5875	40	2.5	1.134	12.21	26.91
17	9/160	.05625	1.42875	36	2.25	1.021	10.99	24.22
18	1/20	.05	1.27	32	2	.9072	9.765	21.53
19	7/160	.04375	1.11125	28	1.75	.7938	8.544	18.84
20	3/80	.0375	.9525	24	1.50	.6804	7.324	16.15
21	11/320	.034375	.873125	22	1.375	.6237	6.713	14.80
22	1/32	.03125	.793750	20	1.25	.567	6.103	13.46
23	9/320	.028125	.714375	18	1.125	.5103	5.493	12.11
24	1/40	.025	.635	16	1	.4536	4.882	10.76
25	7/320	.021875	.555625	14	.875	.3969	4.272	9.42

26	3/160	.01875	.47625	12	.75	.3402	3.662	8.07
27	11/640	.0171875	.4365625	11	.6875	.3119	3.357	7.40
28	1/64	.015625	.396875	10	.625	.2835	3.052	6.73
29	9/640	.0140625	.3571875	9	.5625	.2551	2.746	6.05
30	1/80	.0125	.3175	8	.5	.2268	2.441	5.38
31	7/640	.0109375	.2778125	7	.4375	.1984	2.136	4.71
32	13/1280	.01015625	.25796875	6 1/2	.40625	.1843	1.983	4.37
33	3/320	.009375	.238125	6	.375	.1701	1.831	4.04
34	11/1280	.00859375	.21828125	5 1/2	.34375	.1559	1.678	3.70
35	5/640	.0078125	.1984375	5	.3125	.1417	1.526	3.36
36	9/1280	.00703125	.17859375	4 1/2	.28125	.1276	1.373	3.03
37	17/2560	.006640625	.168671875	4 1/4	.265625	.1205	1.297	2.87
38	1/160	.00625	.15875	4	.25	.1134	1.221	2.69

Reformatted from: **http://www4.law.cornell.edu/uscode/15/206.html**

INDEX

Notes: